AF775463

A Comprehensive Guide to the NIST Cybersecurity Framework 2.0

A Comprehensive Guide to the NIST Cybersecurity Framework 2.0

Strategies, Implementation, and Best Practice

Dr. Jason Edwards
Texas, USA

WILEY

This edition first published 2025
© 2025 John Wiley & Sons Ltd

All rights reserved, including rights for text and data mining and training of artificial technologies or similar technologies. No part of this publication may be reproduced, stored in a retrieval system, or transmitted, in any form or by any means, electronic, mechanical, photocopying, recording or otherwise, except as permitted by law. Advice on how to obtain permission to reuse material from this title is available at http://www.wiley.com/go/permissions.

The right of Jason Edwards to be identified as the author of this work has been asserted in accordance with law.

Registered Offices
John Wiley & Sons, Inc., 111 River Street, Hoboken, NJ 07030, USA
John Wiley & Sons Ltd, The Atrium, Southern Gate, Chichester, West Sussex, PO19 8SQ, UK

For details of our global editorial offices, customer services, and more information about Wiley products visit us at www.wiley.com.

Wiley also publishes its books in a variety of electronic formats and by print-on-demand. Some content that appears in standard print versions of this book may not be available in other formats.

Trademarks: Wiley and the Wiley logo are trademarks or registered trademarks of John Wiley & Sons, Inc. and/or its affiliates in the United States and other countries and may not be used without written permission. All other trademarks are the property of their respective owners. John Wiley & Sons, Inc. is not associated with any product or vendor mentioned in this book.

Limit of Liability/Disclaimer of Warranty
While the publisher and authors have used their best efforts in preparing this work, they make no representations or warranties with respect to the accuracy or completeness of the contents of this work and specifically disclaim all warranties, including without limitation any implied warranties of merchantability or fitness for a particular purpose. No warranty may be created or extended by sales representatives, written sales materials or promotional statements for this work. This work is sold with the understanding that the publisher is not engaged in rendering professional services. The advice and strategies contained herein may not be suitable for your situation. You should consult with a specialist where appropriate. The fact that an organization, website, or product is referred to in this work as a citation and/or potential source of further information does not mean that the publisher and authors endorse the information or services the organization, website, or product may provide or recommendations it may make. Further, readers should be aware that websites listed in this work may have changed or disappeared between when this work was written and when it is read. Neither the publisher nor authors shall be liable for any loss of profit or any other commercial damages, including but not limited to special, incidental, consequential, or other damages.

Library of Congress Cataloging-in-Publication Data Applied for:
Hardback ISBN: 9781394280360

Cover Design: Wiley
Cover Image: © Yuichiro Chino/Getty Images

Set in 9.5/12.5pt STIXTwoText by Straive, Chennai, India

Printed and bound by CPI Group (UK) Ltd, Croydon, CR0 4YY

C9781394280360_220824

To my beloved family. To my mom and dad, who are no longer with us but whose love and guidance continue to inspire me every day. To my wife, Selda, and my children, Michelle, Chris, Ceylin, and Mayra, for your endless love, patience, and encouragement. To my close family members: Derek, Meltem, Nilos, and Ken. And to my sisters, Robin, Kelly, and Lynn. This book is dedicated to you, with all my love and gratitude.

Contents

Preface

In today's digital age, cybersecurity has become indispensable to our personal and professional lives. **As a cybersecurity professor and professional with over two decades of experience, I have witnessed firsthand the evolution of cyber threats and the relentless efforts to mitigate them.** This book is born out of my passion for cybersecurity and a deep understanding of organizations' complexities in safeguarding their digital assets.

The National Institute of Standards and Technology (NIST) Cybersecurity Framework (CSF) has emerged as a cornerstone in cybersecurity. Since its inception, it has provided a structured and flexible approach to managing and reducing cybersecurity risks. The NIST CSF 2.0 represents a significant leap forward, incorporating the latest advancements and addressing the dynamic nature of cyber threats.

This book is designed to be a comprehensive resource for understanding and implementing the NIST CSF 2.0. It delves into the intricacies of the framework, offering detailed explanations of its core components, functions, and implementation strategies. Whether you are a cybersecurity professional, a business leader, or an academic, this guide aims to equip you with the knowledge and tools necessary to enhance your organization's cybersecurity posture.

I have endeavored to make this book accessible to a broad audience, from beginners to seasoned professionals. The language is clear and jargon-free, making complex concepts understandable. At the same time, the depth of coverage ensures that even those with advanced knowledge will find valuable insights.

In writing this book, I have drawn upon my extensive experience and the collective wisdom of cybersecurity experts and practitioners. The practical recommendations, case studies, and actionable advice included herein will bridge the gap between theory and practice, helping organizations implement the NIST CSF effectively and efficiently.

Cybersecurity is a continuous journey, and this book is a step toward mastering it. I invite you to embark on this journey with me, armed with the knowledge and confidence to navigate the ever-changing cyber threat landscape.

July, 2024

Jason Edwards
Texas, USA

Acknowledgments

I am deeply grateful to all those who have contributed to the creation of this book. This endeavor would not have been possible without the support, guidance, and encouragement of many individuals.

First and foremost, I would like to express my heartfelt appreciation to my family for their unwavering support and understanding throughout the writing process. To my wife, Selda, and my children, Michelle, Chris, Ceylin, and Mayra, your patience and encouragement have been my anchor. I am also thankful for the love and support from my extended family: Derek, Meltem, Nilos, Ken, and my sisters Robin, Kelly, and Lynn.

I am indebted to the organizations that have been pivotal in my professional development and the success of this book: Hallmark University, ThriveDX, Cybrary, and the LinkedIn subscribers to the Cyberspear newsletter that I operate.

I extend my gratitude to the following individuals for their insights, support, and encouragement:

Griffin Weaver, Seth Jaffe, Kurt Lubelan, Kim Kemp, Rob Fisher, Don Wuebben, Amy Read, Derek Burkes, Wendell Ladd, Wil Bennet, Ken Hamilton, Clarke Cummings, Gary McAlum, Amil Navarro, Joe Dubbs, Brennan Holland, Ann Kurtz, Allison Carrillo, Brady Justice, Jeff Spaeth, Kristyn Lette, Chinho Ko, Subash Poudyal, PhD, Kul Subedi, PhD, Jim Huseman, Gordon Bjorman, Dr. Angela Dogan, Jerry Smith, Dr. Patrick Woods, Jason Raab, Ankit Kalra, Leead Negri, Kesha Lindbergdashwork, Michael Castillo, Kelley Dadah, Luis Guillermo, Christopher Hickernell, Janice Pryor, Brett Wahlin, James Azar, Michelle Nasser, Luis Gutierrez, Marco Bart.

Your contributions have been invaluable, and I am immensely grateful for your support.

1

Introduction

Understanding the National Institute of Standards and Technology (NIST) Cybersecurity Framework (CSF) can be daunting for many. Its complexity often appears cryptic, bordering on magical for some, yet it is neither. The essence of this book is to demystify the NIST CSF, making it accessible and actionable for businesses of all sizes. It is crafted to serve as a bridge between the high-level guidelines of the NIST CSF and the practical needs of organizations striving to enhance their cybersecurity posture.

This guide is designed for a broad audience, from small businesses to large corporations. The scalability and adaptability of the NIST CSF are among its core strengths, allowing organizations to implement cybersecurity practices that align closely with their specific operational needs and risk profiles. This book will give readers insights into customizing and applying the Framework to best serve their unique circumstances.

Flexibility is a central theme throughout this guide. The cybersecurity controls and recommendations provided herein are meant to be adapted and modified to fit different organizations' specific needs and contexts. This approach encourages readers to think critically about how each control can be tailored to support their cybersecurity goals, reinforcing that there is no one-size-fits-all solution in cybersecurity.

The structure of this book mirrors the practical, hands-on approach to cybersecurity management. Each section within the chapters is designed to introduce one to three controls, accompanied by four to six actionable recommendations. These are crafted to provide readers with clear, actionable steps to improve their cybersecurity measures, making the content directly applicable to their daily operations.

It is important to note that the NIST CSF does not maintain a continuous numbering system for its categories and subcategories; some numbers are intentionally skipped due to their deprecation. This should not be seen as an error or oversight but as an intentional aspect of the Framework's design. The missing numbers highlight the Framework's evolution and adaptability to the changing cybersecurity landscape.

This book is intended to be a practical guide, not an exhaustive academic exploration of the NIST CSF. The aim is to equip readers with the knowledge and tools they need to effectively implement and benefit from the Framework rather than to showcase the historical development of the NIST standards or the technical prowess of its authors.

Adherence to any framework, including the NIST CSF, should not be rigid. This guide emphasizes the importance of tailoring the Framework to meet your organization's needs and circumstances. The goal is to use the NIST CSF as a foundation to build an effective and flexible cybersecurity strategy.

A Comprehensive Guide to the NIST Cybersecurity Framework 2.0: Strategies, Implementation, and Best Practice, First Edition. Jason Edwards.
© 2025 John Wiley & Sons Ltd. Published 2025 by John Wiley & Sons Ltd.

Finally, this book advocates for the principle of reiteration rather than duplication. It is designed to serve as a reference guide that can be consulted repeatedly, offering practical advice and insights rather than a novel narrative. The concise and direct nature of the recommendations and controls presented is intentional, aiming to provide clear guidance without unnecessary complexity. This approach underscores the book's overarching message: the value of pragmatic, adaptable cybersecurity practices that can be tailored to each organization's unique needs.

Why This Book?

The cybersecurity threat landscape is evolving at an unprecedented rate, with new vulnerabilities and attack vectors emerging continuously. This dynamic environment poses a significant challenge for organizations across all sectors, necessitating a robust approach to cybersecurity defense mechanisms. This book is a response to these challenges, offering a comprehensive exploration of the NIST Cybersecurity Framework as a pivotal tool in the fight against cyber threats. It aims to provide readers with a deep understanding of the Framework's structure and how it can be effectively implemented to safeguard digital assets.

Bridging knowledge gaps in cybersecurity practices is a core mission of this guide. Many organizations are disadvantaged, not due to a lack of effort or investment in cybersecurity but because of a fundamental misunderstanding or misapplication of critical principles and practices. This book seeks to rectify this issue by delivering clear, concise, and actionable information on the NIST Cybersecurity Framework, ensuring that readers can understand and apply these practices in a way that significantly boosts their cybersecurity posture.

Frameworks play a crucial role in establishing a comprehensive cybersecurity strategy. They provide a structured and systematic approach to managing and mitigating cyber risks. Among these, the NIST Cybersecurity Framework stands out for its flexibility, comprehensiveness, and adaptability to organizations of various sizes and industries. This book highlights the importance of such frameworks in responding to cyber incidents and fostering a culture of proactive cyber hygiene that can significantly reduce the likelihood and impact of cyberattacks.

This book distinguishes itself through its unique approach to discussing the NIST Cybersecurity Framework. Unlike other texts offering a superficial overview or a highly technical analysis, this guide balances depth and accessibility. It is designed to be of value to cybersecurity professionals while remaining approachable for those new to the field. The book's contributions lie in its practical recommendations, detailed breakdown of the Framework's components, and emphasis on real-world applicability.

Addressing the growing complexity of cybersecurity threats requires more than just theoretical knowledge; it demands a practical understanding of how frameworks like the NIST Cybersecurity Framework can be leveraged in real-world scenarios. This book bridges theory and practice, providing readers with the insights needed to transform their cybersecurity strategies from reactive to proactive.

By focusing on the NIST Cybersecurity Framework, this book fills a critical need for authoritative guidance on one of the world's most respected and widely adopted cybersecurity frameworks. The Framework's emphasis on continuous improvement, risk management, and adaptability makes it an invaluable tool for organizations seeking to navigate the complexities of today's cybersecurity landscape.

The practical applications of the NIST Cybersecurity Framework are vast, extending beyond mere compliance to drive meaningful improvements in cybersecurity practices. This book delves into these applications, offering readers a clear path to understanding the Framework and using it to make tangible improvements to their cybersecurity defenses. It underscores the Framework's role as a set of guidelines and a catalyst for change in how organizations approach cybersecurity.

Finally, this book is an invitation to view cybersecurity through the lens of continuous improvement and strategic alignment. The NIST Cybersecurity Framework is not a static set of recommendations but a living document that evolves in response to new threats and technological advancements. By embracing the Framework's principles, organizations can enhance their cybersecurity posture and align their cybersecurity strategies with their broader business objectives, ensuring that cybersecurity is not just a technical necessity but a strategic asset.

Overview of Cybersecurity Challenges

The cybersecurity landscape is increasingly complex, characterized by various sophisticated threats that target every aspect of the digital environment. From advanced persistent threats to ransomware, phishing, and zero-day exploits, the variety and sophistication of these cyber threats pose a constant challenge to organizations. This complexity is further amplified by the rapid pace of technological advancements and the expanding digital footprint of businesses, making securing digital assets an ever-evolving battle.

Businesses and government sectors share common vulnerabilities that cyberattackers frequently exploit. These vulnerabilities often stem from outdated systems, unpatched software, insufficient network security practices, and employees' lack of cybersecurity awareness. The interconnected nature of digital systems means that a breach in one area can quickly escalate, affecting critical infrastructure, data integrity, and the confidentiality of sensitive information. This shared risk landscape underscores the need for comprehensive and adaptable cybersecurity strategies to protect against known and emerging threats.

The cost of cyber incidents to organizations extends beyond the immediate financial losses associated with data breaches or ransom payments. The long-term repercussions include damage to reputation, loss of customer trust, legal penalties, and the significant resources required for incident response and recovery. The intangible costs, such as the impact on employee morale and the loss of competitive advantage, can also be substantial. These factors together highlight the critical importance of implementing robust cybersecurity measures to mitigate the risk of cyber incidents.

Failure to implement a comprehensive cybersecurity framework like the NIST Cybersecurity Framework can severely affect organizations. Without the structured approach and best practices provided by such frameworks, organizations are often ill-prepared to identify, protect against, detect, respond to, and recover from cybersecurity incidents. This lack of preparedness can lead to increased vulnerability to cyberattacks, potentially resulting in devastating data breaches, financial losses, and erosion of stakeholder trust.

A cohesive cybersecurity strategy can impede an organization's ability to manage and respond to cyber incidents effectively. Without clear guidelines and protocols, the response to cybersecurity threats can be slow and disorganized, allowing attackers more time to exploit vulnerabilities and cause damage. This situation highlights the value of the NIST Cybersecurity Framework's structured approach to managing cyber risks and enhancing an organization's cybersecurity posture.

Organizations that neglect to implement the NIST Cybersecurity Framework may also find themselves at a competitive disadvantage. In an era where cybersecurity is a critical concern for customers and business partners, demonstrating a commitment to cybersecurity through adherence to recognized frameworks can be a significant competitive edge. Moreover, regulatory compliance requirements are increasingly mandating the adoption of such frameworks, and failure to comply can result in legal and financial penalties.

The strategic alignment of cybersecurity practices with business objectives is another critical aspect that can be compromised without the guidance of the NIST Cybersecurity Framework. Cybersecurity is not just a technical issue but a business imperative that affects every aspect of an organization's operations. A framework-based approach ensures that cybersecurity measures are effective and aligned with the organization's goals, risk tolerance, and operational requirements.

In conclusion, the challenges the modern cybersecurity landscape poses are significant but not insurmountable. The NIST Cybersecurity Framework offers an adaptable strategic roadmap for organizations to enhance their cybersecurity defenses, manage cyber risks, and maintain resilience against cyber threats. The consequences of not implementing such a framework can be severe, affecting an organization's financial health, reputation, and operational capability. Therefore, adopting and adapting the NIST Cybersecurity Framework is essential for any organization committed to safeguarding its digital assets and maintaining trust in an increasingly digital world.

2

Understanding the NIST Cybersecurity Framework 2.0

The National Institute of Standards and Technology (NIST) Cybersecurity Framework represents a paradigm shift in national and organizational approaches to cybersecurity. Initially conceptualized to improve the cybersecurity postures of critical infrastructure entities, the Framework has universally been recognized for its adaptability and effectiveness across various sectors. This adaptability is rooted in its design, which allows organizations of different sizes, from small local businesses to global enterprises, to apply the Framework tailored to their unique risk environments and cybersecurity challenges.

Understanding the historical context of the NIST Cybersecurity Framework is essential. It was developed in response to increasing and evolving cyber threats facing organizations and national infrastructure. The Framework's purpose was to create a common language and systematic methodology for managing cybersecurity risk, emphasizing the importance of cybersecurity to the national and economic security of the United States and beyond. This emphasis on a unified approach has facilitated better communication and understanding among stakeholders across different sectors.

The NIST Cybersecurity Framework is built on key principles and objectives to enhance organizations' ability to manage and reduce cybersecurity risks. These principles encourage a holistic view of cybersecurity, integrating it into the organization's overall risk management processes. The Framework's objectives serve as guideposts for organizations, aiming to transform reactive, disjointed responses into a proactive, strategic, and coherent cybersecurity posture.

The Framework's scope and applicability are deliberately broad, designed to be relevant across different sectors, and adaptable to varying organizational sizes and complexities. This universal applicability ensures that the Framework can serve as a foundational tool for entities in industries diverse in energy, healthcare, finance, and education, irrespective of their current cybersecurity maturity.

A critical aspect of the NIST Cybersecurity Framework is its ability to complement other standards and practices. This attribute is precious for organizations that have invested in cybersecurity measures or are bound by sector-specific regulations. By providing a high-level, strategic view of cybersecurity risk management, the Framework can help harmonize existing efforts, making them more effective and cohesive.

The Framework encourages organizations to adopt a continuous improvement approach to cybersecurity, emphasizing the dynamic nature of cyber threats and the need for ongoing assessment and adaptation of cybersecurity practices. This approach helps organizations respond to incidents and anticipate and mitigate potential threats before they can impact the organization.

A Comprehensive Guide to the NIST Cybersecurity Framework 2.0: Strategies, Implementation, and Best Practice,
First Edition. Jason Edwards.
© 2025 John Wiley & Sons Ltd. Published 2025 by John Wiley & Sons Ltd.

Integration of the NIST Cybersecurity Framework into organizational processes can dramatically enhance not only the security but also the resilience of organizations. By providing a structured yet flexible approach, the Framework helps entities develop a cybersecurity strategy that aligns with business objectives, addresses relevant risks, and leverages existing practices and investments.

Finally, the widespread adoption and endorsement of the NIST Cybersecurity Framework underscore its significance and utility in improving national and organizational cybersecurity postures. By fostering a shared understanding and approach to managing cybersecurity risk, the Framework has become an essential resource for organizations seeking to navigate the complex and evolving cybersecurity landscape.

Fundamental Changes from Version 1.X

The evolution from NIST CSF 1.X to 2.0 represents a significant leap forward in addressing the complexities of modern cybersecurity landscapes. This update reflects the collective insights and experiences of a broad range of stakeholders, incorporating the latest best practices and addressing emerging threats. The revisions ensure that the Framework remains a cutting-edge tool for managing cybersecurity risk, adaptable to the rapid changes in the cyber domain.

One of the most significant updates in Version 2.0 is refining the core functions and categories. This addressed the evolving nature of cyber threats and the increasing sophistication of cyberattackers. The updates enhance the clarity and applicability of the Framework, ensuring that it aligns with current cybersecurity challenges and technologies. These changes encourage a more intuitive and practical application of the Framework across different organizational contexts.

The modifications to the implementation tiers are particularly noteworthy. These changes are designed to guide organizations in evolving their cybersecurity practices from reactive to proactive stances. The revised tiers help organizations better understand their cybersecurity maturity levels and provide a more straightforward path for advancement, emphasizing risk management and resilience.

Enhancements in communication and information sharing reflect the growing recognition of the importance of collaboration in cybersecurity. The updated Framework underscores the need for effective communication among internal and external stakeholders, facilitating a more coordinated and agile response to cybersecurity threats. This focus on information sharing is critical in an era where threats can evolve rapidly and spread across organizational and national boundaries.

The introduction of new recommendations for supply chain risk management is a response to the increasing interconnectedness of organizations and the recognition that a chain is only as strong as its weakest link. The updates guide assessing and managing the cybersecurity risks associated with suppliers and partners, an area highlighted by several high-profile breaches.

The changes from Version 1.X to 2.0 also include a more explicit emphasis on privacy and civil liberties, reflecting growing concerns about the balance between security and individual rights. This inclusion demonstrates the Framework's adaptability to broader societal and regulatory changes and underscores the importance of considering these factors in cybersecurity practices.

The revision process also introduced new elements to enhance organizations' resilience to cyber incidents. This resilience is not merely about preventing breaches but ensuring that an organization can operate effectively during and after a cyber event. The updated Framework emphasizes not just the technical aspects of cybersecurity but also the need to incorporate business continuity and recovery planning into the overall cybersecurity strategy.

Furthermore, the updated NIST CSF encourages a more integrated cybersecurity and business strategy approach. Organizations can ensure that cybersecurity measures contribute to overall business performance and value creation by aligning cybersecurity objectives with business goals. This alignment is critical in today's environment, where cybersecurity is not just an IT issue but a strategic business concern.

Lastly, the transition to Version 2.0 highlights the importance of adaptability and customization in cybersecurity. The updated Framework provides a more nuanced approach to applying its principles, recognizing that each organization's cybersecurity needs and challenges are unique. It emphasizes the importance of tailoring the Framework to fit an organization's specific context and risk profile, thereby making it a more effective tool for managing cybersecurity risks.

Core Components of the Framework

The NIST Cybersecurity Framework is structured around core components that offer a comprehensive approach to managing cybersecurity risk. These components are Functions, Categories, and Subcategories, which provide a framework for organizing and sustaining an organization's cybersecurity activities. Understanding and implementing these core components is fundamental to leveraging the Framework effectively.

The Functions provide a high-level organizational view of the lifecycle of managing cybersecurity risks. They encompass Identify, Protect, Detect, Respond, and Recover, creating a strategic foundation for developing a comprehensive cybersecurity program. This structure supports a continuous feedback loop for ongoing improvement, ensuring that cybersecurity measures evolve in line with the changing threat landscape and business requirements.

Categories and Subcategories further break down these Functions into more specific objectives, offering a detailed and actionable approach to achieving the broader outcomes defined by the Functions. Each Category addresses a particular aspect of cybersecurity, such as Asset Management or Access Control. At the same time, Subcategories provide specific targets for each Category, such as establishing data classifications or implementing least privilege principles.

Informative References are another crucial component, offering guidance and resources for achieving the objectives in the Subcategories. These references include industry standards, guidelines, and best practices, providing a rich repository of information to help organizations implement the Framework effectively. They serve as a bridge between the high-level guidance provided by the Framework and the specific actions needed to implement it.

Customization is a fundamental feature of the NIST Cybersecurity Framework, allowing it to be adapted to the diverse needs and circumstances of different organizations. The Framework can be tailored to various sectors, risk environments, and business models, making it a versatile tool applicable to multiple entities. This adaptability is crucial for ensuring the Framework remains relevant and practical across different contexts and industries.

The Framework's integration with existing compliance requirements demonstrates its flexibility and utility. Organizations can align their NIST CSF implementation with other regulatory and industry-specific requirements, streamlining compliance efforts and reducing redundancy. This integration ensures that cybersecurity measures are not just about meeting regulatory obligations but are embedded into the organization's overall risk management strategy.

Overall, the core components of the NIST Cybersecurity Framework form a cohesive and comprehensive approach to cybersecurity. By understanding and applying these components, organizations can develop a robust cybersecurity program that protects against threats, supports their business objectives, and fosters resilience in the face of cyber incidents.

The Functions: Govern, Identify, Protect, Detect, Respond, and Recover

The Govern function is essential for establishing and maintaining a framework for managing cybersecurity risk aligned with organizational strategies and objectives. It involves the development of policies, processes, and standards that guide the organization's cybersecurity efforts, ensuring they are integrated with overall governance structures. This function emphasizes the importance of senior leadership commitment and oversight in fostering a culture of cybersecurity throughout the organization.

Effective governance requires clear communication, defined roles and responsibilities, and regular evaluation of cybersecurity policies and practices. It also involves considering cybersecurity in the context of other business risks, ensuring it is an integral part of the organization's overall risk management framework. This holistic approach helps ensure that cybersecurity investments are aligned with business priorities and effectively mitigate risks.

The Identify function is foundational to effective cybersecurity management. It involves understanding the organization's business context, resources, and cybersecurity risks. This function is critical for identifying the assets, systems, and data that need protection and understanding the threat landscape and the organization's vulnerabilities.

Asset management, risk assessment, and risk management strategies are core components of the Identify function. They enable organizations to prioritize their cybersecurity efforts based on specific risks, vulnerabilities, and business imperatives. This targeted approach ensures that resources are allocated effectively and that the organization's cybersecurity measures are commensurate with the level of risk.

The Protect function is about implementing appropriate safeguards to ensure the delivery of critical services. This includes measures to control access to assets, protect data integrity and confidentiality, and maintain secure environments for information processing. The Protect function covers a broad range of activities, from identity management and access control to data encryption and maintenance of security technologies. It is about creating a barrier against threats while ensuring business operations can continue unimpeded. This function also involves employee training and awareness, as human factors are crucial in maintaining a secure environment.

Effective protection strategies require a layered approach, combining physical, technical, and administrative measures. This multifaceted strategy ensures that if one line of defense fails, others are in place to mitigate the risk. Regular updates and patches to security systems and continuous monitoring for anomalies are essential practices within the Protect function to keep defenses robust and responsive to emerging threats.

The Detect function is critical for promptly identifying the occurrence of a cybersecurity event. This function is rooted in the understanding that no defense can be foolproof. Therefore, continuous monitoring and detection processes are essential to identify and mitigate threats as soon as they emerge. This involves deploying advanced threat detection technologies, regular system, and network analysis, and establishing baseline behaviors for anomaly detection.

Timeliness and accuracy are key in the Detect function, as early detection can significantly reduce the impact of a cyber incident. Organizations need to establish and maintain detection processes that are continually refined and updated in response to evolving threats and organizational changes. Employee vigilance and reporting mechanisms also play a vital role in effectively operating this function.

The Respond function outlines how an organization reacts to detected cybersecurity incidents. The focus here is on containing the impact, eradicating the threat, and recovering normal operations as quickly as possible. This function requires a structured and planned approach, typically outlined in an incident response plan, which defines roles, communication protocols, and procedures for analysis and mitigation.

Effective response strategies are characterized by coordination and agility. They require the ability to assess the severity of an incident quickly, communicate effectively within the organization and with external partners, and take decisive action to limit damage. Continuous learning is also crucial, as post-incident analysis provides valuable insights that can strengthen future response efforts and overall cybersecurity posture.

Finally, the Recover function is about restoring services and capabilities that were impaired due to a cybersecurity incident. This involves technical recovery and restoring trust and confidence among customers, stakeholders, and employees. Recovery planning is intertwined with business continuity planning, ensuring that the organization can return to "business as usual" in the shortest possible time and with minimal disruption.

The Recover function emphasizes the importance of resilience and improvement. After a cyber incident, organizations should focus on immediate recovery, identifying lessons learned, and integrating them into future security strategies. This continuous improvement cycle helps organizations build more robust defenses, enhance resilience, and better prepare for future challenges in the cybersecurity landscape.

CSF Organizational Profiles

Organizational Profiles within the NIST Cybersecurity Framework (CSF) epitomize tailored approaches that embody an organization's specific cybersecurity needs, objectives, and requirements. These profiles allow entities to delineate and understand their unique cybersecurity landscape, encompassing current practices, desired outcomes, and strategies for managing cyber risk. The purpose of these profiles is to provide a structured and standardized approach to cybersecurity that is aligned with the organization's business goals and risk management priorities.

Developing an Organizational Profile entails comprehensively evaluating the organization's cybersecurity practices against the CSF's standards. This process involves gathering input from various organizational stakeholders to ensure a holistic view of cybersecurity needs and priorities. By cataloging existing security measures and comparing them against the CSF's guidelines, organizations can identify gaps in their cybersecurity defenses and areas where existing practices can be strengthened or optimized.

Utilizing a Profile effectively requires regular review and updates to ensure it remains aligned with the organization's changing business environment, risk appetite, and cybersecurity landscape. It is a dynamic tool that guides strategic decision-making in cybersecurity investments and initiatives. Regularly updated Profiles can help organizations maintain resilience against evolving threats and adapt to new technologies and business practices.

Comparing Current State vs. Target State Profiles is critical in strategic cybersecurity planning. This comparison provides a clear roadmap for transitioning from the present cybersecurity maturity level to a more robust and resilient state. By identifying specific actions and priorities, organizations can allocate resources more effectively, focusing on initiatives that offer the highest return on investment regarding risk reduction and operational impact.

Case studies and examples of effective Organizational Profiles serve as invaluable resources for understanding how to implement the CSF successfully. These real-world scenarios can demonstrate how similar organizations have navigated their cybersecurity challenges, providing insights into effective practices, common pitfalls, and practical strategies for achieving cybersecurity objectives. By learning from the experiences of others, organizations can avoid common mistakes and adopt proven approaches to enhance their cybersecurity posture.

CSF Tiers

The CSF Tiers reflect the maturity and sophistication of an organization's cybersecurity risk management processes. They are designed to help organizations contextualize their cybersecurity practices within the Framework of their overall risk management and business operations. The Tiers—Partial (Tier 1), Risk-Informed (Tier 2), Repeatable (Tier 3), and Adaptive (Tier 4)—offer a continuum of approaches for managing cybersecurity risk, from informal and reactive practices to agile and risk-driven strategies.

Selecting the appropriate Tier involves carefully analyzing the organization's business context, risk profile, regulatory requirements, and cybersecurity goals. This decision should balance the potential impact of cybersecurity risks against the costs and efforts required to achieve higher maturity levels. Organizations should consider their current cybersecurity capabilities, strategic objectives, and potential benefits of advancing to a higher tier.

Progressing from one Tier to another requires a methodical approach encompassing planning, implementation, evaluation, and refinement. This progression should be driven by a continuous improvement mindset, with regular assessments to gauge progress and identify improvement areas. As organizations evolve, they should seek to integrate cybersecurity risk management more deeply into their business processes and foster a culture of security awareness and collaboration across all levels of the organization.

However, organizations face several challenges and considerations when assessing and implementing Tiers. Resource limitations, both in terms of budget and skilled personnel, can constrain an organization's ability to advance its cybersecurity maturity. Additionally, the need to balance cybersecurity initiatives with other business priorities can lead to difficult trade-offs. Organizational culture and attitudes toward risk and security also play a significant role in the effectiveness of Tier implementation.

Despite these challenges, advancing through the CSF Tiers can significantly enhance an organization's ability to manage cybersecurity risks effectively. Organizations can ensure that their cybersecurity efforts are practical and sustainable by aligning their cybersecurity practices with their business needs and risk appetite. The Tiers provide a framework for continuous improvement, enabling organizations to respond more adeptly to the evolving cybersecurity landscape while supporting their business objectives.

3

Cybersecurity Controls

In writing this book, my primary aim was to ensure that every reader, irrespective of the size of their organization or the complexity of their cybersecurity needs, finds actionable insights to enhance their digital security posture. To achieve this, I have meticulously detailed the NIST Special Publication 800-53 controls in the appendix for those navigating the more intricate aspects of cybersecurity compliance. However, recognizing the unique challenges faced by smaller enterprises and those in less complex compliance environments, I have also introduced what I term Simplified Security Controls (SSC). These SSCs embody my conviction that effective cybersecurity measures should be within reach for everyone, tailored to make cybersecurity more accessible and manageable without compromising comprehensiveness.

The integration of SSCs alongside the rigorous standards outlined in NIST SP 800-53 underscores the book's holistic approach to cybersecurity. Designed with smaller businesses in mind, the SSCs offer a streamlined method for implementing robust security measures, reflecting my dedication to democratizing cybersecurity. This dual approach ensures that the book serves as a comprehensive guide, equipping readers with practical knowledge and tools that cater to both complex and straightforward cybersecurity needs.

By focusing on both the broader NIST framework and the SSCs, the book emphasizes clarity, applicability, and simplification without losing sight of the need for thoroughness. This balance ensures that readers from all backgrounds, whether educators, students, professionals in nontechnical fields, or small business owners, can navigate the cybersecurity landscape more effectively. The book encourages readers to actively engage with the content, adapt the discussed controls to their specific contexts, and take proactive steps toward enhancing their cybersecurity measures.

Each chapter is crafted to guide readers through the customization of cybersecurity controls to address their unique challenges, resources, and objectives. This tailored approach not only increases the efficacy of the controls but also ensures they offer meaningful and sustainable improvements to an organization's cybersecurity defenses. Furthermore, the book delineates the different categories of controls—preventive, detective, and corrective—integrating them within a comprehensive defense mechanism that aligns with the broader NIST Cybersecurity Framework.

In summary, this book offers a balanced blend of rigorous standards and tailored simplifications at its core, aiming to empower readers with a comprehensive yet practical toolkit for fortifying their organizations against digital threats. Through the inclusion of both NIST SP 800-53 controls and Simplified Security Controls, it represents a commitment to making cybersecurity accessible to all, thereby enabling businesses of any scale to navigate the evolving landscape of cyber threats with confidence and clarity.

A Comprehensive Guide to the NIST Cybersecurity Framework 2.0: Strategies, Implementation, and Best Practice, First Edition. Jason Edwards.
© 2025 John Wiley & Sons Ltd. Published 2025 by John Wiley & Sons Ltd.

Delving Deeper into Cybersecurity Measures

Grasping the essence of cybersecurity controls begins with acknowledging their varied nature. This comprehensive approach is paramount as it allows organizations to consider all possible angles of defense against cyber threats. Every policy, procedure, piece of hardware, or software that contributes to an organization's security posture falls under the umbrella of cybersecurity controls. This holistic view is crucial, as it ensures a thorough and all-encompassing defense mechanism, integrating various aspects of cybersecurity to form a unified barrier against potential breaches and attacks.

The line between specific and general cybersecurity controls is both distinct and nuanced. Specific controls are meticulously designed to counteract targeted threats and secure particular components of an IT infrastructure, such as firewalls for network security or encryption for data protection. On the other hand, general controls, such as user awareness training and regular software updates, serve as the backbone of cybersecurity defense applicable across different scenarios and platforms. Striking the right balance between these two types of controls is essential for creating a resilient and adaptive security strategy to address existing and emerging cyber threats.

Understanding cybersecurity control involves recognizing its fundamental characteristics: relevance, specificity, enforceability, and measurability. These controls must directly contribute to preventing, detecting, and responding to cyber incidents. They should be clearly defined, with specific objectives and performance metrics allowing effective implementation and evaluation. This understanding forms the bedrock upon which effective cybersecurity practices are built, ensuring that every control implemented serves a purpose and contributes to the organization's overall security.

Cybersecurity controls are typically segmented into three primary categories: preventive, detective, and corrective. This categorization facilitates a structured approach to cyber defense, allowing for the development of a comprehensive strategy encompassing all aspects of incident management. Preventive controls aim to avert unauthorized access or alterations; detective controls strive to identify and signal ongoing or completed security breaches; and corrective controls focus on restoring systems and mitigating damage following an incident. A well-rounded cybersecurity strategy employs a balanced mix of these controls tailored to the specific risks and requirements of the organization.

Integrating cybersecurity controls into the broader organizational Framework is a task that requires meticulous planning and alignment with business objectives and regulatory requirements. This ensures that cybersecurity measures are not merely add-ons but integral to the organization's overall operations and strategy. Effective integration involves regular communication between IT and other departments, ensuring that cybersecurity is a shared responsibility in the organization's culture.

The evolution of cybersecurity controls is a testament to the dynamic nature of the digital landscape. Initially focused on physical security and essential password protection, cybersecurity has expanded to include advanced measures such as biometrics, machine learning-based anomaly detection, and blockchain technologies. Understanding this evolution helps cybersecurity professionals appreciate the complexities of modern cyber defense and inspires innovation in developing new controls to address the ever-changing threat landscape.

A comprehensive approach to cybersecurity involves recognizing the various measures that constitute controls and understanding their collective impact. It's about seeing the forest for the trees—appreciating how individual measures form a robust defense system. This perspective is vital for ensuring that all potential security gaps are addressed and that the organization's cybersecurity measures are as comprehensive and effective as possible.

In summary, the distinction between specialized and general controls is not merely academic; it is a practical consideration that impacts the effectiveness of an organization's cybersecurity strategy. By understanding this distinction and applying it to their cybersecurity planning, organizations can ensure they are adequately protected against broad-based and targeted cyber threats. Through careful definition, categorization, and integration of cybersecurity controls, organizations can build a robust framework that addresses current security challenges and is adaptable to future threats and technological advancements.

Comprehensive Assessment of Cybersecurity Safeguards

Understanding the rationale for testing cybersecurity controls extends beyond mere regulatory compliance; it is about ensuring that these measures function effectively in real time to thwart potential threats. This necessity stems from the ever-evolving nature of cyber threats and the complexity of modern IT environments. Verifying the effectiveness of controls is paramount, as theoretical effectiveness does not always translate into practical security. Regular testing provides assurance that controls are correctly implemented and performing as intended, offering the necessary defense against actual cyber incidents.

A detailed methodology overview introduces organizations to the spectrum of testing techniques available, each serving distinct purposes and suited to different controls and environments. From static code analysis and dynamic application testing to network penetration testing and social engineering campaigns, the choice of methodology should align with the organization's specific security needs and threat landscape. Understanding and selecting the appropriate testing techniques are crucial in developing an effective cybersecurity evaluation strategy.

Accessible strategies ensure cybersecurity testing is not just a prerogative of large enterprises with vast resources. Small- and medium-sized organizations can also implement basic yet effective testing methods. Routine actions, such as conducting regular security assessments, engaging in patch management exercises, and employing simple penetration testing tools, can significantly enhance an organization's security stance. These accessible strategies serve as a starting point for building a culture of security and continuous assessment within any organization.

Advanced tactics are crucial for organizations operating in complex or highly targeted environments. This includes sector-specific simulations, in-depth threat modeling, and custom scenario testing, which go beyond conventional methods to uncover subtle or sophisticated vulnerabilities. These advanced approaches require specialized knowledge and resources but are integral to understanding the nuanced threats organizations face in sensitive or critical sectors. Organizations can anticipate and mitigate complex attack vectors more effectively by adopting these sophisticated testing procedures.

The significance of documentation in the testing process cannot be overstressed. Comprehensive records serve multiple functions: they provide evidence for compliance audits, facilitate the identification of trends over time, support the troubleshooting of recurring issues, and aid in transferring knowledge within the organization. Proper documentation ensures that every test contributes to the organization's cumulative cybersecurity knowledge, turning individual assessments into a continuous cycle of learning and improvement.

Embracing continuous improvement through iterative testing and refining cybersecurity controls allows organizations to stay ahead of threats. This dynamic approach involves analyzing the outcomes of each test, understanding the implications of findings, and making informed adjustments to security measures. It's a proactive stance that ensures cybersecurity efforts always align with the current threat environment and organizational changes, fostering a resilient and adaptive security posture.

The emphasis on verifying control effectiveness is not just procedural; it reflects a more profound commitment to safeguarding digital assets against an ever-changing threat landscape. By conducting regular, thorough testing, organizations gain a clearer understanding of their security posture, enabling them to make data-driven decisions and allocate resources more effectively to areas of greatest need or vulnerability.

In wrapping up, the testing and evaluation of cybersecurity controls are indispensable components of a comprehensive cybersecurity strategy. This systematic and ongoing process involves a balanced mix of methodologies, from basic checks accessible to all organizations to advanced tactics designed for intricate systems. Coupled with meticulous documentation and a culture of continuous improvement, this rigorous approach ensures that cybersecurity measures are implemented and continuously adapted to meet the evolving demands of the digital world. Through persistent and detailed evaluation, organizations can fortify their defenses, ensuring they remain robust despite the dynamic threats characteristic of today's cybersecurity landscape.

4

Compliance and Implementation

Implementing the NIST Cybersecurity Framework (CSF) within an organization begins with an initial assessment. This foundational step thoroughly evaluates the organization's cybersecurity posture concerning the benchmarks set by NIST CSF 2.0. Companies must catalog their current cybersecurity policies, practices, and procedures, contrasting them with the comprehensive guidelines of the framework. The objective is to create a precise baseline understanding of where the organization currently stands from a cybersecurity standpoint.

Following the initial assessment, the next critical step is conducting a gap analysis. This process involves identifying the discrepancies between the organization's current cybersecurity measures and the requirements outlined in the NIST CSF. By pinpointing these gaps, an organization can clearly understand the areas that require improvement. The gap analysis highlights vulnerabilities and areas of noncompliance and helps set priorities for the subsequent steps in the implementation process.

Once the gap analysis is complete, developing an action plan is essential. This plan should detail the prioritized steps needed to bridge the identified gaps. It must include specific actions, responsible parties, timelines, and resource allocations. The action plan is a roadmap for aligning the organization's cybersecurity practices with the NIST CSF standards. Effective action plans are detailed and flexible, allowing for adjustments throughout the implementation process.

The execution phase is where plans become a reality. Implementing the changes, enhancements, and new practices outlined in the action plan is critical to improving cybersecurity defenses. This step may involve updating policies, deploying new technologies, or reconfiguring existing systems. Execution requires meticulous management and coordination to implement changes effectively and without disrupting business operations.

Training and awareness for all staff members are integral to the framework's adoption. Employees must understand the new processes, recognize the latest cybersecurity threats, and adapt to the expected behaviors to reinforce the framework's adoption. Regular training sessions, workshops, and communication campaigns can help embed cybersecurity awareness into the organizational culture. This ensures that all personnel are equipped to contribute positively to the organization's cybersecurity posture.

Review and adjustment are ongoing components of the NIST CSF implementation process. Cybersecurity is not a static field; as threats evolve and business objectives change, so must the organization's cybersecurity strategies. Regular reviews of the cybersecurity program, in light of the NIST CSF, help ensure it remains effective and aligned with the organization's needs. Based on feedback, audit results, and evolving threat landscapes, adjustments should be made.

A Comprehensive Guide to the NIST Cybersecurity Framework 2.0: Strategies, Implementation, and Best Practice, First Edition. Jason Edwards.
© 2025 John Wiley & Sons Ltd. Published 2025 by John Wiley & Sons Ltd.

The review process also feeds into the cycle of continuous improvement that the NIST CSF promotes. Organizations should continually use the insights from reviews to refine their cybersecurity practices. This involves revisiting the initial assessments, updating the gap analysis, and adjusting the action plan as necessary. By maintaining this cycle of assessment, analysis, planning, execution, and review, organizations can stay ahead of cybersecurity threats while aligning with the NIST CSF standards.

Adhering to the NIST CSF is not a one-time activity but an ongoing commitment to cybersecurity excellence. The steps for implementing the framework—initial assessment, gap analysis, action plan development, execution, training and awareness, review, and adjustment—are not linear but cyclical. This iterative process helps organizations comply with the framework and foster a culture of continuous cybersecurity improvement. By integrating these steps into their cybersecurity practices, organizations can enhance their defenses, mitigate risks, and better protect their critical assets.

Tailoring the Framework to Different Organizations

The NIST CSF is designed to be adaptable, allowing for customization to fit different organizations' unique needs and circumstances. One of the first steps in tailoring the framework involves considering industry specificity. Each sector faces distinct risks and regulatory requirements. For example, the financial industry may prioritize data privacy and transaction security, while a manufacturing firm might focus on safeguarding production processes. Organizations can ensure more targeted and effective cybersecurity measures by customizing the framework to address these sector-specific concerns.

Size considerations are critical in adapting the NIST CSF for individual organizations. The scale and resources of a business significantly impact its cybersecurity strategies. Small- and medium-sized enterprises (SMEs) might not have the same resources as large corporations but still require robust protection. Tailoring the framework for SMEs might involve focusing on the most critical cybersecurity practices that offer maximum impact for minimal investment. In contrast, larger organizations might implement a more comprehensive range of controls and processes.

Cultural adaptation is key to the effective adoption of the NIST CSF. Every organization's unique culture can significantly influence how cybersecurity policies are perceived and enacted. Tailoring the framework to fit this culture involves engaging with organizational stakeholders to ensure buy-in and compliance. For example, a company that values open communication and collaboration might adopt a more inclusive approach to cybersecurity training and awareness programs.

Aligning the framework with the organization's risk appetite is another important consideration. Each organization has different levels of risk tolerance, influenced by factors such as industry, market position, and financial health. Modifying the framework to reflect these tolerance levels ensures that cybersecurity measures are neither overly burdensome nor insufficiently rigorous. This alignment helps prioritize cybersecurity initiatives and allocate resources more effectively.

Integrating existing policies and procedures ensures that the NIST CSF complements and enhances current practices rather than conflicting with them. Organizations likely already have some cybersecurity measures in place. The framework should build on these foundations, filling gaps and reinforcing strengths rather than replacing established procedures wholesale. This approach promotes consistency and efficiency, using existing resources and efforts best.

Adopting a lifecycle approach to the framework acknowledges that organizations and the cybersecurity landscape are continually evolving. As a business grows, enters new markets, or adopts new technologies, its cybersecurity needs will change. Planning for the continuous adaptation of the NIST CSF ensures that cybersecurity measures remain effective and relevant over time. This involves regular reviews and updates to the cybersecurity program in response to new threats, business objectives, and technological advancements.

In summary, tailoring the NIST Cybersecurity Framework to different organizations requires careful consideration of industry-specific risks, organizational size, cultural factors, risk appetite, existing policies, and the evolving nature of the business and cybersecurity fields. Organizations can ensure a more effective and sustainable cybersecurity posture by customizing the framework to address these factors. This tailored approach helps meet regulatory and compliance requirements, supports strategic business goals, and protects against evolving cyber threats.

Compliance Considerations

Understanding the regulatory landscape is paramount for organizations aiming to comply with the NIST Cybersecurity Framework. Different industries and geographical locations come with varied legal and regulatory obligations that can significantly impact an organization's cybersecurity strategies. Leaders must identify and understand these specific obligations to ensure that their cybersecurity measures align with the NIST CSF and all relevant laws and regulations. This understanding forms the foundation for effective compliance, safeguarding the organization against legal ramifications, and enhancing its cybersecurity posture.

Documentation and evidence are critical to demonstrating compliance with the NIST Cybersecurity Framework and other relevant standards. Organizations must keep detailed records of their cybersecurity policies, procedures, and measures. This documentation should include risk assessments, action plans, incident response records, and training materials. By maintaining comprehensive records, an organization can provide tangible proof of its compliance efforts and cybersecurity initiatives, which is invaluable during audits, legal reviews, or after a security incident.

Continuous monitoring is essential for ongoing compliance assessment and reporting. Cybersecurity is not a static field; threats and regulatory requirements evolve. Establishing processes for continuous monitoring allows organizations to stay ahead of changes and ensure that their cybersecurity measures remain effective and compliant over time. This involves regular reviews of security controls, assessments of new threats, and updates to documentation and practices as necessary.

Stakeholder engagement is crucial in the realm of cybersecurity compliance. Involving legal, regulatory, and business stakeholders ensures compliance activities are well-informed and aligned with broader organizational goals. Legal teams can provide insights into regulatory requirements, while business stakeholders can offer perspectives on how cybersecurity measures impact operations. Collaborative engagement facilitates a more cohesive and effective approach to compliance, ensuring that all facets of the organization are considered.

Addressing noncompliance is an inevitable aspect of managing cybersecurity. When gaps in compliance are identified, organizations must develop and implement strategies to mitigate risks and address these gaps promptly. This might involve revising policies, enhancing security controls, or providing additional training to staff. Proactive management of noncompliance helps avoid legal or financial penalties and strengthens the organization's overall cybersecurity framework.

International considerations are increasingly important in the globalized business environment. Organizations operating across different jurisdictions face the complex task of complying with multiple cybersecurity regulations. Navigating this complexity requires a nuanced approach, understanding the legal frameworks of each jurisdiction, and tailoring cybersecurity measures to meet these varied requirements. International compliance demands ongoing attention and adaptation to ensure that global operations do not expose the organization to undue risk.

In the NIST Cybersecurity Framework context, compliance is not merely about adhering to rules but about embedding cybersecurity best practices into the organization's fabric. This approach ensures compliance contributes positively to the organization's overall security posture, rather than being seen as burdensome. By viewing compliance through the lens of strategic cybersecurity management, organizations can turn regulatory requirements into opportunities for enhancing their security measures and business operations.

Finally, an effective compliance program under the NIST Cybersecurity Framework extends beyond mere adherence to external regulations. It involves a commitment to ongoing improvement, stakeholder engagement, and a proactive stance toward cybersecurity challenges. Organizations that approach compliance as an integral part of their cybersecurity strategy can meet regulatory demands and foster a culture of security, resilience, and trust. This holistic approach to compliance underscores the importance of cybersecurity in the modern business landscape and aligns regulatory obligations with strategic business objectives.

Integrating with Other Standards and Frameworks

Integrating the NIST CSF with other standards and frameworks requires a detailed comparative analysis to identify synergies and conflicts. Organizations often operate within a complex regulatory and operational landscape where multiple cybersecurity frameworks and standards apply. Understanding how the NIST CSF aligns with, complements, or diverges from these other frameworks. This analysis should highlight areas of overlap where adherence to the NIST CSF may also fulfill other regulatory requirements and pinpoint areas where additional measures may be needed to address specific framework mandates.

Developing harmonization strategies is essential for organizations seeking to effectively reconcile and leverage different standards. Harmonization involves aligning practices and controls from various frameworks to avoid redundancy and ensure a cohesive cybersecurity strategy. This approach streamlines cybersecurity efforts and reduces compliance costs and complexity. By identifying commonalities between the NIST CSF and other frameworks, organizations can create a unified set of cybersecurity practices that meet multiple requirements efficiently.

Leveraging industry-specific best practices within the framework's structure is another vital integration aspect. Each industry faces unique cybersecurity challenges and risks, leading to the development of specialized best practices and standards. Integrating these practices with the NIST CSF can enhance the framework's effectiveness and relevance for specific sectors. Organizations should seek to understand their industry's cybersecurity norms and standards and incorporate these into their NIST CSF implementation.

Cross-reference mapping is a practical tool for integrating the NIST CSF with other frameworks. Creating a comprehensive map linking NIST CSF controls with those of other frameworks can help organizations understand how different requirements interrelate. This mapping can clarify which controls satisfy multiple requirements and where additional actions might be necessary. It serves as a visual aid and reference point for managing compliance with multiple cybersecurity frameworks simultaneously.

Custom integration plans are crucial for ensuring a cohesive and comprehensive approach to cybersecurity. Every organization's unique situation requires tailored strategies to integrate various standards and frameworks effectively. These plans should consider the organization's specific risk profile, operational needs, and existing cybersecurity practices. A custom integration plan can help ensure that the organization's cybersecurity approach is holistic and aligned with its overall business strategy.

Evolution and adaptation are key to maintaining alignment with changing standards and frameworks. The cybersecurity landscape is dynamic, with new threats and standards evolving to address these challenges. Organizations must keep abreast of these changes to ensure their cybersecurity practices remain current and effective. Regular reviews and updates to the organization's cybersecurity strategy, including how it integrates various standards and frameworks, are essential for staying ahead of potential threats and maintaining compliance.

In summary, integrating the NIST CSF with other standards and frameworks is a multifaceted process that requires careful analysis, strategic planning, and ongoing adaptation. Organizations can ensure a comprehensive and cohesive cybersecurity approach by identifying synergies and conflicts, developing harmonization strategies, leveraging industry-specific best practices, creating cross-reference mappings, and crafting custom integration plans. Additionally, by remaining adaptable to the evolution of cybersecurity standards, organizations can ensure that their cybersecurity measures remain effective and aligned with best practices and regulatory requirements.

Finally, integrating multiple cybersecurity frameworks and standards should not be viewed as an end but as part of a continuous process of improvement and adaptation. Organizations should strive to create a cybersecurity culture that values integration, harmonization, and constant learning. This approach ensures that cybersecurity efforts are aligned with current standards and resilient and adaptable to future changes in the cybersecurity landscape.

5

Organizational Context (GV.OC)

> Cybersecurity is the backbone of organizational integrity, not merely a technical requirement but a strategic asset that enhances trust, resilience, and mission fulfillment when aligned with our mission.

Integrating cybersecurity seamlessly into an organization's mission is essential, emphasizing the need to align cybersecurity strategies with the organization's objectives. This approach underscores the importance of engaging stakeholders actively in the cybersecurity process. Navigating the complex landscape of legal and regulatory requirements is critical for maintaining trust and effectiveness in cybersecurity efforts. Understanding the organization's dependencies on various services and capabilities is also highlighted, advocating for a proactive stance in protecting these key elements. Such insights position cybersecurity not merely as a technical necessity but as a strategic asset that is vital for the organization's integrity, resilience, and the fulfillment of its mission.

GV.OC-01: The Organizational Mission Is Understood and Informs Cybersecurity Risk Management

The symbiotic relationship between an organization's mission and cybersecurity strategies is foundational to understanding and implementing effective risk management. Cybersecurity is not merely a technical challenge but a strategic one intrinsically linked to the organizational mission. This section begins by breaking down the components of the mission statement relevant to cybersecurity. Each element of the mission statement, whether it pertains to serving a specific customer base, achieving excellence in a particular domain, or upholding specific values, has implications for cybersecurity. By dissecting the mission statement, organizations can identify areas where cybersecurity efforts must be concentrated.

Analysis of the cybersecurity implications for each component of the mission statement reveals potential vulnerabilities and opportunities. For instance, the mission statement's commitment to customer privacy necessitates stringent data protection measures. Strategies for embedding cybersecurity considerations into the mission's execution involve integrating cybersecurity goals with business objectives, ensuring that robust cybersecurity practices support each facet of the mission.

Detailed interpretation of the organizational mission statement through a cybersecurity lens further emphasizes the importance of alignment. By mapping organizational goals to specific

A Comprehensive Guide to the NIST Cybersecurity Framework 2.0: Strategies, Implementation, and Best Practice, First Edition. Jason Edwards.
© 2025 John Wiley & Sons Ltd. Published 2025 by John Wiley & Sons Ltd.

cybersecurity initiatives, organizations can ensure their cybersecurity efforts are protective and strategic. The prioritization of cybersecurity projects based on their impact on organizational objectives enables a targeted approach to risk management. Integrating cybersecurity milestones into business plans ensures that cybersecurity is continuously considered in operational and strategic decisions.

Leadership plays a critical role in integrating the mission into cybersecurity policy. Top management is responsible for championing cybersecurity and developing and endorsing policies that reflect the organizational mission. Leadership must also communicate the importance of cybersecurity to the organization's mission, ensuring that all employees understand their role in safeguarding the organization. This involves allocating resources and supporting cybersecurity initiatives, demonstrating the organization's commitment to security as a component of its mission.

Mission-centric risk assessment and prioritization are vital to effective cybersecurity planning. Techniques for identifying and assessing risks to the organizational mission allow for a focused approach to cybersecurity, where the most significant threats are addressed first. This involves integrating mission-centric risk assessments into the cybersecurity strategy, ensuring mission-critical risks are continuously monitored and reassessed.

Communicating the importance of the mission in cybersecurity to all employees is essential for fostering a culture of security. Developing communication plans and training modules that emphasize each employee's role in mission-centric cybersecurity helps engage and educate the workforce. Engagement strategies and feedback mechanisms enhance employee understanding and commitment to cybersecurity practices.

Incorporating mission considerations into cybersecurity training and awareness programs ensures that training is relevant and impactful. Customization of training programs to highlight mission-specific cybersecurity aspects, incorporation of mission-driven scenarios in awareness initiatives, and continuous evaluation of training content keep the workforce informed and prepared for emerging cybersecurity challenges.

Finally, continuous mission review to identify emerging cybersecurity needs ensures that the organization's cybersecurity strategies evolve with its mission. Regular review processes, identification of emerging threats, and mechanisms for updating cybersecurity strategies in response to mission evolution are crucial for maintaining resilience against evolving cyber threats. Feedback from various organizational levels plays a key role in this continuous review process, ensuring that the organization's cybersecurity strategies remain aligned with its mission and values.

Recommendations

- **Embed Cybersecurity in Mission Statements**: Organizations should explicitly incorporate cybersecurity considerations into their mission statements to underscore their commitment to protecting assets and stakeholder interests. This can foster a culture where security is perceived as integral to the organization's success.
- **Enhance Leadership Engagement**: Encourage leadership to participate actively in cybersecurity discussions and decision-making processes. Their involvement can drive home the importance of security as a fundamental component of the organizational mission, ensuring adequate resources and focus on cybersecurity initiatives.
- **Integrate Cybersecurity and Business Objectives**: Develop a framework for aligning cybersecurity initiatives with business goals, ensuring that security efforts directly support the mission. This can include setting security objectives in harmony with business strategies and evaluating cybersecurity projects through their contribution to the mission.

- **Implement Mission-Centric Security Training**: Tailor cybersecurity training programs to reflect the organization's mission, incorporating scenarios and examples that highlight the impact of cybersecurity on achieving mission objectives. This approach can enhance employee engagement and understanding of their role in protecting the organization.
- **Continuously Review and Adapt Cybersecurity Strategies**: Establish a regular review cycle for assessing the alignment of cybersecurity strategies with the evolving organizational mission. This should include mechanisms for rapidly adapting to new threats, technologies, and business priorities, ensuring that cybersecurity remains a dynamic and integral part of the mission's execution.

NIST 800-53 Controls

(Refer to our appendix for a comprehensive description of the NIST 800-53 Controls, essential for organizations adhering to strict regulatory standards.)

- PM-1

Simplified Security Controls (SSC)

This section outlines our simplified security controls (SSC), tailored for environments beyond the scope of NIST compliance. These controls offer streamlined, effective security solutions suitable for any organization seeking to enhance its cybersecurity defenses with clear, actionable measures.

Control 1: Cybersecurity Mission Integration Framework

Control Objective: To systematically incorporate cybersecurity considerations into the organizational mission and daily operations, ensuring that cybersecurity becomes an integral part of the organizational culture and mission execution.

Implementation Steps:

1.1 **Development of an Integration Framework**: Create a comprehensive framework that outlines how cybersecurity considerations will be integrated into the organizational mission and daily operations. This includes identifying key mission components and determining how cybersecurity supports each element.

1.2 **Stakeholder Engagement**: Engage with stakeholders from various departments (e.g., IT, HR, marketing, and operations) to understand their roles and responsibilities in mission execution and how cybersecurity impacts their activities.

1.3 **Framework Implementation and Monitoring**: Implement the integration framework across the organization, with regular monitoring and reporting mechanisms, to assess its effectiveness and identify areas for improvement.

Expected Outcome: The Cybersecurity Mission Integration Framework ensures that cybersecurity is seamlessly integrated into all aspects of the organizational mission and operations, fostering a culture of security awareness and resilience. This leads to more effective cybersecurity practices aligned with and supportive of the organizational goals, enhancing overall mission success.

Control 2: Mission-Aligned Cybersecurity Leadership Development Program

Control Objective: To cultivate a leadership team well-versed in cybersecurity principles and their importance to the organizational mission, ensuring that leaders can effectively champion cybersecurity initiatives and integrate them with business strategies.

Implementation Steps:

2.1 **Curriculum Development**: Develop a leadership development program focused on cybersecurity, covering topics such as the symbiotic relationship between cybersecurity and the organizational mission, leadership roles in cybersecurity policy development, and strategies for communicating the importance of cybersecurity.

2.2 **Program Delivery**: Deliver the program to current and emerging leaders through workshops, seminars, and executive coaching, ensuring that leaders at all levels have the knowledge and skills to advocate for and integrate cybersecurity into their operations.

2.3 **Evaluation and Continuous Improvement**: Regularly evaluate the effectiveness of the leadership development program in fostering a culture of cybersecurity awareness among leaders, making adjustments as necessary based on feedback and changing cybersecurity landscapes.

Expected Outcome: The Mission-Aligned Cybersecurity Leadership Development Program ensures that organizational leaders are prepared to effectively integrate cybersecurity considerations into their strategic decision-making and daily operations. This leadership competency in cybersecurity supports a proactive security posture and aligns cybersecurity initiatives with the organizational mission.

Control 3: Dynamic Cybersecurity Risk Management Process

Control Objective: To implement a flexible and responsive cybersecurity risk management process that prioritizes risks based on their potential impact on the organizational mission, ensuring timely and effective mitigation strategies.

Implementation Steps:

3.1 **Establishment of a Mission-Centric Risk Assessment Methodology**: Develop a risk assessment methodology that prioritizes cybersecurity risks based on their impact on the organizational mission, considering mission criticality, data sensitivity, and operational importance.

3.2 **Regular Risk Assessments and Prioritization**: Conduct regular risk assessments to identify and evaluate new and existing cybersecurity risks, prioritizing them based on the established methodology.

3.3 **Implementation of Mitigation Strategies and Continuous Monitoring**: Implement appropriate mitigation strategies for prioritized risks and establish a continuous monitoring program to identify and respond to evolving threats and vulnerabilities quickly.

Expected Outcome: The Dynamic Cybersecurity Risk Management Process ensures that cybersecurity risks are managed to align with the organization's mission priorities. By focusing on the most significant risks to the mission, the organization can allocate resources more effectively and enhance its overall security posture, thereby protecting mission-critical operations and assets.

GV.OC-02: Internal and External Stakeholders are Understood, and Their Needs and Expectations Regarding Cybersecurity Risk Management Are Understood and Considered

The influence of stakeholders on cybersecurity risk management is profound and multifaceted. Stakeholders, including employees, customers, partners, and regulatory bodies, each bring unique perspectives and requirements to the organization's cybersecurity landscape. Understanding these perspectives is crucial for developing a cybersecurity strategy that meets regulatory compliance and aligns with business objectives and stakeholder expectations. This alignment is challenging due to various stakeholder groups' diverse and sometimes conflicting needs.

Mapping and profiling internal and external stakeholders are foundational to reconciling these diverse expectations with an organization's cybersecurity strategies. Organizations can categorize stakeholders effectively by identifying them based on their influence and interest, developing detailed profiles, and highlighting each group's cybersecurity expectations and needs. Stakeholder maps and profiles are invaluable tools for guiding communication and engagement efforts, ensuring that strategies are developed with a comprehensive understanding of stakeholder perspectives.

Assessing stakeholder needs and cybersecurity expectations involves collecting and analyzing input from various stakeholder groups. This process is vital for integrating stakeholder feedback into cybersecurity risk assessments and policies. Prioritizing stakeholder needs in cybersecurity policy development ensures that the organization's cybersecurity measures are responsive to stakeholder concerns, thereby enhancing trust and cooperation. Continuous stakeholder engagement is essential for refining cybersecurity strategies to accommodate evolving expectations and emerging risks.

The strategic inclusion of stakeholder needs in cybersecurity policy formation is a delicate balance. It requires a process that incorporates diverse stakeholder inputs while maintaining the integrity and coherence of cybersecurity policies. Stakeholder consultation and feedback during policy drafting foster inclusivity and transparency, leading to policies comprehensively addressing key cybersecurity expectations across stakeholder groups.

Communication plays a pivotal role in aligning cybersecurity practices with stakeholder expectations. Developing targeted communication plans for different stakeholder groups, employing techniques to convey cybersecurity risks and measures effectively, and utilizing engagement strategies to build support are critical for ensuring that cybersecurity initiatives are well-received and supported. Feedback mechanisms are also crucial for gauging stakeholder satisfaction and adjusting communication approaches.

Another critical aspect of stakeholder management is educating stakeholders about their impact on cybersecurity risks. Initiatives informing stakeholders about their cybersecurity roles and tools and resources to identify and mitigate risks empower stakeholders to contribute positively to the organization's cybersecurity posture. Case studies and strategies for encouraging proactive participation are effective ways to underscore the importance of every stakeholder's actions.

Establishing feedback mechanisms for ongoing stakeholder engagement ensures that the organization remains responsive to stakeholder needs and expectations. Regularly incorporating stakeholder feedback into cybersecurity improvements fosters a culture of continuous learning and adaptation. Evaluating the effectiveness of these mechanisms and adapting engagement strategies based on stakeholder feedback is essential for maintaining robust and dynamic cybersecurity practices.

Finally, adapting cybersecurity strategies to accommodate stakeholder feedback is essential for responsive and resilient cybersecurity governance. Frameworks for integrating feedback into planning and case examples of strategy adjustments based on input highlight the dynamic nature of cybersecurity management. Regular reviews and updates of cybersecurity strategies, informed by stakeholder feedback, ensure that the organization's cybersecurity measures remain effective and aligned with stakeholder needs and objectives.

Recommendations

- **Develop Comprehensive Stakeholder Maps and Profiles**: Create detailed maps and profiles of internal and external stakeholders, focusing on their cybersecurity expectations and needs.

This will aid in developing targeted strategies that effectively address diverse stakeholder concerns.

- **Incorporate Stakeholder Feedback into Cybersecurity Policies**: Establish mechanisms for regular collection and analysis of stakeholder feedback. Integrate this feedback into formulating and continuously improving cybersecurity policies, ensuring policies remain relevant and effective in meeting stakeholder expectations.
- **Enhance Stakeholder Communication and Engagement**: Implement targeted communication and engagement strategies for different stakeholder groups. Use clear, jargon-free language to convey cybersecurity risks and initiatives, building support and fostering a shared commitment to cybersecurity.
- **Educate Stakeholders on Cybersecurity Impact**: Launch educational initiatives to increase stakeholder awareness of their cybersecurity roles. Provide resources and tools to help stakeholders identify and mitigate cybersecurity risks, emphasizing the collective responsibility for cybersecurity.
- **Establish Continuous Feedback and Adaptation Mechanisms**: Set up ongoing channels for stakeholder feedback and regularly review cybersecurity strategies in response to this input. Adapt cybersecurity policies and practices to reflect changing stakeholder needs and emerging cybersecurity trends, ensuring the organization's cybersecurity posture is resilient and dynamic.

NIST 800-53 Controls

(Refer to our appendix for a comprehensive description of the NIST 800-53 Controls, essential for organizations adhering to strict regulatory standards.)

- PM-09,PM-18,PM-30,SR-03,SR-05,SR-06,SR-08

Simplified Security Controls (SSC)

This section outlines our SSC, tailored for environments beyond the scope of NIST compliance. These controls offer streamlined, effective security solutions suitable for any organization seeking to enhance its cybersecurity defenses with clear, actionable measures.

Control 1: Stakeholder Engagement and Analysis Program

Control Objective: To systematically identify, engage, and understand the needs and expectations of both internal and external stakeholders regarding cybersecurity risk management, ensuring that cybersecurity strategies and policies are inclusive and reflective of stakeholder perspectives.

Implementation Steps:

1.1 **Identification and Categorization of Stakeholders**: Utilize techniques such as stakeholder mapping to identify and categorize stakeholders based on their level of influence, interest in cybersecurity, and potential impact on the cybersecurity strategy.

1.2 **Conduct Stakeholder Needs Assessment**: Through surveys, interviews, and workshops, collect detailed information on stakeholder expectations, needs, and perceptions related to cybersecurity risk management.

1.3 **Analysis and Integration of Stakeholder Feedback**: Analyze collected data to understand stakeholder concerns and expectations. Integrate these insights into the cybersecurity risk management process, tailoring strategies and policies to address key stakeholder needs.

Expected Outcome: Establishing a comprehensive Stakeholder Engagement and Analysis Program ensures that the organization's cybersecurity risk management strategy is developed with a clear understanding of stakeholder needs and expectations. This leads to more effective and inclusive cybersecurity policies, increased stakeholder buy-in, and enhanced protection against cybersecurity threats.

Control 2: Cybersecurity Policy Co-Development Initiative

Control Objective: To facilitate the inclusive development of cybersecurity policies by actively involving stakeholders in the drafting process, ensuring that policies are comprehensive, practical, and aligned with organizational and stakeholder needs.

Implementation Steps:

2.1 **Establishment of a Policy Co-Development Framework**: Create a framework that outlines the process for stakeholder involvement in policy development, including mechanisms for consultation, feedback, and collaborative drafting.

2.2 **Stakeholder Consultation and Drafting Sessions**: Organize sessions where stakeholders can contribute their insights and feedback on policy drafts, leveraging their diverse perspectives to enhance policy relevance and effectiveness.

2.3 **Policy Review and Finalization**: Incorporate stakeholder feedback into the policy drafts, conduct a final review to ensure all concerns are addressed, and then formalize the policies for implementation.

Expected Outcome: The Cybersecurity Policy Co-Development Initiative ensures that cybersecurity policies reflect organizational objectives and consider the needs and expectations of a broad spectrum of stakeholders. This collaborative approach promotes greater policy adherence, reduces the risk of oversight, and strengthens the organization's cybersecurity posture.

Control 3: Continuous Stakeholder Feedback and Cybersecurity Adaptation Process

Control Objective: To establish a dynamic process for collecting and incorporating stakeholder feedback into ongoing cybersecurity strategy and policy adaptation, ensuring that cybersecurity measures remain relevant and effective in meeting organizational and stakeholder needs.

Implementation Steps:

3.1 **Implementation of a Continuous Feedback Mechanism**: Develop mechanisms (e.g., online platforms, regular meetings, and suggestion boxes) for stakeholders to continuously provide feedback on cybersecurity practices and policies.

3.2 **Regular Review and Analysis of Stakeholder Feedback**: Schedule regular sessions to review and analyze stakeholder feedback, identifying trends, concerns, and opportunities for improvement in cybersecurity measures.

3.3 **Adaptation and Communication of Changes**: Based on feedback analysis, adapt cybersecurity strategies and policies as needed. Communicate changes and updates to stakeholders, ensuring transparency and maintaining trust.

Expected Outcome: The Continuous Stakeholder Feedback and Cybersecurity Adaptation Process ensures that the organization's cybersecurity measures are agile and responsive to stakeholder input. This ongoing engagement fosters a culture of collaboration and trust, enhances stakeholder satisfaction, and ensures that cybersecurity strategies evolve in alignment with both internal and external expectations and needs.

GV.OC-03: Legal, Regulatory, and Contractual Requirements Regarding Cybersecurity—Including Privacy and Civil Liberties Obligations—Are Understood and Managed

The evolving legal and regulatory landscape presents opportunities and challenges for organizations pursuing cybersecurity excellence. As cybersecurity threats continue to escalate in complexity and volume, so too do the legal, regulatory, and contractual requirements aimed at mitigating these threats. Recognizing the importance of compliance is not merely about adhering to laws; it's a strategic imperative that enhances an organization's cybersecurity posture and fosters trust among customers, partners, and regulators. Navigating these requirements, however, demands a nuanced understanding of the legal landscape and a commitment to integrating these obligations into the core of cybersecurity strategies.

Keeping abreast of the current and emerging cybersecurity laws and regulations is a critical first step for organizations. This task involves developing strategies for monitoring legal developments and analyzing how these changes impact organizational operations. The applicability of specific regulations can vary widely depending on factors such as the industry sector, data types handled, and geographic location of data storage or processing. Techniques for interpreting these legal requirements in the cybersecurity context are crucial for ensuring that organizations not only comply with the letter of the law but also embrace the spirit of these regulations to enhance their security posture.

Integrating compliance requirements into cybersecurity policies and procedures requires a structured approach. Organizations should develop frameworks that seamlessly incorporate legal and regulatory requirements into their cybersecurity practices. This involves drafting policies that clearly communicate compliance obligations to all employees and stakeholders while establishing procedures that support compliance and security objectives. Continuous improvement of these policies and procedures is necessary to adapt to the changing compliance landscape, ensuring that organizations remain agile in the face of evolving legal obligations.

Establishing a legal review process for cybersecurity operations ensures that cybersecurity decisions and activities align with legal and regulatory expectations. This process should involve legal experts in the review of cybersecurity practices, incident responses, and data breach management. Such involvement ensures that organizations can swiftly adapt their cybersecurity strategies based on legal advice, minimizing legal risks and enhancing compliance.

Employee education on compliance obligations in cybersecurity contexts is another key component. Organizations must develop targeted training programs that address legal and regulatory compliance, ensuring that employees understand their role in maintaining this compliance. Incorporating compliance topics into ongoing cybersecurity awareness initiatives helps reinforce this knowledge, creating a compliance culture permeating every level of the organization.

Incident response protocols must also consider legal and regulatory aspects to ensure comprehensive and compliant responses to cybersecurity incidents. This includes integrating legal requirements into planning and execution, coordinating with legal teams on regulatory notifications and disclosures, and maintaining documentation practices for potential legal proceedings. Post-incident analysis is crucial for identifying areas for compliance improvement and reducing future legal risks.

Developing a privacy and civil liberties protection framework is critical in today's data-driven world. This framework should establish privacy principles that guide cybersecurity measures, implement data protection controls, and ensure compliance with national and international privacy regulations. Engaging with stakeholders on privacy matters promotes transparency and accountability, bolstering trust and enhancing the organization's reputation.

Regular updates and revisions of compliance measures are necessary to keep pace with the dynamic legal and regulatory environment. This entails setting up processes for monitoring developments, scheduling regular reviews of compliance policies, and implementing rapid changes in response to new legal requirements. Audits and assessments are vital in verifying compliance, identifying gaps, and providing a feedback loop for continuous improvement.

Recommendations

- **Leverage Legal and Compliance Expertise**: Establish a cross-functional team comprising legal, compliance, and cybersecurity experts to ensure a holistic approach to managing legal and regulatory requirements. This team should be responsible for monitoring legal developments, interpreting their implications for the organization, and guiding the integration of compliance measures into cybersecurity strategies.
- **Enhance Employee Compliance Education**: Develop comprehensive training programs focused on legal and regulatory compliance within the cybersecurity context. Regularly update these programs to reflect changes in the legal landscape and assess employee comprehension to reinforce a culture of compliance.
- **Implement Dynamic Compliance Frameworks**: Create adaptable compliance frameworks that quickly incorporate changes in legal and regulatory requirements. These frameworks should facilitate the alignment of cybersecurity policies and procedures with evolving legal obligations, ensuring resilience against compliance risks.
- **Prioritize Privacy and Civil Liberties**: Integrate privacy principles and civil liberties protections into all cybersecurity operations. Adopt a privacy-by-design approach that embeds data protection controls from the outset, aligning with organizational values and regulatory expectations.
- **Engage in Continuous Improvement**: Regularly review and update compliance policies, procedures, and practices to address the changing legal and regulatory landscape. Employ audits and assessments to identify compliance gaps and opportunities for enhancement, ensuring ongoing compliance and cybersecurity effectiveness.

NIST 800-53 Controls

(Refer to our appendix for a comprehensive description of the NIST 800-53 Controls, essential for organizations adhering to strict regulatory standards.)

- AC-01,AT-01,AU-01,CA-01,CM-01,CP-01,IA-01,IR-01,MA-01,MP-01,PE-01,PL-01,PM-01,PS-1, PT-01,RA-01,SA-01,SC-01,SI-01,SR-01,PM-28

Simplified Security Controls (SSC)

This section outlines our SSC, tailored for environments beyond the scope of NIST compliance. These controls offer streamlined, effective security solutions suitable for any organization seeking to enhance its cybersecurity defenses with clear, actionable measures.

Control 1: Compliance Mapping and Integration System

Control Objective: To create a comprehensive system that maps out all relevant legal, regulatory, and contractual cybersecurity obligations and integrates these requirements into the organization's cybersecurity policies and procedures.

Implementation Steps:

1.1 **Identification and Documentation of Requirements**: Utilize a dedicated team to identify and document all legal, regulatory, and contractual obligations related to cybersecurity, focusing on those relevant to the organization's operations and industry.

1.2 **Development of a Compliance Mapping Framework**: Develop a framework that maps each identified obligation to specific cybersecurity policies, procedures, and controls, ensuring complete coverage of compliance requirements.

1.3 **Integration and Operationalization of Requirements**: Integrate the compliance mapping framework into the organization's cybersecurity policy development process, ensuring that all new and revised policies and procedures reflect current compliance obligations.

Expected Outcome: The Compliance Mapping and Integration System ensures that the organization's cybersecurity policies and procedures align with legal, regulatory, and contractual obligations. This systematic approach facilitates compliance, reduces the risk of legal penalties, and enhances the organization's cybersecurity posture by ensuring that all relevant requirements are met and maintained.

Control 2: Legal and Compliance Review Board

Control Objective: To establish a cross-functional board that oversees the review of cybersecurity practices, decisions, and incidents from a legal and compliance perspective, ensuring that the organization's cybersecurity operations adhere to legal requirements and best practices.

Implementation Steps:

2.1 **Formation of the Review Board**: Create a board comprising legal experts, cybersecurity professionals, compliance officers, and representatives from affected business units, ensuring diverse perspectives on cybersecurity and legal matters.

2.2 **Regular Review Schedule**: Set a regular schedule for the board to review ongoing cybersecurity practices, recent security incidents, and legal and regulatory changes impacting cybersecurity.

2.3 **Guidance and Recommendations**: Provide guidance on legal compliance issues, recommend adjustments to cybersecurity practices based on legal advice, and ensure that incident response activities are conducted to consider legal obligations and protect the organization.

Expected Outcome: The Legal and Compliance Review Board ensures that cybersecurity operations are consistently reviewed for legal and regulatory compliance, reducing the risk of legal penalties and enhancing the organization's ability to respond to incidents in a legally sound manner.

Control 3: Cybersecurity Compliance Training Program

Control Objective: To develop and implement a training program focused on cybersecurity-related legal, regulatory, and contractual obligations to enhance employee understanding and compliance.

Implementation Steps:

3.1 **Program Development**: Develop a comprehensive training program that covers essential legal and regulatory requirements related to cybersecurity, tailored to different roles within the organization.

3.2 **Mandatory Participation**: Make participation in the training program mandatory for all employees, with specialized modules for those in roles with significant cybersecurity responsibilities.

3.3 **Continuous Update and Assessment**: Regularly update the training content to reflect changes in the legal and regulatory landscape and assess employee comprehension and application of the material through tests and practical exercises.

Expected Outcome: The Cybersecurity Compliance Training Program ensures that all employees know their roles and responsibilities in maintaining legal and regulatory compliance in cybersecurity activities. This widespread understanding reduces non-compliance risk and enhances the organization's overall cybersecurity posture by promoting compliant behavior across all levels.

GV.OC-04: Critical Objectives, Capabilities, and Services that Stakeholders Depend on or Expect from the Organization are Understood and Communicated

The alignment of cybersecurity with critical organizational objectives is pivotal for ensuring that an organization's essential functions and services are secured against potential cyber threats. This alignment starts with a deep understanding of the interplay between organizational objectives and cybersecurity, highlighting the importance of identifying which objectives require robust cybersecurity support. Strategies to ensure cybersecurity initiatives contribute effectively to achieving these objectives are vital, although challenges in aligning these efforts with broader organizational goals can be complex. Addressing these challenges requires a nuanced approach, balancing cybersecurity efforts with the organization's strategic aims.

Identifying and documenting the essential services and capabilities critical to stakeholders form the cornerstone of effective cybersecurity planning. This process involves determining which services are vital from the stakeholder's perspective and documenting their cybersecurity needs. Engaging stakeholders in identifying and validating these critical services ensures that cybersecurity planning is grounded in the operational realities and expectations of those the organization serves. The documentation of these services plays a crucial role in cybersecurity planning, allocating resources, and prioritizing cybersecurity efforts.

Evaluating cybersecurity risks relative to critical objectives and capabilities involves a structured approach to risk assessment. Frameworks for assessing these risks must consider the potential impact on essential services and the broader organizational goals. This evaluation prioritizes cybersecurity measures, ensuring that resources are allocated effectively to protect against threats with the highest potential to impact critical services. Integrating risk evaluation into ongoing cybersecurity strategy development allows organizations to remain agile, adapting their strategies to mitigate evolving threats.

Engaging stakeholders to validate and refine cybersecurity objectives is essential for ensuring cybersecurity measures align with stakeholder needs and expectations. Strategies for involving stakeholders in this process include setting and reviewing cybersecurity objectives, incorporating their feedback into cybersecurity priorities, and aligning expectations with the organization's practical cybersecurity capabilities. Transparency and effective communication are critical in these engagement efforts, building trust and fostering a collaborative approach to cybersecurity.

Customizing cybersecurity controls to protect identified critical services ensures that protective measures are tailored to the specific needs of these services. This customization involves applying best practices and industry standards, continuously evaluating and adjusting controls to address evolving threats, and prioritizing cybersecurity resilience to ensure the uninterrupted delivery of critical services. The role of cybersecurity in maintaining the continuity and reliability of these services cannot be overstated, necessitating a dedicated focus on protecting these vital functions.

Developing communication plans for cyber risks related to critical organizational services involves crafting strategies to inform stakeholders about potential cyber threats and the importance of cybersecurity measures in mitigating these risks. These plans should utilize clear and accessible language, engage stakeholders in discussions about risk mitigation strategies, and provide regular updates on the status of cybersecurity efforts. Effective communication is fundamental to managing stakeholder expectations and fostering a shared understanding of the importance of cybersecurity in protecting critical services.

Cross-functional coordination is critical for maintaining and improving the protection of essential services. Establishing collaborative frameworks between cybersecurity teams and other functional areas facilitates information sharing and coordination of actions to safeguard these vital services. Leadership plays a crucial role in fostering a culture of cross-functional collaboration, with regular evaluations of coordination effectiveness and identifying improvement opportunities to enhance the overall cybersecurity posture.

Finally, establishing metrics for cybersecurity performance regarding critical services involves developing key performance indicators (KPIs) that measure the effectiveness of cybersecurity initiatives. These metrics should focus on the impact of cybersecurity on the availability and reliability of critical services, guiding improvements and resource allocation decisions. Communicating these performance metrics to stakeholders demonstrates the value of cybersecurity efforts and reinforces the organization's commitment to protecting the services upon which stakeholders depend.

Recommendations

- **Integrate Cybersecurity with Business Strategy**: Ensure cybersecurity initiatives align closely with the organization's critical objectives and services. This involves regular collaboration between cybersecurity leaders and business executives to ensure mutual understanding and support.
- **Enhance Stakeholder Engagement**: Develop ongoing communication and feedback mechanisms with stakeholders to understand their needs better and align cybersecurity efforts with their expectations. This should include regular updates on cybersecurity risks and mitigation measures.
- **Implement Dynamic Risk Assessment Processes**: Adopt a dynamic approach to evaluating risks to critical services and capabilities, ensuring that risk assessments are regularly updated to reflect the evolving threat landscape and changes in organizational objectives.
- **Customize Cybersecurity Controls**: Tailor cybersecurity measures to address specific needs and risks for critical services and capabilities. This customization should be based on thoroughly analyzing each service's importance and vulnerability to disruption.
- **Foster a Culture of Cyber Resilience**: Promote organizational resilience by incorporating cybersecurity considerations into the planning and delivery all critical services. This includes developing incident response plans and business continuity strategies that ensure the reliability and continuity of these services.

NIST 800-53 Controls

(Refer to our appendix for a comprehensive description of the NIST 800-53 Controls, essential for organizations adhering to strict regulatory standards.)

- PM-08,PM-11,CP-02,PM-30,RA-09

Simplified Security Controls (SSC)

This section outlines our SSC, tailored for environments beyond the scope of NIST compliance. These controls offer streamlined, effective security solutions suitable for any organization seeking to enhance its cybersecurity defenses with clear, actionable measures.

Control 1: Critical Service Identification and Documentation Process

Control Objective: To systematically identify, document, and maintain a comprehensive record of all critical services and capabilities that stakeholders depend on, including their specific cybersecurity requirements.

Implementation Steps:

1.1 **Stakeholder Consultation for Service Identification**: Engage with stakeholders through interviews, surveys, and workshops to identify services and capabilities critical to the organization's operations and stakeholder satisfaction.

1.2 **Documentation of Critical Services**: Develop a detailed documentation process that captures critical services, their importance, cybersecurity requirements, and specific compliance or regulatory obligations.

1.3 **Review and Update Mechanism**: Establish a regular review schedule to update the documentation of critical services, ensuring it reflects changes in organizational objectives, stakeholder needs, and the cybersecurity landscape.

Expected Outcome: Establishing a comprehensive and regularly updated record of critical services and their cybersecurity requirements enhances the organization's ability to prioritize cybersecurity efforts. It ensures that cybersecurity measures are aligned with protecting the most vital services to stakeholder satisfaction and organizational success.

Control 2: Cross-Functional Cybersecurity Risk Assessment Team

Control Objective: To evaluate cybersecurity risks relative to critical objectives and capabilities through a cross-functional team approach, ensuring a comprehensive understanding of risks and prioritization of cybersecurity measures.

Implementation Steps:

2.1 **Formation of the Risk Assessment Team**: Form a cross-functional team including representatives from cybersecurity, IT, business units, and other relevant departments to ensure a holistic view of cybersecurity risks to critical services.

2.2 **Development of a Risk Assessment Framework**: Develop a framework that aligns cybersecurity risk assessments with critical organizational objectives and services, focusing on potential impacts and vulnerabilities.

2.3 **Continuous Risk Evaluation and Reporting**: Implement a continuous risk evaluation process to identify and assess new and evolving threats to critical services. Regularly report findings to senior management and adjust cybersecurity strategies as necessary.

Expected Outcome: A cross-functional cybersecurity risk assessment team ensures that cybersecurity risks are evaluated from multiple perspectives, leading to a prioritization of measures that protect critical services and support organizational objectives. This approach facilitates informed decision-making and resource allocation toward areas of highest impact.

Control 3: Cybersecurity Performance Metrics for Critical Services

Control Objective: To establish KPIs that measure the effectiveness of cybersecurity practices in protecting critical services, ensuring continuous improvement and accountability.

Implementation Steps:

3.1 **Development of Cybersecurity KPIs**: In consultation with stakeholders, develop specific KPIs that measure the effectiveness, efficiency, and impact of cybersecurity practices on the availability, integrity, and confidentiality of critical services.

3.2 **Implementation of Measurement Tools**: Deploy tools and processes to regularly measure these KPIs, providing ongoing visibility into the performance of cybersecurity controls and their alignment with critical service protection.

3.3 **Review and Adjustment of Cybersecurity Strategies**: Use the insights gained from KPI measurement to guide the continuous improvement of cybersecurity strategies and controls, ensuring they remain effective in protecting critical services against evolving threats.

Expected Outcome: Establishing and monitoring cybersecurity performance metrics for critical services ensures that the organization can quantitatively assess and demonstrate the value of cybersecurity investments. This data-driven approach facilitates targeted improvements, better resource allocation, and enhanced protection of services critical to stakeholders, leading to increased trust and reliability.

GV.OC-05: Outcomes, Capabilities, and Services that the Organization Depends on Are Understood and Communicated

The foundational role of cybersecurity in safeguarding the key outcomes, capabilities, and services an organization depends on cannot be overstated. This relationship underscores the critical nature of recognizing and protecting dependencies within cybersecurity planning. The repercussions of cybersecurity incidents on these organizational aspects are significant, potentially derailing operations and compromising service delivery. Employing strategies to enhance resilience and minimize vulnerabilities associated with these critical dependencies is essential for maintaining organizational integrity and continuity.

Cataloging organizational dependencies and understanding their cybersecurity implications is vital in this process. It involves thoroughly identifying and documenting these dependencies, followed by analyzing the cybersecurity risks associated with each. Prioritizing these dependencies based on their risk and impact on the organization enables focused cybersecurity efforts. Engaging stakeholders in this process ensures the catalog is comprehensive and reflects the organization's operational realities.

Analyzing the impact of cybersecurity on organizational capabilities and services involves assessing how issues in this domain can affect the organization's essential functions. This analysis benefits from reviewing case studies of past cybersecurity incidents to understand their consequences and employing methodologies to quantify potential impacts. Strategies developed from this analysis aim to mitigate risks to the organization's critical dependencies, preserving operational capacity and service delivery.

Communication strategies are crucial for raising awareness of these dependencies and the associated cyber risks. Developing targeted communication plans and utilizing various platforms and media can effectively convey the importance of cybersecurity in protecting these organizational aspects. Engagement practices and feedback mechanisms are instrumental in ensuring the message resonates across the organization, leading to informed actions and enhanced cybersecurity posture.

Prioritizing cybersecurity efforts based on the criticality of organizational dependencies ensures that resources are allocated effectively. This process involves setting criteria for prioritizing initiatives and dynamically adjusting these priorities in response to the changing risk landscape. Regular evaluations confirm that cybersecurity efforts are aligned with organizational priorities, safeguarding the most critical dependencies.

Integrating dependency considerations into cybersecurity risk assessments enables a comprehensive understanding of the risks associated with key organizational aspects. This inclusion enriches the overall cybersecurity strategy, guiding decision-making and ensuring a focused approach to mitigating dependency risks. Continuous review and updates of risk assessments reflect changes in organizational dependencies, maintaining the relevance of cybersecurity strategies.

Facilitating cross-departmental understanding of cybersecurity dependencies fosters a collaborative approach to protecting these critical aspects of the organization. Strategies for promoting collaboration, alongside tailored training and awareness programs, highlight the interconnected nature of organizational operations and the importance of a unified cybersecurity stance. Exercises and simulations further emphasize the significance of safeguarding dependencies and enhancing organizational resilience.

Regular reviews of organizational dependencies and adjustments to the cybersecurity posture based on these reviews are essential for maintaining alignment with organizational needs. Leadership is critical in endorsing and implementing changes that address dependency risks, ensuring that cybersecurity strategies are integrated into strategic planning and operational practices. This ongoing process supports the organization's ability to adapt to evolving threats and maintain the integrity of critical services and capabilities.

Recommendations

- **Implement Comprehensive Dependency Mapping**: Develop and regularly update a detailed map of organizational dependencies, focusing on those critical to operations and service delivery. This map should guide cybersecurity strategy and resource allocation.
- **Enhance Stakeholder Engagement**: Actively involve stakeholders in identifying and assessing organizational dependencies. This engagement should include regular communication on the status of cybersecurity efforts and how they protect these critical areas.
- **Prioritize Based on Dependency Criticality**: Establish a prioritization framework for cybersecurity initiatives that are directly informed by the criticality of organizational dependencies. This approach ensures focused protection where it is most needed.
- **Regularly Update Cybersecurity Strategies**: Incorporate findings from dependency analyses into ongoing cybersecurity strategy development. This ensures the organization's cybersecurity efforts align with its operational and service delivery priorities.
- **Foster a Culture of Continuous Learning**: Develop targeted training initiatives to manage organizational dependencies within the cybersecurity context. Include these topics in certification paths and use interactive learning techniques to enhance training effectiveness.

NIST 800-53 Controls

(Refer to our appendix for a comprehensive description of the NIST 800-53 Controls, essential for organizations adhering to strict regulatory standards.)

- PM-11,PM-30,RA-07,SA-09,SR-05

Simplified Security Controls (SSC)

This section outlines our SSC, tailored for environments beyond the scope of NIST compliance. These controls offer streamlined, effective security solutions suitable for any organization seeking to enhance its cybersecurity defenses with clear, actionable measures.

Control 1: Dependency Mapping and Cybersecurity Implication Analysis

Control Objective: To systematically identify and document all critical organizational dependencies, analyzing their cybersecurity implications to prioritize protection efforts effectively.

Implementation Steps:

1.1 **Conduct a Comprehensive Dependency Mapping Exercise**: Engage teams across the organization to identify and map out all critical services, processes, and technologies the organization depends on for its operational success.

1.2 **Analyze Cybersecurity Implications for Each Dependency**: For each identified dependency, assess the cybersecurity risks associated with its disruption, compromise, or failure, considering both direct and indirect impacts.

1.3 **Prioritize Dependencies Based on Cybersecurity Risk**: Utilize the risk analysis to prioritize dependencies, focusing cybersecurity efforts on those with the highest potential impact on organizational objectives and stakeholder value.

Expected Outcome: A clear and prioritized map of organizational dependencies and associated cybersecurity risks, enabling targeted protection efforts where needed. This approach ensures that cybersecurity resources are allocated efficiently, enhancing resilience against threats to critical dependencies.

Control 2: Cybersecurity Dependency Awareness and Communication Program

Control Objective: To develop and implement a comprehensive awareness and communication program highlighting the importance of protecting critical dependencies from cybersecurity threats and fostering a culture of cybersecurity awareness across the organization.

Implementation Steps:

2.1 **Develop Targeted Communication Plans**: Create communication plans that address different segments of the organization, tailored to each group's specific roles and responsibilities in protecting critical dependencies.

2.2 **Utilize Diverse Communication Channels**: Employ a mix of communication channels (e.g., internal webinars, newsletters, and intranet posts) to disseminate information about the importance of cybersecurity in safeguarding organizational dependencies.

2.3 **Incorporate Feedback Mechanisms**: Implement mechanisms to gather feedback on the effectiveness of communication strategies, adjusting approaches based on stakeholder input to ensure maximum engagement and comprehension.

Expected Outcome: A well-informed organization where all members understand their role in protecting critical dependencies against cybersecurity threats. This program aims to enhance the cybersecurity culture across the organization, leading to proactive behaviors that support the overall cybersecurity posture and resilience.

Control 3: Integration of Dependency Risks into Cybersecurity Strategy

Control Objective: To ensure that the organization's cybersecurity strategy explicitly addresses the risks to critical dependencies, incorporating dependency considerations into risk assessments, resource allocation, and strategic planning.

Implementation Steps:

3.1 **Incorporate Dependency Analysis into Risk Assessments**: Ensure that cybersecurity risk assessments explicitly consider risks to critical dependencies, including the potential impact of their compromise on organizational objectives.

3.2 **Align Cybersecurity Strategy with Dependency Protection Needs**: Adjust the cybersecurity strategy to prioritize the protection of identified critical dependencies, allocating resources based on each dependency's criticality and risk level.

3.3 **Establish a Continuous Review Process**: Implement a process for regularly reviewing dependency risks and cybersecurity strategy effectiveness, incorporating changes in the threat landscape, dependency landscape, and organizational objectives.

Expected Outcome: A cybersecurity strategy closely aligns with the organization's critical dependencies, ensuring that efforts and resources protect the most impactful areas. This strategic alignment enhances the organization's ability to withstand and recover from cybersecurity incidents, safeguarding essential services and capabilities upon which the organization and its stakeholders rely.

6

Risk Management Strategy (GV.RM)

> In the fabric of organizational resilience, each thread of cybersecurity risk management we weave strengthens our defense and our position in the digital frontier.

In today's digital-first environment, establishing robust cybersecurity risk management protocols is beneficial and imperative for organizational sustainability and resilience. This necessitates a comprehensive approach where objectives for risk management are clearly defined and aligned with stakeholder expectations, ensuring a coherent understanding of risk appetite and tolerance. It integrates cybersecurity risk considerations into the broader enterprise risk management (ERM) framework, fostering informed strategic decision-making regarding risk responses. Furthermore, it emphasizes the importance of clear communication channels within and with external entities to identify, assess, and prioritize cybersecurity threats and opportunities. By adopting a standardized methodology for evaluating cybersecurity risks and recognizing strategic opportunities that cybersecurity can present, organizations can better navigate the complexities of the digital landscape, turning potential vulnerabilities into competitive advantages.

GV.RM-01: Risk Management Objectives are Established and Agreed to by Organizational Stakeholders

The foundation of a robust cybersecurity strategy begins with a clear definition of risk management objectives. These objectives guide organizational efforts toward prioritizing and mitigating risks in alignment with the overall business goals. Establishing such objectives involves a deep understanding of the organization's risk appetite, the landscape of potential cyber threats, and the critical assets that must be protected. Effective risk management objectives are about identifying what to protect and setting measurable goals for how protection is implemented and maintained. This ensures that cybersecurity efforts are strategic, focused, and capable of evolving with the changing nature of cyber threats.

Setting cybersecurity risk management objectives necessitates the active involvement of key stakeholders across the organization. This collaboration ensures that the objectives are comprehensive and reflect the concerns and priorities of various departments. Engaging stakeholders from the outset facilitates a broader understanding of the organization's risk profile and fosters a security awareness and accountability culture. By incorporating diverse perspectives,

A Comprehensive Guide to the NIST Cybersecurity Framework 2.0: Strategies, Implementation, and Best Practice, First Edition. Jason Edwards.
© 2025 John Wiley & Sons Ltd. Published 2025 by John Wiley & Sons Ltd.

organizations can craft objectives that are aligned with their strategic vision and adaptable to the dynamic cybersecurity landscape.

Aligning cybersecurity risk management objectives with business goals is crucial for ensuring cybersecurity initiatives bolster rather than hinder organizational growth. This alignment emphasizes cybersecurity as a strategic enabler, safeguarding and facilitating business operations. It involves translating the technical aspects of cyber risk management into business impacts, enabling decision-makers to understand the importance of cybersecurity investments. Ultimately, this linkage ensures that the organization's risk management efforts directly contribute to achieving its overarching goals.

Establishing measurable objectives is key to effective cybersecurity risk management, allowing organizations to track progress and adapt performance-based strategies. These objectives should be specific, achievable, relevant, and time-bound (SMART), providing clear targets for risk reduction efforts. Measurable objectives facilitate the quantification of risk management success, making it easier to communicate value and make informed decisions about resource allocation and strategic direction in cybersecurity efforts.

Incorporating industry standards and best practices into cybersecurity risk management objectives is essential for ensuring that an organization's efforts are comprehensive and compliant with regulatory requirements. This alignment helps benchmark an organization's security posture against established frameworks such as the NIST Cybersecurity Framework, thereby identifying areas for improvement. Adapting these standards to fit the organization's unique context ensures that risk management objectives are not only ambitious but also attainable and relevant to the specific cybersecurity challenges faced.

Effective communication and endorsement of risk management objectives across the organization are critical for ensuring all team members are aligned and motivated to achieve these goals. This involves clear messaging from leadership and regular updates on progress and challenges. By fostering an organizational culture that prioritizes cybersecurity, organizations can ensure that risk management objectives are understood, accepted, and integrated into daily operations, enhancing overall security posture.

The cybersecurity landscape is constantly evolving, requiring regularly reviewing and adjusting risk management objectives. This iterative process ensures that objectives remain relevant and responsive to new threats and organizational changes. Incorporating feedback from these reviews allows for the continuous improvement of the cybersecurity strategy, ensuring that risk management efforts are proactive and aligned with current and future security challenges.

Risk management objectives must be seamlessly integrated into the overall business strategy for cybersecurity efforts to be truly effective. This integration ensures that cybersecurity is not an afterthought but a fundamental aspect of strategic planning and decision-making. By embedding cybersecurity into the strategic core of the organization, businesses can ensure that risk management supports and enhances their operational and strategic goals, thereby comprehensively securing their assets and reputation.

Assigning clear roles and responsibilities is vital for successfully implementing and maintaining cybersecurity risk management objectives. This accountability ensures that individuals and teams are aware of their specific duties and empowered to take action to safeguard the organization's digital assets. Leadership plays a crucial role in fostering a culture of accountability, where everyone understands the importance of their contributions to the organization's cybersecurity posture and is committed to achieving the set objectives.

Recommendations

- **Engage Broadly with Stakeholders**: Ensure the involvement of stakeholders from various departments in formulating risk management objectives to capture a wide range of insights and ensure alignment with organizational goals.
- **Align Objectives with Business Strategy**: Regularly review and align risk management objectives with the overall business strategy to ensure that cybersecurity efforts support and enable business operations.
- **Establish Clear Metrics**: Define clear, measurable objectives and establish metrics and KPIs to track progress, enabling timely adjustments to the cybersecurity strategy.
- **Incorporate Industry Standards**: Leverage industry standards and best practices to benchmark and guide the development of risk management objectives, ensuring compliance and enhancing security posture.
- **Foster a Culture of Accountability**: Promote accountability and ownership across all levels of the organization, with clear responsibilities and regular training to ensure that everyone contributes to the cybersecurity efforts.

NIST 800-53 Controls

(Refer to our appendix for a comprehensive description of the NIST 800-53 Controls, essential for organizations adhering to strict regulatory standards.)

- PM-09,RA-07,SR-02

Simplified Security Controls (SSC)

This section outlines our Simplified Security Controls (SSC), tailored for environments beyond the scope of NIST compliance. These controls offer streamlined, effective security solutions suitable for any organization seeking to enhance its cybersecurity defenses with clear, actionable measures.

Control 1: Stakeholder Engagement and Consensus Building Program

Control Objective: To ensure that risk management objectives are collaboratively developed with input from all relevant organizational stakeholders, reflecting a comprehensive understanding of cybersecurity risks and business priorities.

Implementation Steps:

1.1 **Identify and Map Stakeholders**: Identify key stakeholders across the organization, including IT, business units, compliance, legal, and executive leadership, to ensure diverse perspectives are incorporated.

1.2 **Conduct Stakeholder Engagement Workshops**: Organize workshops and meetings to discuss and gather input on risk management objectives, facilitating a collaborative environment where stakeholders can voice concerns and suggestions.

1.3 **Develop Consensus on Risk Objectives**: Utilize the information gathered from stakeholder engagements to develop a set of risk management objectives that balance diverse needs and priorities, achieving consensus among stakeholders.

Expected Outcome: A set of risk management objectives widely supported across the organization, reflecting a balanced consideration of cybersecurity risks and business priorities. This consensus-building approach fosters organizational alignment on risk priorities and enhances the effectiveness of cybersecurity risk management efforts.

Control 2: Integration of Cybersecurity Risk Objectives with Business Strategy

Control Objective: To ensure that cybersecurity risk management objectives fully align with and support the organization's overarching business goals and strategic priorities.

Implementation Steps:

2.1 **Analyze Business Goals and Strategies**: Work closely with strategic planning and business units to understand current and future business goals, identifying where cybersecurity risk management can support these objectives.

2.2 **Align Risk Objectives with Business Goals**: Develop cybersecurity risk management objectives that directly support business goals, ensuring that risk management activities enhance rather than hinder organizational objectives.

2.3 **Communicate and Endorse Alignment**: Secure executive endorsement and communicate the alignment between cybersecurity risk objectives and business goals across the organization, ensuring clear understanding and support.

Expected Outcome: Cybersecurity risk management objectives clearly align with and directly support the organization's business goals, ensuring that cybersecurity efforts are perceived as value-adding rather than merely compliance-driven. This alignment promotes a culture of integrating cybersecurity into strategic planning and decision-making processes.

Control 3: Dynamic Review and Adjustment of Cybersecurity Risk Management Objectives

Control Objective: To establish a process for the regular review and dynamic adjustment of cybersecurity risk management objectives, ensuring they remain relevant in changing business needs, technological advancements, and evolving threat landscapes.

Implementation Steps:

3.1 **Establish a Regular Review Schedule**: Set a regular schedule for reviewing cybersecurity risk management objectives, involving key stakeholders in the review process to ensure objectives remain aligned with business goals and the risk environment.

3.2 **Incorporate Feedback and Environmental Changes**: Analyze feedback from stakeholders, lessons learned from incident responses, changes in business strategies, technological advancements, and evolving threats to adjust risk management objectives as necessary.

3.3 **Communicate Adjustments and Rationale**: Communicate any adjustments to risk management objectives and the rationale behind them across the organization, ensuring transparency and maintaining stakeholder engagement and support.

Expected Outcome: A set of cybersecurity risk management objectives that are dynamic and responsive to the changing business, technological, and threat environments. This approach ensures that the organization's risk management efforts remain effective, relevant, and aligned with internal priorities and external requirements, enhancing the organization's overall security posture and resilience.

GV.RM-02: Risk Appetite and Risk Tolerance Statements are Established, Communicated, and Maintained

Understanding and managing risk is a foundational aspect of cybersecurity. The concepts of risk appetite and risk tolerance are at the heart of this endeavor. Risk appetite refers to the risk an organization accepts to pursue its objectives. In contrast, risk tolerance defines the levels of variation an organization is willing to endure within its risk appetite. These concepts are theoretical and practical tools for guiding an organization's risk management strategy. Establishing clear statements for both is crucial for effective cybersecurity governance.

The process of articulating risk appetite and tolerance begins with a thorough understanding of an organization's objectives, operating environment, and potential risks. Developing risk appetite and tolerance statements requires a collaborative effort involving leadership and key stakeholders. This ensures that the statements accurately reflect the organization's strategic vision and capacity to manage risk. Leadership involvement is particularly important as it signifies an organizational commitment to risk management.

Quantifying risk appetite and tolerance can be challenging. It involves translating qualitative risk assessments into quantitative measures, which can then be used to guide decision-making. This quantification process is crucial for aligning the risk statements with organizational goals and culture. It enables organizations to decide which risks to accept, mitigate, or transfer.

Balancing organizational growth with cybersecurity risk exposure is a delicate act. Growth strategies often entail taking on new risks, and as such, an organization's risk appetite and tolerance must evolve to accommodate these changes. Case studies of organizations that successfully balance growth and risk can provide valuable insights. These examples highlight the importance of regularly adjusting risk appetite and tolerance in response to strategic shifts.

Communicating risk appetite and tolerance throughout an organization is vital for ensuring that all employees understand the boundaries of acceptable risk. This is often achieved through targeted training programs. Developing effective training materials and tailoring them to different roles within the organization can reinforce the importance of risk management. Moreover, evaluating the effectiveness of these training programs is essential for ensuring that the risk management message is received and understood.

Documenting and regularly updating risk appetite statements ensures they remain relevant and reflect the current risk landscape and organizational objectives. Best practices in documentation and communication ensure that changes to these statements are clearly articulated and disseminated throughout the organization. This ongoing maintenance is crucial for the resilience of an organization's risk management framework.

Ensuring daily operations and decision-making processes align with defined risk appetite and tolerance is critical for risk consistency. Employing tools and frameworks that support this alignment can help organizations maintain a risk-aware culture. Leadership plays a crucial role in modeling risk-consistent behaviors, highlighting the importance of executive endorsement in effective risk management.

Employing robust systems and tools for tracking risk decisions helps organizations identify when they are deviating from their defined risk parameters. Indicators and metrics that measure adherence can inform adjustments to risk management strategies, ensuring that they remain effective and aligned with organizational objectives.

Recommendations

- **Engage Leadership at All Levels**: Ensure organizational leaders understand and endorse the risk appetite and tolerance statements. Their active participation in risk management processes reinforces the importance of these concepts to the entire organization.
- **Integrate Risk Management into Strategic Planning**: Make risk management a central component of strategic planning processes. This integration ensures that risk considerations are factored into decision-making, aligning operational activities with the organization's risk appetite and tolerance.

- **Foster Open Communication about Risk**: Encourage a culture where employees feel comfortable discussing risk and reporting potential issues. Open communication channels can help identify and address risks before they escalate into significant threats.
- **Implement Continuous Risk Education**: Develop ongoing training programs to inform employees about risk management practices, risk appetite and tolerance changes, and their roles in supporting these initiatives. Continuous education is key to maintaining a risk-aware culture.
- **Utilize Technology to Enhance Risk Visibility**: Leverage technology solutions that provide real-time insights into risk exposure and compliance with risk appetite and tolerance. Enhanced visibility into risk can help organizations respond more effectively to emerging threats.

NIST 800-53 Controls

(Refer to our appendix for a comprehensive description of the NIST 800-53 Controls, essential for organizations adhering to strict regulatory standards.)

- PM-09

Simplified Security Controls (SSC)

This section outlines our SSC, tailored for environments beyond the scope of NIST compliance. These controls offer streamlined, effective security solutions suitable for any organization seeking to enhance its cybersecurity defenses with clear, actionable measures.

Control 1: Development and Documentation of Risk Appetite and Tolerance Statements

Control Objective: To formally establish, document, and communicate the organization's risk appetite and tolerance levels, providing a clear framework for making risk-based decisions that support organizational objectives.

Implementation Steps:

1.1 **Collaborative Development of Statements**: Engage key stakeholders from across the organization, including executive leadership, cybersecurity teams, business units, and risk management, to collaboratively develop risk appetite and tolerance statements.

1.2 **Formal Documentation**: Document the agreed-upon risk appetite and tolerance levels in a formal statement, clearly outlining the types and levels of risks the organization is willing to accept in pursuit of its strategic objectives.

1.3 **Approval and Communication**: Obtain formal approval of senior leadership's risk appetite and tolerance statements and communicate these statements across the organization through various channels to ensure widespread understanding and alignment.

Expected Outcome: A clearly defined, documented, and communicated set of risk appetite and tolerance statements that guide the organization's risk management practices and decision-making processes, ensuring consistency and alignment with strategic objectives.

Control 2: Integration of Risk Appetite in Strategic Decision-Making

Control Objective: To ensure the organization's risk appetite and tolerance levels are integrated into strategic decision-making processes, aligning cybersecurity risk management with business strategy and growth objectives.

Implementation Steps:

2.1 **Risk Assessment Alignment**: Align risk assessment methodologies and processes with the defined risk appetite and tolerance levels, ensuring that risk evaluations consider organizational willingness to accept risk.

2.2 **Decision-Making Frameworks**: Incorporate risk appetite and tolerance considerations into strategic decision-making frameworks, requiring that all major decisions are evaluated against these criteria to ensure alignment.

2.3 **Monitoring and Reporting**: Implement monitoring and reporting mechanisms to assess how well strategic decisions and risk management practices align with the established risk appetite and tolerance levels.

Expected Outcome: Strategic decisions and risk management practices consistently aligned with the organization's risk appetite and tolerance, promoting balanced organizational growth with controlled cybersecurity risk exposure.

Control 3: Training and Awareness Programs on Risk Appetite and Tolerance

Control Objective: To develop and deliver comprehensive training and awareness programs aimed at embedding an understanding of the organization's risk appetite and tolerance across all levels of the organization.

Implementation Steps:

3.1 **Curriculum Development**: Develop a training curriculum that includes detailed information on the organization's risk appetite and tolerance, their importance, and how they should influence day-to-day decision-making and risk management practices.

3.2 **Program Delivery**: Deliver the training through various formats (e.g., workshops, e-learning modules, and presentations) to ensure broad and effective dissemination across the organization.

3.3 **Feedback and Iteration**: Collect feedback on the training programs to assess understanding and applicability, making necessary adjustments to improve future training effectiveness.

Expected Outcome: An organization-wide understanding and awareness of risk appetite and tolerance, ensuring that these concepts are consistently considered in daily operations, risk management practices, and decision-making processes, fostering a risk-informed organizational culture.

GV.RM-03: Cybersecurity Risk Management Activities and Outcomes Are Included in Enterprise Risk Management Processes

Integrating cybersecurity risk management into ERM processes represents a pivotal shift toward recognizing digital threats as a core component of organizational risk. This chapter delves into the intricate process of embedding cybersecurity considerations into the broader ERM framework, underlining the importance of cybersecurity in today's digitally driven world. Businesses face many cyber threats that can compromise sensitive data, disrupt operations, and inflict significant financial and reputational damage. Hence, including cybersecurity within ERM is not merely a strategic choice but a necessity.

Effective integration begins with understanding the pivotal role of cybersecurity within the enterprise risk framework. Cybersecurity is no longer viewed as a technical issue to be managed by IT departments alone but as a critical business function that supports overall organizational resilience and strategic objectives. Establishing clear cybersecurity risk management objectives that align with enterprise goals facilitates a more holistic approach to risk management. This alignment ensures that cyber risk mitigation efforts are not siloed but integral to the enterprise's risk management strategy.

Developing and communicating key cybersecurity metrics is essential for reporting and evaluating the effectiveness of cyber risk management strategies. These metrics should be carefully selected to provide insight into the organization's cybersecurity posture and the effectiveness of implemented controls. By integrating cybersecurity data into enterprise risk dashboards and reports, organizations can provide a comprehensive view of their risk landscape to stakeholders, enabling informed decision-making.

Aligning cyber risk management activities with enterprise risk objectives requires a coordinated effort across the organization. It involves identifying and prioritizing cyber risks based on their potential impact on the organization's strategic goals. Through case studies and real-world examples, this section demonstrates how aligning cybersecurity efforts with broader enterprise objectives has enabled organizations to mitigate risks more effectively and capitalize on strategic opportunities.

Cross-functional teams play a crucial role in enhancing cyber risk management. Organizations can identify and address cyber risks more comprehensively by leveraging diverse perspectives from various departments. Effective collaboration and communication among these teams are critical for identifying vulnerabilities, developing mitigation strategies, and implementing robust cybersecurity measures.

Communicating cybersecurity risk management outcomes to key stakeholders is another vital component of integrating cybersecurity into ERM processes. This involves developing tailored messages that resonate with different stakeholder groups and utilizing various communication channels to ensure broad reach and impact. Evaluating the effectiveness of these communication efforts is crucial for ensuring that stakeholders are well-informed and engaged in the organization's cyber risk management practices.

Embedding cyber risk considerations into business decision-making processes underscores the importance of cybersecurity in strategic planning. Incorporating cyber risk information into decision frameworks enables organizations to make more informed choices, balancing risk and opportunity in pursuit of strategic objectives. This section explores strategies for integrating cyber risk analysis into decision-making and highlights case examples where this integration has significantly influenced business outcomes.

Continuous improvement is the linchpin of effective cyber risk management. By establishing frameworks for assessing and enhancing cyber risk management processes, organizations can adapt to the evolving threat landscape. This involves regular reviews, feedback loops, and implementing changes based on lessons learned. The chapter concludes by examining how continuous improvement processes contribute to organizational resilience, supported by case studies that illustrate the positive impact of comprehensive cyber risk management.

Recommendations

- **Establish a Unified Risk Management Framework**: Implement a governance framework that integrates cybersecurity risks into the broader ERM process, ensuring alignment with organizational objectives and facilitating comprehensive risk oversight.
- **Develop Robust Communication Strategies**: Create and implement detailed communication plans that effectively convey the importance of cybersecurity risk management to all stakeholders, utilizing a variety of channels to reach different audiences within and outside the organization.

- **Leverage Cross-Functional Expertise**: Form cross-functional teams comprising members from IT, cybersecurity, business units, and risk management to foster a collaborative approach to identifying, assessing, and mitigating cyber risks.
- **Implement Continuous Monitoring and Improvement**: Adopt continuous monitoring tools and regular review cycles to assess the effectiveness of cybersecurity measures, making adjustments based on emerging threats and evolving business objectives.
- **Foster a Culture of Cyber Risk Awareness**: Cultivate a risk-aware culture through ongoing education and training programs for employees at all levels, emphasizing the role of every individual in maintaining cybersecurity and enhancing organizational resilience.

NIST 800-53 Controls

(Refer to our appendix for a comprehensive description of the NIST 800-53 Controls, essential for organizations adhering to strict regulatory standards.)

- PM-03,PM-09,PM-30,RA-07,SR-02

Simplified Security Controls (SSC)

This section outlines our SSC, tailored for environments beyond the scope of NIST compliance. These controls offer streamlined, effective security solutions suitable for any organization seeking to enhance its cybersecurity defenses with clear, actionable measures.

Control 1: Formal Integration of Cybersecurity into Enterprise Risk Management Framework

Control Objective: To ensure cybersecurity risks are identified, assessed, and managed within the broader ERM framework, aligning cybersecurity risk management activities with overall enterprise risk objectives.

Implementation Steps:

1.1 **Develop Integration Guidelines**: Establish clear guidelines for integrating cybersecurity risk management into the existing ERM framework, detailing roles, responsibilities, and processes.

1.2 **Cyber Risk Identification and Assessment**: Implement processes to systematically identify and assess cybersecurity risks, using criteria that align with the broader enterprise risk assessment methodologies.

1.3 **Incorporate Cyber Risks into ERM Reporting**: Ensure that cybersecurity risks and their management outcomes are included in enterprise risk reporting, providing a holistic view of organizational risks to senior management and relevant stakeholders.

Expected Outcome: Cybersecurity risks are effectively managed within the broader context of enterprise risks, ensuring that strategic decisions reflect a comprehensive understanding of all risks facing the organization, including those related to cybersecurity.

Control 2: Cross-Functional Cyber Risk Management Teams

Control Objective: To leverage cross-functional expertise in managing cybersecurity risks by establishing teams that include members from cybersecurity, IT, business units, and risk management, enhancing the identification, assessment, and mitigation of cyber risks.

Implementation Steps:

2.1 **Team Formation**: Form cross-functional teams with representatives from cybersecurity, IT, risk management, business units, and other relevant departments to ensure a diverse understanding of cybersecurity risks and their impacts.

2.2 **Regular Risk Management Meetings**: Conduct meetings to discuss emerging cybersecurity threats, vulnerabilities, and incidents and their potential impact on the organization's risk posture.

2.3 **Actionable Risk Mitigation Plans**: Develop and implement risk mitigation plans informed by cross-functional insights, ensuring that cybersecurity risk management activities are aligned with ERM objectives.

Expected Outcome: Enhanced identification, assessment, and mitigation of cybersecurity risks through the utilization of cross-functional expertise, leading to more effective and comprehensive cybersecurity risk management practices that are integrated with overall ERM efforts.

Control 3: Cybersecurity Risk Considerations in Business Decision-Making

Control Objective: To embed cybersecurity risk considerations into all significant business decision-making processes, ensuring that decisions reflect an understanding of potential cybersecurity risks and their implications.

Implementation Steps:

3.1 **Define Risk Consideration Processes**: Establish formal processes for considering cybersecurity risks in business decisions, including criteria for when and how cybersecurity risk assessments should influence decision-making.

3.2 **Training and Awareness for Decision-Makers**: Provide training and awareness programs for key decision-makers on the importance of cybersecurity risks and how to integrate risk considerations into business decisions.

3.3 **Review and Feedback Loop**: Implement a review process to evaluate the effectiveness of incorporating cybersecurity risk considerations into decision-making, providing feedback and adjustments as necessary.

Expected Outcome: Business decisions across the organization are made with a clear understanding of cybersecurity risks and their potential impacts. This ensures that such considerations are integral to strategic planning and operational decision-making, enhancing organizational resilience against cyber threats.

GV.RM-04: Strategic Direction That Describes Appropriate Risk Response Options Is Established and Communicated

The essence of strategic risk response planning in cybersecurity lies in its ability to address imminent threats and foresee and prepare for potential future vulnerabilities. This process is crucial in the digital age, where the complexity and frequency of cyberattacks continue to rise. An effective cyber risk response strategy encompasses a thorough understanding of potential threats, evaluating their impact on business objectives, and developing a coherent plan to address these risks in alignment with the organization's strategic goals. This chapter explores the intricacies of formulating, implementing, and continually refining risk response strategies within the NIST Cybersecurity Framework 2.0 framework.

Formulating a risk response strategy that aligns with business objectives begins with a deep dive into the organization's strategic goals and the potential cyber risks that could impede these goals. The process requires a collaborative effort across departments to ensure the risk response strategy supports business objectives without stifling innovation. Criteria for selecting risk response options include cost-effectiveness, impact on business operations, and alignment with overall business strategy. Real-world examples illustrate how organizations have successfully aligned their risk response strategies with business priorities, emphasizing the need for flexibility as business objectives evolve.

Identifying a range of cyber risk response options is critical to developing a comprehensive risk management plan. Traditional risk responses such as accepting, avoiding, transferring, or mitigating risk must be considered alongside innovative approaches tailored to the cyber domain. The selection of appropriate risk response options is based on a thorough risk assessment, understanding the organization's risk appetite, and balancing the cost of response measures against the potential impact of risks. However, challenges arise in identifying and selecting the most influential risk response options, necessitating a systematic approach to evaluate the benefits and drawbacks of each option.

Stakeholder involvement in selecting risk response strategies is imperative for ensuring that the chosen approaches are viable and supported across the organization. Engaging stakeholders in the risk response planning process helps balance diverse perspectives and ensures that the selected strategies meet the needs of different business areas. Techniques for effective stakeholder engagement include workshops, surveys, and stakeholder meetings, although challenges such as conflicting interests and communication barriers may arise. Overcoming these challenges requires transparent communication and a commitment to finding consensus.

Training is a cornerstone for the effective implementation of risk response plans. Tailored training programs must be developed to prepare employees for executing the strategy, ensuring they understand the rationale behind risk response measures and their responsibilities in implementing them. The effectiveness of risk response training programs is enhanced by ongoing education and regular updates to reflect changes in the risk landscape and business objectives.

Communication protocols for risk response measures are essential for ensuring all stakeholders know their roles and responsibilities in risk mitigation efforts. Effective communication strategies include using intranet sites, emails, and regular meetings to discuss risk response actions. However, challenges in communicating complex risk information can arise, necessitating clear, accessible language and multiple communication channels to reach diverse audiences.

Testing and validating selected risk response options before full implementation is crucial to ensure their effectiveness. This involves conducting simulations, tabletop exercises, and other validation activities to evaluate how the risk response measures perform under various scenarios. Based on the outcomes of these tests, adjustments may be necessary to refine the risk response strategies.

Integrating risk responses into incident management plans ensures that the organization is prepared to respond to cyber incidents when they occur effectively. This integration requires a clear understanding of the roles and responsibilities of all parties involved in incident response and establishing processes that align incident management efforts with the overall risk response strategy. Challenges in integration can stem from differing objectives and priorities between incident management and risk management teams, necessitating a coordinated approach to address these challenges.

Regular review and adaptation of risk response strategies are essential for maintaining their relevance and effectiveness in a rapidly evolving cyber threat landscape. This process involves periodic reassessment of the risk environment, stakeholder feedback, and the performance of risk response measures. Flexibility and adaptability are key in ensuring the risk response strategy remains aligned with changing business needs and risk profiles.

Recommendations

- **Develop Dynamic Risk Response Frameworks**: Create adaptable strategies that quickly adjust to new threats and business objectives. This includes regular updates to risk assessments and response plans to reflect the current threat landscape.
- **Foster a Culture of Collaboration**: Enhance collaboration between cybersecurity teams, business units, and stakeholders to ensure comprehensive risk response strategies align with organizational goals. Promote an environment where communication is open, and feedback is actively sought and incorporated.
- **Prioritize Training and Awareness**: Invest in continuous training and awareness programs to ensure employees understand their risk response and incident management roles. Tailor training initiatives to address specific risk scenarios and response procedures.
- **Leverage Technology for Risk Management**: Utilize advanced technologies and cybersecurity tools to support the implementation of risk response strategies. This includes automated risk analysis tools, incident response platforms, and continuous monitoring solutions.
- **Engage in Continuous Learning**: Adopt a posture of continuous learning from past incidents and evolving threats. Incorporate lessons learned into risk response plans and use them as opportunities for improving cybersecurity practices and strengthening organizational resilience.

NIST 800-53 Controls

(Refer to our appendix for a comprehensive description of the NIST 800-53 Controls, essential for organizations adhering to strict regulatory standards.)

- PM-09,PM-28,PM-30,SR-02

Simplified Security Controls (SSC)

This section outlines our SSC, tailored for environments beyond the scope of NIST compliance. These controls offer streamlined, effective security solutions suitable for any organization seeking to enhance its cybersecurity defenses with clear, actionable measures.

Control 1: Development of a Comprehensive Risk Response Strategy

Control Objective: To establish a strategic direction that defines appropriate risk response options, aligning cybersecurity risk management efforts with organizational business objectives and risk appetite.

Implementation Steps:

1.1 **Strategic Risk Assessment**: Conduct a comprehensive assessment to identify and prioritize cybersecurity risks based on their potential impact on the organization's business objectives and risk appetite.

1.2 **Formulation of Risk Response Options**: Develop a range of risk response options, including avoidance, mitigation, transfer, and acceptance, tailored to the organization's strategic goals and risk-tolerance levels.

1.3 **Strategy Documentation and Approval**: Document the risk response strategy in a formal plan and obtain approval from senior management, ensuring that the strategy aligns with overall business objectives and risk management policies.

Expected Outcome: A clearly defined and approved risk response strategy that guides the organization in effectively managing cybersecurity risks in alignment with its business objectives and risk appetite, enhancing its resilience and strategic decision-making capabilities.

Control 2: Stakeholder Engagement in Risk Response Selection

Control Objective: To involve key stakeholders in selecting risk response strategies, ensuring that the chosen responses are viable, supported across the organization, and aligned with business priorities.

Implementation Steps:

2.1 **Identification of Key Stakeholders**: Identify and map key stakeholders who will be impacted by or have influence over cybersecurity risk response decisions, including business unit leaders, IT, legal, and compliance teams.

2.2 **Stakeholder Consultation Process**: Facilitate workshops or meetings with identified stakeholders to discuss potential risk response options, gathering input and preferences to inform decision-making.

2.3 **Consensus Building and Strategy Finalization**: Build consensus among stakeholders on preferred risk response strategies and finalize decisions, ensuring that responses are practical and aligned with organizational capabilities and goals.

Expected Outcome: Risk response strategies developed with comprehensive stakeholder involvement, leading to greater organizational buy-in, more effective implementation of risk responses, and alignment of cybersecurity initiatives with business priorities.

Control 3: Training and Simulation Exercises for Risk Response Implementation

Control Objective: To prepare the organization to effectively implement risk response plans through targeted training programs and simulation exercises, ensuring readiness and capability to act on risk responses as needed.

Implementation Steps:

3.1 **Development of Training Programs**: Create targeted training programs for relevant personnel, focusing on the roles, responsibilities, and actions required under different risk response scenarios.

3.2 **Conduct Simulation Exercises**: Organize simulation exercises to practice implementing risk response strategies in a controlled environment, allowing teams to familiarize themselves with response procedures and identify areas for improvement.

3.3 **Review and Feedback**: Conduct reviews after training and simulation exercises to gather feedback, assess effectiveness, and make necessary adjustments to risk response plans and training materials.

Expected Outcome: The organization is well-prepared to implement risk response strategies effectively, with personnel trained and familiar with their roles in risk responses. Simulation exercises enhance the organization's readiness to manage cybersecurity risks dynamically, reducing the potential impact of incidents and aligning with strategic risk management objectives.

GV.RM-05: Lines of Communication Across the Organization Are Established for Cybersecurity Risks, Including Risks from Suppliers and Other Third Parties

Effective communication is the cornerstone of managing cybersecurity risks within any organization. It is crucial for disseminating vital risk information, raising awareness, and guiding the response to cyber threats. This chapter underscores the importance of establishing robust communication strategies to handle cybersecurity risks, emphasizing formal and informal channels to ensure comprehensive coverage and understanding across the organization. Clear and consistent communication cannot be overstated, as it directly influences the organization's ability to effectively respond to and mitigate cyber threats, enhancing overall risk awareness and organizational resilience.

Establishing formal and informal communication channels for risk information is a foundation for effective cybersecurity risk management. Formal channels, such as emails, intranet sites, and dedicated risk management platforms, provide structured and authoritative sources of information. Conversely, informal channels, such as discussions in social areas, team meetings, and even casual conversations, play a vital role in disseminating risk information quickly and fostering a culture of openness and vigilance. Balancing these methods ensures that risk information is accessible to all, enhancing clarity and ensuring critical messages reach the intended audiences immediately.

Clarity and consistency in risk-related messaging are paramount. Developing clear and concise risk messages involves understanding the audience, simplifying complex information, and emphasizing actionable insights. Maintaining consistency across all communication channels reinforces the message, aids in comprehension, and builds trust among stakeholders. However, achieving this clarity and consistency poses challenges, such as overcoming technical jargon and aligning messages with the diverse understanding levels of the audience. Addressing these challenges requires a concerted effort and a clear strategy, including regular training and relatable examples.

Facilitating cross-departmental collaboration on cyber risk communication amplifies the effectiveness of the communication strategy. Encouraging departments to share information and collaborate on risk assessments and response plans fosters a unified approach to cybersecurity. Strategies to enhance collaboration include establishing cross-functional teams, regular inter-departmental meetings, and shared platforms for information exchange. Overcoming barriers to collaboration, such as departmental silos and varying priorities, is essential for a cohesive and effective risk communication strategy.

In today's interconnected business environment, involving suppliers and third parties in risk communication processes is essential. Establishing communication protocols with external partners ensures they know and can respond to cybersecurity risks affecting the organization. This involvement also ensures that suppliers and third parties adhere to the organization's cybersecurity standards, thereby mitigating risks in the supply chain. Effective communication with external entities requires clear agreements, regular updates, and feedback and incident reporting mechanisms.

Leveraging technology enhances the effectiveness of risk communication. Tools and platforms that support real-time communication, automated alerts, and interactive training can significantly improve the dissemination and reception of risk information. However, selecting the right technology solutions and ensuring their effective implementation pose challenges, including integrating existing systems and providing user adoption. Highlighting successful implementations can serve as a guide for organizations looking to enhance their risk communication through technology.

Educating employees on their roles in risk communication is critical. Developing targeted training programs that cater to different roles within the organization ensures that employees are aware of the risks and understand their part in mitigating them. Such education fosters a proactive cybersecurity culture and empowers employees to act as the first defense against cyber threats. Regular assessments of training effectiveness and updates to the curriculum ensure that the training remains relevant and impactful.

Crisis communication planning for cybersecurity incidents prepares the organization to respond swiftly and effectively in the event of a cyberattack. An effective crisis communication plan outlines the protocols for internal and external communication, ensuring that accurate information is disseminated quickly to minimize panic and confusion. Preparing for potential challenges in crisis communication, such as misinformation and rapid escalation, is essential for maintaining control during an incident.

Recommendations

- **Implement Regular Training and Awareness Programs**: Continuously educate employees on cybersecurity risks and their roles in communication, emphasizing the importance of vigilance and reporting.
- **Establish Clear Protocols for External Communication**: Develop and maintain clear guidelines for communicating with suppliers, third parties, and the public during regular operations and in the event of a cybersecurity incident.
- **Leverage Advanced Communication Technologies**: Adopt and integrate advanced communication tools that facilitate real-time alerts, automated notifications, and interactive training to enhance the reach and effectiveness of risk communication.
- **Foster a Culture of Open Communication**: Encourage an organizational culture where employees feel comfortable reporting cybersecurity concerns and incidents without fear of reprisal, ensuring timely and effective responses to risks.
- **Conduct Regular Reviews and Simulations**: Regularly review and update communication strategies, conduct simulations and drills to test the effectiveness of crisis communication plans, and make necessary adjustments based on lessons learned and feedback.

NIST 800-53 Controls

(Refer to our appendix for a comprehensive description of the NIST 800-53 Controls, essential for organizations adhering to strict regulatory standards.)

- PM-09,PM-30

Simplified Security Controls (SSC)

This section outlines our SSC, tailored for environments beyond the scope of NIST compliance. These controls offer streamlined, effective security solutions suitable for any organization seeking to enhance its cybersecurity defenses with clear, actionable measures.

Control 1: Establishment of a Cybersecurity Communication Framework

Control Objective: To create a structured framework for communicating cybersecurity risks across the organization, ensuring that all stakeholders, including suppliers and third parties, are informed and engaged in a timely and effective manner.

Implementation Steps:

1.1 **Framework Development**: Develop a comprehensive communication framework that identifies key cybersecurity risk information, stakeholders, communication channels, frequency, and protocols for regular and emergency communications.

1.2 **Integration of Suppliers and Third Parties**: Explicitly include suppliers and third-party partners in the communication framework, establishing clear guidelines and channels for sharing risk information and collaborating on risk mitigation.

1.3 **Framework Implementation and Training**: Implement the framework across the organization and conduct training sessions for employees, management, and third parties to ensure understanding and adherence to communication protocols.

Expected Outcome: A well-established cybersecurity communication framework enhances the organization's ability to disseminate risk information effectively, ensuring all stakeholders are informed and respond appropriately to cybersecurity risks. This framework fosters a culture of transparency and collaboration, contributing to more effective risk management.

Control 2: Cross-Departmental Cyber Risk Communication Committees

Control Objective: To facilitate ongoing cross-departmental collaboration and communication on cybersecurity risks, ensuring comprehensive risk understanding and coordinated response efforts across the organization.

Implementation Steps:

2.1 **Committee Formation**: Establish cross-departmental committees comprising representatives from IT, cybersecurity, legal, HR, operations, and other relevant departments, along with supplier and third-party representatives as appropriate.

2.2 **Regular Meetings and Updates**: Schedule regular meetings for these committees to share information on current cybersecurity risks, discuss risk impacts across departments, and coordinate response strategies.

2.3 **Action Plan Development**: Develop action plans based on committee discussions, outlining specific tasks, responsibilities, and timelines for addressing identified cybersecurity risks and enhancing organizational resilience.

Expected Outcome: Cross-departmental cyber risk communication committees ensure that cybersecurity risks are understood and managed from a holistic organizational perspective, promoting a unified response to emerging threats and vulnerabilities. This collaborative approach enhances risk awareness and response coordination across all levels of the organization.

Control 3: Crisis Communication Plans for Cybersecurity Incidents

Control Objective: To prepare the organization for effective communication during a cybersecurity crisis, minimizing confusion and reputational damage while ensuring stakeholders are accurately informed throughout the incident response process.

Implementation Steps:

3.1 **Crisis Communication Plan Development**: Develop comprehensive crisis communication plans that include predefined messages, communication channels, spokespersons, and escalation protocols for different cybersecurity incidents.

3.2 **Simulation and Training**: Conduct simulation exercises to test the crisis communication plans and train the crisis communication team and spokespersons to execute the plan under various scenarios.

3.3 **Continuous Review and Improvement**: Regularly review and update the crisis communication plans based on lessons learned from simulations and actual incidents, ensuring the plans remain relevant and practical.

Expected Outcome: The organization possesses a robust crisis communication plan that enables it to respond quickly and effectively to cybersecurity incidents, maintaining stakeholder trust by providing clear, accurate, and timely information. This preparedness reduces the potential impact of incidents on the organization's reputation and operations.

GV.RM-06: A Standardized Method for Calculating, Documenting, Categorizing, and Prioritizing Cybersecurity Risks Is Established and Communicated

Establishing standardized methods for assessing cybersecurity risks is crucial in today's complex and evolving threat landscape. This chapter delves into the significance of adopting uniform approaches to cybersecurity risk assessment, emphasizing the role of standardization in enhancing the clarity, consistency, and comparability of risk management efforts. Standardized methods facilitate a common understanding among stakeholders, enable benchmarking cybersecurity practices, and foster informed decision-making across the organization.

Developing consistent risk calculation metrics and formulas is the first step toward effective cybersecurity risk management. The selection of appropriate metrics and formulas is guided by the nature of the organization's operations, its risk appetite, and the regulatory environment in which it operates. Organizations can ensure a systematic approach to identifying, assessing, and prioritizing risks by establishing a set of standardized risk calculation methods. These methods must be adaptable to reflect the organization's risk profile and the changing cyber threat environment.

Effective risk documentation is pivotal in the risk management process. It maintains detailed records of identified risks, their assessment, and response strategies. Best risk documentation practices include standardized templates, clear and concise language, and a centralized repository for easy access and reference. This not only aids in the ongoing management of risks but also supports regulatory compliance and audit processes. However, maintaining up-to-date and accessible risk documentation presents challenges, including the need for regular updates and the management of sensitive information.

Categorizing risks based on their impact and likelihood is essential for prioritizing risk management efforts. This involves evaluating the potential consequences of each risk and the probability of its occurrence. Effective risk categorization helps organizations focus their resources on the most significant threats, ensuring an efficient and targeted risk response. Nevertheless, accurately categorizing risks can be challenging due to the subjective nature of impact and likelihood assessments, requiring a balanced approach and, where possible, quantitative data to inform decision-making.

Prioritizing cybersecurity risks based on standardized criteria ensures that risk management efforts are aligned with the organization's strategic objectives and risk tolerance. This process involves developing criteria that consider factors such as the severity of impact, regulatory requirements, and business objectives. Prioritization enables organizations to allocate resources effectively, focusing on mitigating high-priority risks that pose the greatest threat to achieving business goals.

Communicating the risk assessment process and its results ensures stakeholder buy-in and facilitates a coordinated risk response. Effective communication strategies involve tailoring messages to the needs and understanding of different stakeholder groups and utilizing various communication platforms to ensure widespread dissemination of risk information. Despite its importance, risk communication faces challenges, including overcoming stakeholder biases and clarifying complex risk information.

Integrating standardized risk assessment methods into continuous risk monitoring is critical for maintaining an up-to-date view of the organization's risk landscape. Continuous risk monitoring allows for the timely detection of new or evolving risks, enabling prompt and effective responses. This integration relies on establishing mechanisms for regular risk assessments and updates to the risk management plan, ensuring that risk mitigation strategies remain relevant and practical.

Training stakeholders on standardized risk assessment techniques is crucial for building a risk-aware culture within the organization. Tailored training programs enhance stakeholders' understanding of risk assessment processes and their role in managing cybersecurity risks. Ongoing education and training are essential for keeping pace with the dynamic nature of cyber threats and the evolving regulatory landscape.

Utilizing technology to support standardized risk management processes offers significant benefits, including efficiency gains, improved accuracy, and enhanced analytical capabilities. Technology tools can automate risk calculation, documentation, and prioritization, streamlining risk management. However, selecting and implementing the right technology solutions requires careful consideration of the organization's specific needs, integration capabilities, and the potential for scalability.

Recommendations

- **Adopt a Continuous Improvement Approach**: Regularly review and update risk assessment methodologies to incorporate the latest threat intelligence, industry best practices, and lessons learned from past incidents.
- **Enhance Stakeholder Engagement**: Foster an organizational culture that values open communication and collaboration on cybersecurity risk management. Engage various stakeholders in the risk assessment to gather diverse perspectives and insights.
- **Leverage Quantitative Data Where Possible**: Incorporate quantitative data into the risk assessment process to support objective decision-making. Use data analytics and threat modeling to enhance the accuracy of risk calculations and prioritizations.
- **Strengthen Documentation Practices**: Implement robust documentation protocols to ensure comprehensive, accurate, and accessible records of risk assessments, decisions, and actions. Utilize document management systems to organize and secure risk documentation effectively.
- **Integrate Advanced Technologies**: Explore and integrate advanced technologies such as artificial intelligence and machine learning to automate routine tasks, enhance risk analysis, and support dynamic risk management efforts.

NIST 800-53 Controls

(Refer to our appendix for a comprehensive description of the NIST 800-53 Controls, essential for organizations adhering to strict regulatory standards.)

- PM-09,PM-18,PM-28,PM-30,RA-03

Simplified Security Controls (SSC)

This section outlines our SSC, tailored for environments beyond the scope of NIST compliance. These controls offer streamlined, effective security solutions suitable for any organization seeking to enhance its cybersecurity defenses with clear, actionable measures.

Control 1: Establishment of a Unified Cybersecurity Risk Assessment Framework

Control Objective: To create and implement a standardized framework for calculating, documenting, categorizing, and prioritizing cybersecurity risks, ensuring consistency and comprehensiveness in risk management activities across the organization.

Implementation Steps:

1.1 **Framework Development**: Develop a comprehensive cybersecurity risk assessment framework that includes standardized metrics, calculation formulas, and risk categorization and prioritization criteria.

1.2 **Documentation and Training**: Document the framework in detailed guidelines and train relevant stakeholders, including IT, security teams, and department heads, to ensure understanding and proper implementation.

1.3 **Integration into Risk Management Processes**: Integrate the standardized framework into existing risk management processes, ensuring that all cybersecurity risks are assessed, documented, and managed according to the established guidelines.

Expected Outcome: A consistent and practical approach to cybersecurity risk management across the organization, enabling accurate risk assessment, clear documentation, and appropriate prioritization of cybersecurity risks based on their impact and likelihood.

Control 2: Continuous Education Program on Standardized Risk Assessment

Control Objective: To ensure that all stakeholders involved in cybersecurity risk management are continuously educated on standardized risk assessment techniques, maintaining the efficacy and accuracy of risk assessments.

Implementation Steps:

2.1 **Development of Education Materials**: Create comprehensive training materials covering standardized risk assessment methods, including case studies and best practices.

2.2 **Regular Training Sessions**: Conduct regular training sessions for all stakeholders involved in risk assessment processes, including new employees and those in changing roles.

2.3 **Feedback and Update Mechanism**: Implement a feedback mechanism to gather insights on the training program's effectiveness and areas for improvement, updating the education materials and training approach accordingly.

Expected Outcome: Stakeholders across the organization understand the standardized risk assessment methods, ensuring that risk management activities are performed consistently and effectively. Continuous education supports the adaptation to new threats and changes in the cybersecurity landscape.

Control 3: Utilization of Technology for Risk Management Standardization

Control Objective: To leverage technology solutions to support the standardized calculation, documentation, categorization, and prioritization of cybersecurity risks, enhancing the efficiency and accuracy of risk management processes.

Implementation Steps:

3.1 **Technology Selection**: Identify and select technology solutions, such as risk management software, that support the standardized risk assessment framework and can be integrated with existing systems.

3.2 **Customization and Implementation**: Customize the selected technology solutions to fit the organization's specific risk assessment framework, ensuring they facilitate the calculation, documentation, categorization, and prioritization of risks defined in the framework.

3.3 **Training and Adoption**: Train relevant stakeholders on using the technology solutions effectively and monitor the adoption and usage to ensure they enhance the risk management process.

Expected Outcome: Enhanced efficiency and accuracy in the organization's risk management activities through the effective use of technology, supporting the standardized approach to cybersecurity risk assessment, and ensuring high consistency and comprehensiveness in managing cybersecurity risks.

GV.RM-07: Strategic Opportunities (i.e., Positive Risks) Are Characterized and Are Included in Organizational Cybersecurity Risk Discussions

Risk is often associated with adverse outcomes in cybersecurity, such as data breaches or system failures. However, an evolving perspective emphasizes identifying and leveraging positive cybersecurity risks or strategic opportunities that can drive innovation and growth. Recognizing and capitalizing on these opportunities requires a shift from viewing cybersecurity solely as a protective measure to seeing it as a catalyst for business enhancement and competitive advantage.

The concept of positive risks in cybersecurity introduces the idea that not all risks lead to negative outcomes; some can present opportunities for significant benefits. Identifying these opportunities involves assessing potential cybersecurity risks for their ability to contribute positively to the organization's goals, such as enabling new technological advancements or market differentiators. The challenge lies in distinguishing between purely negative risks and those that can yield positive outcomes if managed effectively.

Aligning positive risk opportunities with business innovation strategies is crucial for organizations seeking to harness cybersecurity as a growth driver. This alignment integrates risk management processes with business planning and innovation initiatives, ensuring cybersecurity considerations support and enhance strategic objectives. Through case studies and examples, we see organizations that have successfully leveraged cybersecurity risks to their advantage, demonstrating the potential for these risks to contribute to business innovation.

Communicating the value of positive risks to stakeholders is essential in exploiting strategic opportunities. This communication must articulate the benefits of pursuing certain cybersecurity risks, framing them within the context of strategic business objectives and growth opportunities. Developing plans for exploiting identified opportunities then involves detailed strategies for risk management, resource allocation, and execution. However, gaining support for these initiatives can be challenging, as stakeholders often must embrace a level of risk outside their comfort zones.

A balanced view of risks and opportunities is critical for effective cybersecurity planning. Integrating both aspects into the cybersecurity strategy enables organizations to manage risks prudently while pursuing opportunities for innovation and competitive advantage. Fostering an organizational culture that values risk management and opportunity exploitation is essential for maintaining this balance, encouraging a proactive approach to cybersecurity that supports strategic business objectives.

Establishing criteria for evaluating and prioritizing positive risks is fundamental to effective risk management. These criteria should consider the potential impact of the risk on the organization's strategic goals, the feasibility of exploiting the opportunity, and the resources required. Dynamic prioritization allows the organization to adapt its focus as internal and external factors change, ensuring that resources are allocated to opportunities with the most significant potential benefit.

Incentivizing, identifying, and reporting positive risks is a strategy to encourage employees and other stakeholders to contribute to the organization's risk management efforts. Developing incentive mechanisms, such as recognition programs or rewards for innovative risk identification, can help cultivate a culture that values the discovery and exploration of strategic opportunities. However, creating such a culture presents challenges, including overcoming resistance to new ideas and ensuring that the potential benefits of positive risks are well understood.

Incorporating positive risk management into the overall risk framework ensures a comprehensive risk approach that includes protective measures and strategic opportunity exploitation. This integration requires adjustments to risk management policies and practices to accommodate the pursuit of positive risks, supported by leadership's commitment to a balanced risk management approach.

Recommendations

- **Foster a Culture of Innovation**: Encourage a corporate culture that views cybersecurity as a defense mechanism and a springboard for innovation and competitive differentiation.

- **Enhance Stakeholder Engagement**: Develop robust engagement strategies to ensure stakeholders understand and support the pursuit of positive cybersecurity risks as part of the organization's broader strategic objectives.
- **Implement Dynamic Risk Assessment Processes**: Adopt flexible risk assessment methodologies that can quickly adapt to changes in the business environment, identifying threats and opportunities with equal proficiency.
- **Create Incentives for Opportunity Identification**: Establish reward systems that recognize and encourage identifying and exploring positive risks, fostering an environment where innovative thinking is valued.
- **Integrate Positive Risk Management into Strategic Planning**: Ensure that positive risk management is integral to strategic planning processes, allowing for the seamless alignment of cybersecurity initiatives with business innovation efforts.

NIST 800-53 Controls

(Refer to our appendix for a comprehensive description of the NIST 800-53 Controls, essential for organizations adhering to strict regulatory standards.)

- PM-09,PM-18,PM-28,PM-30,RA-03

Simplified Security Controls (SSC)

This section outlines our SSC, tailored for environments beyond the scope of NIST compliance. These controls offer streamlined, effective security solutions suitable for any organization seeking to enhance its cybersecurity defenses with clear, actionable measures.

Control 1: Framework for Identifying and Evaluating Positive Cybersecurity Risks

Control Objective: To systematically identify, evaluate, and prioritize positive cybersecurity risks (opportunities) that can contribute to organizational innovation and strategic objectives.

Implementation Steps:

1.1 **Development of Identification Criteria**: Establish criteria for identifying potential positive risks in cybersecurity initiatives, focusing on opportunities for innovation, efficiency improvements, and competitive advantages.

1.2 **Evaluation and Prioritization Process**: Implement a process for evaluating identified positive risks against predefined criteria such as potential impact on business goals, feasibility, and alignment with the organization's risk appetite.

1.3 **Integration into Cybersecurity Planning**: Incorporate the prioritization of positive risks into regular cybersecurity planning and strategy discussions, ensuring a balanced view of risks and opportunities in decision-making.

Expected Outcome: An established process for recognizing and leveraging positive cybersecurity risks, facilitating a proactive approach to finding strategic opportunities within cybersecurity operations and planning, and enhancing organizational growth and competitive positioning.

Control 2: Communication Protocol for Strategic Cybersecurity Opportunities

Control Objective: To develop and implement a formal communication protocol that ensures strategic opportunities identified within cybersecurity risks are effectively communicated and exploited across the organization.

Implementation Steps:

2.1 **Establishment of Communication Channels**: Identify and formalize communication channels for sharing information about positive cybersecurity risks with key stakeholders, including executive leadership, IT, business units, and innovation teams.

2.2 **Development of Messaging Guidelines**: Create guidelines for the content and presentation of opportunities to ensure clarity, consistency, and compelling arguments for their strategic value and potential impact.

2.3 **Regular Review and Feedback Sessions**: Schedule sessions to discuss identified opportunities, gather feedback, and decide on actionable steps to exploit these opportunities for organizational benefit.

Expected Outcome: Effective and structured communication of cybersecurity-derived strategic opportunities across the organization, leading to informed decision-making and the adoption of initiatives that drive innovation and strategic advantage.

Control 3: Incentive Program for Positive Risk Identification

Control Objective: To encourage the proactive identification and reporting of positive cybersecurity risks by establishing an incentive program that rewards individuals and teams for contributions that lead to realizing strategic opportunities.

Implementation Steps:

3.1 **Program Design**: Design an incentive program that recognizes and rewards employees who successfully identify and articulate positive risks with potential strategic benefits.

3.2 **Implementation and Promotion**: Roll out the incentive program across the organization, promoting it through internal communications to ensure widespread awareness and participation.

3.3 **Monitoring and Recognition**: Monitor the program's effectiveness in generating actionable positive risk opportunities, recognizing and rewarding contributions in a way that highlights their value to the organization.

Expected Outcome: Increased engagement and participation in identifying positive cybersecurity risks, leading to a culture that actively seeks and exploits opportunities for innovation and strategic growth within the cybersecurity domain.

7

Roles, Responsibilities, and Authorities (GV.RR)

> In the chess game of cybersecurity, the leaders set the board, defining the strategy and culture that guide our defenses against ever-evolving threats.

In today's rapidly evolving cybersecurity landscape, the role of organizational leadership in shaping and upholding a robust cybersecurity posture cannot be overstated. As cyber threats become increasingly sophisticated, the need for a comprehensive approach to managing these risks has led to the development of practices and frameworks designed to safeguard digital assets effectively. Among these, the strategic involvement of organizational leaders in cybersecurity risk management stands out as a critical factor for success. Leadership's responsibility encompasses not just the establishment of a risk-aware and ethical culture but also entails accountability for the continuous improvement of cybersecurity measures within the organization. This section delves into the multifaceted roles of leadership in cybersecurity, from defining accountability frameworks and promoting a culture of risk awareness to developing and enforcing cybersecurity policies and fostering an environment where cybersecurity initiatives are supported and effectively integrated into the organization's fabric. Through a detailed exploration of leadership strategies, ethical considerations, policy development, and the cultivation of support for cybersecurity initiatives, this narrative underscores the pivotal role of leadership in navigating the complex cybersecurity domain, ultimately providing actionable recommendations to strengthen this crucial aspect of organizational risk management.

GV.RR-01: Organizational Leadership Is Responsible and Accountable for Cybersecurity Risk and Fosters a Culture That Is Risk-Aware, Ethical, and Continually Improving

The significance of organizational leadership in managing cybersecurity risks cannot be overstated. Leaders play a pivotal role in setting the tone at the top, directly influencing an organization's cybersecurity posture and culture. This section delves into the multifaceted responsibilities of leadership in cybersecurity risk management, emphasizing the importance of a comprehensive accountability framework, strategic approaches to fostering a risk-aware culture, ethical considerations in decision-making, the development and review of cybersecurity policies, effective communication, the implementation of continuous improvement models, and the cultivation of leadership support for cybersecurity initiatives.

A Comprehensive Guide to the NIST Cybersecurity Framework 2.0: Strategies, Implementation, and Best Practice, First Edition. Jason Edwards.
© 2025 John Wiley & Sons Ltd. Published 2025 by John Wiley & Sons Ltd.

At the heart of effective cybersecurity risk management lies the introduction to the leadership's role. Leaders are not just figureheads; their actions and decisions profoundly impact the organization's approach to managing cybersecurity risks. They ensure cybersecurity measures align with the organization's goals and objectives, safeguarding its assets, reputation, and stakeholders. The accountability framework defines the roles and responsibilities of leaders in cybersecurity, ensuring that they are clearly understood and effectively implemented across all levels of the organization.

Leadership strategies for promoting a risk-aware culture are essential for embedding cybersecurity awareness into the organizational fabric. Leaders must champion cybersecurity initiatives, demonstrating their importance through their actions and communications. By prioritizing cybersecurity in their strategic planning and resource allocation, leaders can encourage a culture that values and practices diligent risk management. This culture is crucial for the early detection and mitigation of cybersecurity threats.

Ethical considerations play a critical role in cybersecurity decision-making. Leaders must navigate the complex landscape of cybersecurity ethics, making decisions that protect the organization while respecting privacy, data protection laws, and ethical norms. This requires a deep understanding of the moral implications of cybersecurity actions and developing guidelines that ensure ethical practices are followed consistently.

The involvement of leadership in developing and reviewing cybersecurity policies is indispensable. Policies are the backbone of an organization's cybersecurity framework, providing the guidelines and standards for behavior and decision-making. Leaders must be actively involved in creating, reviewing, and updating these policies to ensure they remain relevant and effective in the face of evolving cybersecurity threats.

Communicating the importance of cybersecurity from the top down is crucial for ensuring that all employees understand its significance and their role in safeguarding the organization. Leadership must be clear, consistent, and compelling in their communications, embedding cybersecurity as a core value of the organization.

Implementing a continuous improvement model for cybersecurity reflects the dynamic nature of cyber threats and the need for organizations to adapt continually. Leaders should advocate for adopting models that promote ongoing assessment, learning, and enhancement of cybersecurity measures. This approach improves the organization's cybersecurity posture and fosters a culture of innovation and resilience.

Cultivating leadership support for cybersecurity initiatives is fundamental to their success. Leaders must be visible champions of these initiatives, providing the necessary resources and support to achieve their objectives. This involves engaging with cybersecurity teams, understanding their challenges, and facilitating cross-functional collaboration to drive comprehensive cybersecurity solutions.

Assessing the effectiveness of leadership in cybersecurity risk oversight involves regular evaluations of how well leaders are fulfilling their responsibilities in this area. This assessment can help identify areas for improvement, ensuring that leadership remains a strong pillar in the organization's cybersecurity efforts.

Recommendations

- **Establish Clear Cybersecurity Accountability Structures:** Organizations should define and communicate the roles and responsibilities of their leaders in cybersecurity, ensuring that accountability is understood and accepted at all levels.

- **Promote Continuous Education and Awareness:** Leaders should engage in and promote ongoing cybersecurity education and awareness programs, staying informed about the latest cybersecurity threats and best practices.
- **Foster an Ethical Cybersecurity Culture:** Develop and implement ethical guidelines for cybersecurity practices, ensuring decisions reflect organizational values and legal requirements.
- **Regularly Review and Update Cybersecurity Policies:** Leadership should be actively involved in periodically reviewing and updating cybersecurity policies to address new threats and incorporate best practices.
- **Implement and Support Continuous Improvement Models:** Encourage the adoption of continuous improvement models in cybersecurity practices, supporting initiatives that enhance the organization's cybersecurity posture over time.

NIST 800-53 Controls

(Refer to our appendix for a comprehensive description of the NIST 800-53 Controls, essential for organizations adhering to strict regulatory standards.)

- PM-02,PM-19,PM-23,PM-24,PM-29

Simplified Security Controls (SSC)

Control 1: Leadership Accountability and Responsibility Charter

Control Objective: To formalize the roles, responsibilities, and accountability of organizational leadership in governing and overseeing cybersecurity risk management, ensuring a top-down approach to fostering a risk-aware culture.

Implementation Steps:

1.1 **Charter Development:** Develop a Leadership Accountability and Responsibility Charter that defines the cybersecurity governance roles, responsibilities, and accountability mechanisms for senior leaders and board members.

1.2 **Approval and Adoption:** Present the charter for approval to the board and senior leadership team, ensuring alignment and commitment to the defined roles and responsibilities.

1.3 **Communication and Training:** Communicate the charter throughout the organization and provide training for leaders and employees on the importance of leadership accountability in cybersecurity risk management.

Expected Outcome: A clear and formalized understanding across the organization of leadership's role and accountability in cybersecurity, fostering a culture of risk awareness, ethical decision-making, and continuous improvement driven from the top.

Control 2: Ethical Decision-Making Framework in Cybersecurity

Control Objective: To embed ethical considerations into cybersecurity decision-making processes, ensuring that decisions reflect the organization's values and commitment to ethical practices.

Implementation Steps:

2.1 **Framework Development:** Develop an Ethical Decision-Making Framework specific to cybersecurity, outlining ethical principles, guidelines, and a decision-making process that considers legal, moral, and social implications.

2.2 **Integration with Cybersecurity Policies:** Integrate the ethical decision-making framework into existing cybersecurity policies and decision-making processes, ensuring that moral considerations are a standard part of cybersecurity governance.

2.3 **Awareness and Training Programs:** Implement awareness and training programs for all employees, focusing on the importance of ethics in cybersecurity, how to apply the ethical decision-making framework, and the impact of ethical considerations on organizational integrity and trust.

Expected Outcome: An organizational culture that prioritizes ethical considerations in all cybersecurity decisions, enhancing the integrity, trustworthiness, and social responsibility of the organization's cybersecurity practices.

Control 3: Continuous Improvement Program for Cybersecurity Practices

Control Objective: To establish a continuous improvement program for leadership-driven cybersecurity practices to ensure the organization's cybersecurity posture evolves in line with emerging threats and technologies.

Implementation Steps:

3.1 **Program Development:** Develop a continuous improvement program that includes mechanisms for regular assessment of cybersecurity practices, identifying areas for improvement, and implementing enhancements.

3.2 **Leadership Involvement and Support:** Ensure active involvement of organizational leadership in the program, including regular reviews of program outcomes and provision of necessary resources for implementing improvements.

3.3 **Feedback Loops and Adaptation:** Create feedback loops that allow for the collection and incorporation of insights from across the organization and external cybersecurity developments into the continuous improvement process.

Expected Outcome: A dynamic and responsive cybersecurity practice that continually adapts and improves, supported by leadership commitment and resources, ensuring the organization stays ahead of cybersecurity risks and aligns with best practices and technological advancements.

GV.RR-02: Roles, Responsibilities, and Authorities Related to Cybersecurity Risk Management Are Established, Communicated, Understood, and Enforced

Establishing, communicating, understanding, and enforcing roles, responsibilities, and authorities within the realm of cybersecurity risk management constitute the foundation of an effective cybersecurity posture. This section delves into the critical process of structuring cybersecurity roles and responsibilities, the establishment of clear cybersecurity authorities and reporting lines, methods for communicating roles across the organization, educational programs aimed at ensuring a thorough understanding of cybersecurity responsibilities, enforcement mechanisms, regular reviews for alignment with cybersecurity goals, addressing noncompliance, integrating cybersecurity roles into the organizational structure, and the crucial role of internal audit in assessing compliance with these responsibilities.

The introduction to structuring cybersecurity roles and responsibilities underscores the imperative of having a clear, organized framework that delineates each role's specific duties and authorities. Such a framework facilitates the efficient management of cybersecurity risks and ensures that every member of the organization understands their part in the collective effort to safeguard digital assets.

Establishing clear cybersecurity authorities and reporting lines is essential for creating an environment where decisions and actions related to cybersecurity are promptly and effectively communicated. This structure ensures a clear chain of command and accountability, which is crucial for rapidly responding to cybersecurity incidents and the strategic management of cybersecurity initiatives.

Effective processes for communicating cybersecurity roles across the organization are vital. These processes must be designed to reach every corner of the organization, ensuring everyone, from the executive board to the newest employees, knows their cybersecurity responsibilities. Communication strategies can include regular meetings, training sessions, newsletters, and intranet postings.

Educational programs play a crucial role in ensuring an understanding of cybersecurity responsibilities. These programs should provide initial training upon role assignment and offer ongoing education to keep pace with the evolving cyber threat landscape. The goal is to create a culture of continuous learning and adaptation, reinforcing the importance of cybersecurity in the daily activities of all employees.

Enforcement mechanisms for cybersecurity policies and roles are necessary to ensure compliance and to address deviations proactively. These mechanisms can include audits, reviews, and disciplinary actions for noncompliance. Effective enforcement fosters a culture of accountability and reinforces the seriousness with which the organization views its cybersecurity obligations.

Regular reviews of roles and responsibilities for alignment with cybersecurity goals ensure that the organization's cybersecurity strategy remains effective and responsive to new threats. These reviews allow for the timely adjustment of roles and responsibilities in response to changes in the cybersecurity landscape, technological advancements, and shifts in the organization's structure or business model.

Addressing noncompliance and enhancing accountability in cybersecurity roles are critical for maintaining the integrity of an organization's cybersecurity framework. Mechanisms for addressing noncompliance should be clear, fair, and consistently applied, ensuring all employees understand the consequences of failing to meet their cybersecurity responsibilities.

Integrating cybersecurity roles into the organizational structure is crucial for embedding cybersecurity into the organization's fabric. This integration ensures that cybersecurity considerations are part of strategic decision-making processes and that cybersecurity is not siloed but is a cross-functional priority.

The role of internal audits in assessing compliance with cybersecurity responsibilities cannot be overstated. Internal audits provide an independent evaluation of how well the organization adheres to established cybersecurity policies and procedures, offering insights into improvement areas and helping enforce accountability.

Recommendations

- **Develop Comprehensive Role Definitions:** Organizations should meticulously define and document cybersecurity roles and responsibilities, ensuring clarity and comprehensibility. This effort should encompass all levels of the organization, from top management to operational staff.
- **Implement Robust Communication Strategies:** Establish ongoing and multifaceted communication strategies to inform all employees about their cybersecurity responsibilities and any changes to cybersecurity policies or procedures.
- **Prioritize Cybersecurity Education and Training:** Invest in comprehensive, continuous education and training programs that cover the basics of cybersecurity and address the specific responsibilities associated with each role within the organization.
- **Enforce Policies Through Clear Mechanisms:** Create and communicate clear mechanisms for enforcing cybersecurity policies, including consequences for noncompliance. Ensure that these mechanisms are applied consistently across the organization.

- **Conduct Regular Role and Responsibility Reviews:** Review and update roles and responsibilities to ensure they align with current cybersecurity goals and the threat landscape. Engage stakeholders from various parts of the organization in this process to provide comprehensive coverage and buy-in.

NIST 800-53 Controls

(Refer to our appendix for a comprehensive description of the NIST 800-53 Controls, essential for organizations adhering to strict regulatory standards.)

- PM-02,PM-13,PM-19,PM-23,PM-24,PM-29

Simplified Security Controls (SSC)

Control 1: Comprehensive Cybersecurity Role Definition and Documentation

Control Objective: To clearly define, document, and maintain up-to-date descriptions of roles, responsibilities, and authorities for all positions involved in cybersecurity risk management, ensuring clarity and accountability across the organization.

Implementation Steps:

1.1 **Role Definition Workshop:** Organize workshops with key cybersecurity stakeholders to define and review roles, responsibilities, and authorities related to cybersecurity risk management.

1.2 **Documentation and Approval:** Document the defined roles and responsibilities in an official company policy or framework and obtain approval from senior management to ensure alignment and commitment.

1.3 **Dissemination and Acknowledgment:** Communicate the documented roles and responsibilities across the organization through multiple channels (e.g., intranet, email, and meetings) and require acknowledgment from all employees to ensure understanding and acceptance.

Expected Outcome: Clearly defined and widely understood cybersecurity roles and responsibilities across the organization, leading to enhanced accountability, efficient risk management processes, and a strong cybersecurity posture.

Control 2: Cybersecurity Role and Responsibility Education Programs

Control Objective: To ensure all employees understand their roles and responsibilities in cybersecurity risk management through comprehensive and ongoing educational programs.

Implementation Steps:

2.1 **Program Development:** Develop an educational program tailored to different roles within the organization, focusing on the importance of cybersecurity, specific role-based responsibilities, and how individuals can contribute to risk management efforts.

2.2 **Regular Training Sessions:** Schedule and conduct regular training sessions, leveraging a variety of formats (e.g., online courses, workshops, and seminars) to accommodate different learning preferences and schedules.

2.3 **Assessment and Feedback:** Implement assessment mechanisms to measure program effectiveness and gather feedback for continuous improvement, ensuring the program remains relevant and engaging over time.

Expected Outcome: A well-informed workforce that understands individual and collective responsibilities in cybersecurity risk management, leading to a culture of accountability and proactive risk management behaviors across the organization.

Control 3: Accountability and Enforcement Mechanism for Cybersecurity Roles

Control Objective: To establish and enforce accountability for cybersecurity roles and responsibilities, ensuring compliance with defined policies and procedures.

Implementation Steps:

3.1 **Establishment of Enforcement Policies:** Develop and communicate clear policies regarding enforcing cybersecurity roles and responsibilities, including potential consequences for noncompliance.

3.2 **Monitoring and Reporting System:** Implement a system for monitoring compliance with defined cybersecurity roles and responsibilities and for reporting instances of noncompliance to relevant authorities within the organization.

3.3 **Review and Adjustment Process:** Establish a regular review process for assessing enforcement mechanisms' effectiveness and making necessary adjustments based on feedback and observed compliance rates.

Expected Outcome: A strong culture of accountability for cybersecurity, where individuals understand the importance of their roles and are motivated to comply with organizational policies and procedures. Noncompliance is promptly addressed, reinforcing the seriousness with which the organization treats cybersecurity risk management.

GV.RR-03: Adequate Resources Are Allocated Commensurate with the Cybersecurity Risk Strategy, Roles, Responsibilities, and Policies

Allocating adequate resources in line with an organization's cybersecurity risk strategy is critical to effective cybersecurity risk management. This involves financial investments and the commitment of human capital and technological resources. This section discusses the introduction to resource allocation for cybersecurity risk management, strategic planning for cybersecurity resource requirements, budgeting for cybersecurity in line with organizational risk appetite, ensuring sufficient human capital investment for cybersecurity initiatives, allocation of technological resources to support cybersecurity measures, integration of resource planning with cybersecurity policy implementation, periodic review of resource allocation effectiveness, adjusting resources in response to the evolving cybersecurity landscape, and training and development as part of cybersecurity resource allocation.

The resource allocation process for cybersecurity risk management begins with a comprehensive understanding of the organization's current cybersecurity posture and strategic goals. Recognizing the dynamic nature of cyber threats, organizations need to adopt a flexible approach to resource allocation that can adapt to changing circumstances. This approach should be underpinned by a strategic planning process that assesses current and future resource requirements in light of the organization's cybersecurity risk strategy.

Budgeting for cybersecurity must be aligned with the organization's risk appetite, ensuring that investments are proportional to the identified risks and potential impacts. This necessitates a clear understanding of the organization's tolerance for risk and the allocation of resources accordingly. The budget should cover all aspects of cybersecurity, including prevention, detection, response, and recovery efforts and ongoing maintenance and improvement of cybersecurity measures.

Investment in human capital is paramount to the success of cybersecurity initiatives. Organizations must ensure that they have a sufficient number of skilled cybersecurity professionals who are equipped to manage and mitigate risks effectively. This includes hiring new talent and investing in the ongoing development of current staff to keep pace with the rapidly evolving threat landscape.

Technological resources play a crucial role in supporting cybersecurity measures. This encompasses the procurement and implementation of security tools and technologies and the maintenance and upgrading of existing systems to ensure they remain effective against new threats. The allocation of technological resources should be guided by thoroughly evaluating the organization's specific needs and vulnerabilities.

Integrating resource planning with cybersecurity policy implementation ensures that the allocation of resources is directly linked to the organization's cybersecurity objectives and policies. This integration facilitates a cohesive approach to cybersecurity, where resources are strategically deployed to support the most critical areas of need.

The effectiveness of resource allocation must be periodically reviewed to ensure that investments are delivering the desired outcomes. This review process should consider the financial aspects and the efficiency and effectiveness of human capital and technological investments in improving the organization's cybersecurity posture.

Adjusting resources in response to the evolving cybersecurity landscape is crucial. As cyber threats grow in sophistication and scale, organizations must remain vigilant and flexible, ready to reallocate resources as necessary to address new and emerging risks. This may involve increasing investments in certain areas, reallocating resources from one initiative to another, or adopting new technologies and approaches.

Training and development are integral components of cybersecurity resource allocation. By investing in the continuous education and training of their cybersecurity workforce, organizations can ensure that their teams are equipped with the latest knowledge and skills to defend against cyber threats effectively.

Recommendations

- **Conduct a Comprehensive Cybersecurity Resource Assessment:** Regularly evaluate your organization's cybersecurity needs to ensure resource allocation aligns with the current risk landscape and organizational objectives.
- **Align Budgeting with Risk Appetite:** Develop a cybersecurity budget that reflects your organization's risk tolerance, ensuring that investments are focused on areas of greatest need and potential impact.
- **Invest in Human Capital:** Prioritize the recruitment and development of skilled cybersecurity professionals and foster a culture of continuous learning to maintain a knowledgeable and responsive cybersecurity team.
- **Leverage Technology Strategically:** Implement and maintain technological solutions directly supporting your cybersecurity objectives, ensuring your technological investments are scalable and adaptable to evolving threats.
- **Review and Adjust Resources Regularly:** Establish a process for the ongoing review of cybersecurity resource allocation, adjusting your strategy to respond to new challenges and opportunities in the cybersecurity landscape.

NIST 800-53 Controls

(Refer to our appendix for a comprehensive description of the NIST 800-53 Controls, essential for organizations adhering to strict regulatory standards.)

- PM-3

Simplified Security Controls (SSC)

Control 1: Cybersecurity Resource Assessment and Planning Process

Control Objective: To ensure that cybersecurity resources (financial, human, and technological) are strategically planned and allocated in line with the organizational cybersecurity risk strategy, roles, responsibilities, and policies.

Implementation Steps:

1.1 **Conduct a Comprehensive Resource Needs Assessment:** Annually assess the resources required for effective cybersecurity risk management, considering current risks, strategic objectives, and the cybersecurity maturity model.

1.2 **Develop a Resource Allocation Plan:** Based on the assessment, develop a detailed plan outlining the required budget, staffing levels, and technology investments necessary to meet cybersecurity objectives.

1.3 **Secure Approval and Implement Allocation:** Present the resource allocation plan to senior management for approval. Once approved, systematically allocate resources according to the plan, ensuring that all cybersecurity initiatives are adequately supported.

Expected Outcome: A well-defined and executed process for cybersecurity resource planning and allocation that aligns with the organization's risk management strategy and objectives, ensuring that all cybersecurity efforts are adequately resourced for effective implementation and sustained operations.

Control 2: Periodic Review and Adjustment of Cybersecurity Resources

Control Objective: To regularly review the allocation and effectiveness of resources dedicated to cybersecurity, making adjustments in response to changes in the cybersecurity landscape, emerging threats, and organizational priorities.

Implementation Steps:

2.1 **Establish a Review Schedule:** Set a regular schedule for reviewing the effectiveness and adequacy of allocated cybersecurity resources, including budget, personnel, and technology.

2.2 **Conduct Resource Effectiveness Assessments:** Evaluate the impact of allocated resources on cybersecurity posture improvements, identifying areas where resources are either insufficient or could be optimized.

2.3 **Adjust Resources as Necessary:** Based on the assessment outcomes, reallocate or adjust resources to address gaps, inefficiencies, or emerging needs, ensuring ongoing alignment with cybersecurity objectives and risk strategies.

Expected Outcome: A dynamic and responsive approach to cybersecurity resource allocation ensures that resources are continually aligned with current and emerging cybersecurity needs, enhancing the organization's ability to manage and mitigate risks effectively.

Control 3: Training and Development for Cybersecurity Capacity Building

Control Objective: To prioritize training and development as critical components of cybersecurity resource allocation, ensuring that the organization's human capital has the knowledge and skills necessary to address current and future cybersecurity challenges.

Implementation Steps:

3.1 **Identify Training Needs:** Regularly assess the training and development needs of the cybersecurity team and other relevant personnel, identifying gaps in skills and knowledge that could impact cybersecurity risk management.

3.2 **Develop and Implement Training Programs:** Develop comprehensive training and professional development programs that address identified needs, leveraging a mix of in-house, external, and online training resources.

3.3 **Evaluate Training Effectiveness:** Implement mechanisms to evaluate the effectiveness of training programs in improving cybersecurity capabilities and adjust the programs based on feedback and evolving cybersecurity requirements.

Expected Outcome: A well-trained and skilled cybersecurity workforce capable of effectively managing and mitigating cybersecurity risks, supported by ongoing investment in professional development and training programs that ensure personnel remain at the forefront of cybersecurity practices and technologies.

GV.RR-04: Cybersecurity Is Included in Human Resource Practices

Integrating cybersecurity into human resource (HR) practices represents a critical strategy for enhancing an organization's overall cybersecurity posture. This section explores the necessity of incorporating cybersecurity considerations throughout the HR management lifecycle, including recruitment and hiring practices, employee onboarding and training programs, performance management processes, job descriptions, and career development plans, employee awareness programs, incident response training, offboarding processes, and the role of HR in communicating cybersecurity expectations. Organizations can foster a culture prioritizing cybersecurity at every level through a comprehensive approach.

The introduction to integrating cybersecurity into HR management highlights the importance of embedding cybersecurity awareness and competencies across all employee-related processes. By aligning HR practices with cybersecurity strategies, organizations can ensure that every employee contributes to the security of information assets from the moment they are hired.

Cybersecurity considerations in recruitment and hiring practices are crucial for identifying candidates who have the necessary technical skills and demonstrate a strong understanding of and commitment to cybersecurity principles. This includes evaluating candidates' awareness of cybersecurity issues and their approach to handling sensitive information.

Including cybersecurity in employee onboarding and training programs ensures new hires have the knowledge and tools to protect the organization from day one. Comprehensive onboarding should cover the organization's cybersecurity policies, the importance of data protection, and the individual's role in maintaining a secure information environment.

Performance management processes that reflect cybersecurity competencies allow organizations to assess and encourage the development of cybersecurity skills among their workforce. By incorporating cybersecurity goals and objectives into performance evaluations, employers can reinforce the importance of security-aware behavior and recognize individuals who contribute to the organization's cybersecurity efforts.

Incorporating cybersecurity criteria into job descriptions and career development plans helps to clarify the security-related expectations for each role within the organization. This clarity ensures employees understand how their positions contribute to the organization's cybersecurity posture. It encourages them to acquire the necessary skills to advance their careers while enhancing organizational security.

Employee awareness programs addressing cybersecurity best practices are essential for maintaining a vigilant workforce. Regular, engaging training sessions can help to keep cybersecurity at the forefront of employees' minds, encouraging them to adopt secure behaviors and report potential security incidents promptly.

As part of employee development, incident response training prepares staff to act effectively in the event of a cybersecurity incident. This training should be tailored to different employees' specific roles and responsibilities, ensuring that everyone knows what to do in the event of an attack or breach.

Offboarding processes that ensure cybersecurity risk mitigation are critical for protecting organizational information when employees leave. Procedures should include revoking access rights, returning company-owned devices, and debriefing sessions to identify potential security concerns.

The role of HR in communicating cybersecurity expectations to employees underscores the importance of clear, consistent messaging about the organization's commitment to cybersecurity. HR can play a pivotal role in reinforcing the message that everyone has a part to play in protecting the organization's digital assets.

Recommendations

- **Embed Cybersecurity Criteria in Hiring Practices:** Ensure that cybersecurity awareness and competencies are key criteria in recruiting and selecting all employees, not just those in IT roles.
- **Mandate Ongoing Cybersecurity Training:** Implement comprehensive cybersecurity training for all employees to foster a culture of security awareness and vigilance.
- **Integrate Cybersecurity into Performance Evaluations:** Incorporate cybersecurity goals and achievements into performance management processes to incentivize and reward security-conscious behavior.
- **Develop Role-Specific Cybersecurity Training:** Tailor cybersecurity training and development programs to employees' specific roles and responsibilities, ensuring that everyone is equipped to contribute to the organization's cybersecurity efforts.
- **Establish Secure Offboarding Procedures:** Implement thorough offboarding processes that address cybersecurity risks, ensuring the secure transition of roles and responsibilities when employees exit the organization.

NIST 800-53 Controls

(Refer to our appendix for a comprehensive description of the NIST 800-53 Controls, essential for organizations adhering to strict regulatory standards.)

- PM-13,PS-01,PS-07,PS-09

Simplified Security Controls (SSC)

Control 1: Integration of Cybersecurity Criteria in Recruitment and Hiring

Control Objective: To incorporate cybersecurity competencies and awareness into the recruitment and hiring processes, ensuring that new hires possess the necessary cybersecurity knowledge or demonstrate the potential to develop such competencies.

Implementation Steps:

1.1 **Develop Cybersecurity Competency Criteria:** Collaborate with cybersecurity and HR teams to define essential cybersecurity competencies and awareness levels for various roles within the organization.

1.2 **Incorporate Criteria into Job Descriptions:** Integrate the defined cybersecurity competency criteria into job descriptions and hiring materials, ensuring that expectations are clear to potential candidates.

1.3 **Assess Cybersecurity Competencies During Hiring:** Implement assessment mechanisms (e.g., interview questions and practical tests) to evaluate candidates' cybersecurity competencies and awareness as part of the hiring process.

Expected Outcome: Enhanced cybersecurity posture of the organization through the recruitment of individuals who possess critical cybersecurity knowledge and competencies or demonstrate the ability and willingness to develop them, contributing to a more robust cybersecurity culture.

Control 2: Cybersecurity Onboarding and Continuous Training Programs

Control Objective: To ensure that all employees, regardless of their role, receive comprehensive cybersecurity training as part of their onboarding process and continue to receive updated training throughout their employment.

Implementation Steps:

2.1 **Develop a Comprehensive Onboarding Program:** Create an onboarding program that includes cybersecurity training modules tailored to the new hire's role within the organization.

2.2 **Implement Continuous Cybersecurity Training:** Establish a continuous training program that provides regular, updated training sessions on current cybersecurity threats, best practices, and organizational policies.

2.3 **Monitor and Evaluate Training Effectiveness:** Implement mechanisms to monitor participation and assess the effectiveness of the cybersecurity training programs, making adjustments based on feedback and the evolving cybersecurity landscape.

Expected Outcome: A workforce that is consistently aware of and trained in the latest cybersecurity practices and threats, contributing to the overall resilience of the organization against cyberattacks.

Control 3: Performance Management Reflecting Cybersecurity Competencies

Control Objective: To integrate cybersecurity competencies into employee performance management processes, ensuring that employees are evaluated and recognized for their contributions to cybersecurity within their roles.

Implementation Steps:

3.1 **Define Cybersecurity Performance Metrics:** In collaboration with cybersecurity experts, define specific metrics or indicators that reflect an employee's contribution to cybersecurity initiatives and practices.

3.2 **Integrate Metrics into Performance Reviews:** Incorporate the defined cybersecurity performance metrics into the regular performance review process for all employees.

3.3 **Provide Feedback and Development Opportunities:** Use performance reviews to provide feedback on cybersecurity practices and identify opportunities for further development and training in cybersecurity competencies.

Expected Outcome: A performance management process that encourages and rewards contributions to the organization's cybersecurity efforts, fostering a culture of continuous improvement and accountability in cybersecurity practices across all roles.

8

Policy (GV.PO)

> Crafting cybersecurity policies is not just about defense; it's about creating a dynamic blueprint that evolves with our digital landscape, ensuring our resilience and alignment with our core mission.

In the intricate realm of cybersecurity, the creation and rigorous upkeep of policies tailored to an organization's unique landscape stand as a crucial defense against digital threats. Such policies must be intricately designed to mesh seamlessly with the organization's overarching goals and strategic direction, ensuring comprehensive protection and alignment with its mission. The process involves a symbiotic collaboration between the leadership and the cybersecurity teams, fostering a security culture permeating every level of the organization. Organizations can significantly bolster their cybersecurity posture by employing innovative communication and training techniques and leveraging cutting-edge technology for enforcement and monitoring. This approach underscores the necessity for policies to be fluid and capable of evolving in tandem with new technological advancements, emerging threats, and changes in the regulatory environment, thus maintaining their relevance and effectiveness in a rapidly changing digital world.

GV.PO-01: Policy for Managing Cybersecurity Risks Is Established Based on Organizational Context, Cybersecurity Strategy, and Priorities and Is Communicated and Enforced

Introducing a policy framework for managing cybersecurity risks marks the first step in a comprehensive security strategy. It sets a formal foundation, defining an organization's cybersecurity efforts' scope, objectives, and expectations. Such a framework not only delineates the roles and responsibilities across the organization but also establishes a benchmark against which the effectiveness of cybersecurity measures can be assessed. This initial phase is critical; it requires a clear understanding of the organization's mission, values, and operational landscape, ensuring the policies developed are applicable and enforceable.

Customizing cybersecurity policies to fit the organizational context is not merely a recommendation but a necessity. A one-size-fits-all approach to cybersecurity is ineffective, given the unique risks, resources, and regulatory requirements different organizations face. Tailoring policies involves a thorough risk assessment, identifying vulnerabilities, and prioritizing actions based on potential impact. This customization ensures that the policies are relevant, realistic, and

A Comprehensive Guide to the NIST Cybersecurity Framework 2.0: Strategies, Implementation, and Best Practice, First Edition. Jason Edwards.
© 2025 John Wiley & Sons Ltd. Published 2025 by John Wiley & Sons Ltd.

aligned with the organization's needs, enhancing their acceptance and implementation across departments.

Aligning policies with the organizational cybersecurity strategy and priorities is another critical component. This alignment ensures that the policies support broader security objectives and are integrated with other risk management efforts. It involves a strategic view of the organization's long-term goals, technology landscape, and cybersecurity threats. Such alignment helps create a cohesive and proactive cybersecurity environment that can adapt to threats, technologies, and business objectives.

The engagement and approval process of cybersecurity policies by leadership cannot be overstated. Leadership's support and commitment are essential for the legitimacy and success of cybersecurity policies. This process involves presenting the proposed policies to the leadership for review, adjustments, and approval. It is an opportunity to ensure that the policies have the backing necessary for enforcement and that they align with the organization's overall strategic direction.

Once approved, disseminating cybersecurity policies across the organization is the next crucial step. Effective communication strategies ensure that all employees know the policies, understand their roles and responsibilities, and recognize the importance of compliance. This dissemination can take various forms, including training sessions, intranet posts, and policy manuals tailored to suit the organization's culture and communication preferences.

Training programs play a pivotal role in supporting the understanding and enforcement of policies. By educating employees about the policies, their purpose, and the specific actions required for compliance, organizations can foster a culture of cybersecurity awareness. Training should be ongoing, reflecting updates to the policies and emerging cyber threats, to maintain a high level of vigilance and preparedness among staff.

Implementing enforcement mechanisms for compliance with cybersecurity policies is essential for their effectiveness. These mechanisms may include regular audits, monitoring tools, and disciplinary actions for noncompliance. They serve as both a deterrent against policy violations and a means to quickly identify and rectify lapses in compliance, ensuring that cybersecurity measures are consistently applied.

Monitoring and reporting on policy adherence is crucial for maintaining and improving the cybersecurity posture. This involves regular assessments of how well the policies are being followed and the effectiveness of the enforcement mechanisms. Insights gained from these evaluations can guide updates to the policies and enforcement strategies, ensuring they remain relevant and effective in the face of changing cybersecurity threats and business needs.

Leveraging technology to aid in policy communication and enforcement can significantly enhance the effectiveness of the cybersecurity policy framework. Tools such as policy management software, automated compliance checks, and real-time monitoring solutions can streamline policy implementation and maintenance, reduce human error risk, and provide actionable intelligence for decision-making.

Recommendations

- **Engage All Stakeholders Early**: Involve representatives from various departments in the policy development process. This inclusivity ensures that the policies are comprehensive, practical, and have the buy-in necessary for effective implementation.
- **Continuous Risk Assessment**: Regularly update the cybersecurity risk assessment to reflect new threats, technologies, and business processes. This ongoing assessment informs the need for policy adjustments and ensures the policies align with the organization's risk posture.

- **Clear Communication and Training**: Develop clear, concise, and accessible policy documents. Complement these with robust training programs that are updated regularly to keep pace with changes in cybersecurity practices and threats.
- **Enforcement and Accountability**: Establish clear consequences for noncompliance and enforce them consistently. However, it also creates an environment where employees feel comfortable reporting security incidents or policy violations without fear of reprisal.
- **Leverage Technology**: Utilize technology to facilitate the communication, enforcement, and monitoring of cybersecurity policies. Automation and cybersecurity tools can provide critical support in maintaining a solid security posture.

NIST 800-53 Controls

(Refer to our appendix for a comprehensive description of the NIST 800-53 Controls, essential for organizations adhering to strict regulatory standards.)

- AC-01,AT-01,AU-01,CA-01,CM-01,CP-01,IA-01,IR-01,MA-01,MP-01,PE-01,PL-01,PM-01,PS-01, PT-01,RA-01,SA-01,SC-01,SI-01,SR-01

Simplified Security Controls (SSC)

Control 1: Development and Customization of Cybersecurity Policies

Control Objective: To establish a comprehensive set of cybersecurity policies tailored to the organizational context, align with the cybersecurity strategy and priorities, and provide clear guidance on managing cybersecurity risks.

Implementation Steps:

1.1 **Policy Development Workshop**: Organize workshops with key stakeholders from IT, cybersecurity, legal, HR, and business units to draft cybersecurity policies that address the specific needs and context of the organization.

1.2 **Customization to Organizational Context**: Ensure policies are customized to reflect the organization's specific cybersecurity risks, technological environment, business processes, and regulatory obligations.

1.3 **Review and Approval**: Submit the drafted policies for review and approval by executive leadership to ensure alignment with overall organizational strategy and priorities.

Expected Outcome: A set of approved cybersecurity policies customized to the organization's specific context, providing a clear framework for managing cybersecurity risks and aligning with strategic objectives.

Control 2: Comprehensive Dissemination and Training on Cybersecurity Policies

Control Objective: To ensure that all employees are informed about the cybersecurity policies, understand their implications, and are trained on their responsibilities within these policies.

Implementation Steps:

2.1 **Communication Plan**: Develop and execute a communication plan that uses multiple channels to disseminate cybersecurity policies across the organization, ensuring all employees are reached.

2.2 **Tailored Training Programs**: Implement training programs tailored to different roles within the organization, focusing on the relevance of the cybersecurity policies to each role and the specific actions required for compliance.

2.3 **Feedback and Q&A Sessions**: Organize feedback and Q&A sessions to address employees' questions or concerns about the policies, ensuring they are fully understood.

Expected Outcome: All employees are aware of, understand, and are equipped to comply with the cybersecurity policies, contributing to a culture of cybersecurity awareness and risk management across the organization.

Control 3: Monitoring, Reporting, and Enforcement of Policy Adherence

Control Objective: To establish mechanisms for monitoring compliance with cybersecurity policies, reporting on adherence levels, and enforcing policies across the organization.

Implementation Steps:

3.1 **Monitoring Mechanisms**: Implement monitoring mechanisms, leveraging manual audits and automated tools, to assess compliance with cybersecurity policies regularly.

3.2 **Regular Reporting**: Develop a regular reporting schedule to update leadership on compliance levels, highlighting areas of concern and recommending corrective actions.

3.3 **Enforcement Procedures**: Establish clear procedures for addressing noncompliance with cybersecurity policies, including corrective actions and, where necessary, disciplinary measures.

Expected Outcome: Effective monitoring and enforcement mechanisms ensure high levels of adherence to cybersecurity policies, with regular reporting keeping leadership informed and enabling prompt action to address any areas of noncompliance.

GV.PO-02: Policy for Managing Cybersecurity Risks Is Reviewed, Updated, Communicated, and Enforced to Reflect Changes in Requirements, Threats, Technology, and Organizational Mission

The realm of cybersecurity is in a constant state of evolution, with new threats emerging, technologies advancing, and regulatory landscapes shifting. In this dynamic environment, the static application of cybersecurity policies is inadequate for safeguarding an organization's digital assets. Introducing the dynamic nature of cybersecurity policy management sets the stage for understanding the critical need for policies that can adapt to changing circumstances.

Cybersecurity is not a static field; the threats that organizations face are continually evolving, as are the technologies at their disposal and the regulatory requirements they must adhere to. Therefore, a robust cybersecurity policy framework must be living documents—flexible and responsive to these changes to remain relevant and effective.

Establishing a scheduled review process for cybersecurity policies is fundamental to this adaptive approach. Organizations should implement regular intervals—annually, bi-annually, or following significant changes in the operational or threat landscape—for policy review. This scheduled process ensures that policies do not become outdated but evolve with the organization's external and internal environments. The review process should be thorough, involving stakeholders from across the organization to ensure comprehensive coverage of all areas of cybersecurity.

Adapting policies to reflect evolving legal and regulatory requirements is critical to policy management. Organizations must adjust their policies accordingly to remain compliant as governments and industries introduce new regulations to address emerging cybersecurity threats and concerns. This adaptation involves staying abreast of legal changes and understanding their implications for the organization's operations, which often requires the expertise of legal professionals and cybersecurity experts.

Incorporating emerging threat intelligence into policy updates is another essential component. As cyber adversaries continually develop new techniques and tactics, organizations must ensure that their cybersecurity policies reflect the latest threat intelligence. This incorporation can involve revising incident response plans, updating security protocols, and adjusting monitoring strategies to effectively address new types of cyber threats.

Adjusting policies to keep pace with advancements in technology is equally important. Organizations adopting new technologies—such as cloud services, IoT devices, and AI applications—must reassess their cybersecurity policies to ensure they address the unique risks and vulnerabilities these technologies introduce. This adjustment ensures that policies remain relevant and effective in protecting against threats that target or exploit these new technologies.

Ensuring policies remain aligned with the shifting organizational mission is crucial. As organizations grow, diversify, or change direction, their cybersecurity policies must evolve to support the new mission and objectives. This alignment ensures that cybersecurity efforts are fully integrated into the organization's overall strategy, supporting its goals rather than hindering them.

Stakeholder involvement in the policy review and update process is key to the success of dynamic policy management. By engaging stakeholders from different departments and levels of the organization, policymakers can gain valuable insights into the practical implications of policies, identify potential gaps or issues, and foster a culture of cybersecurity awareness and responsibility throughout the organization.

Communication strategies for newly updated policies are critical in ensuring that all organization members know and understand the latest cybersecurity expectations and requirements. Effective communication can involve training sessions, internal newsletters, or dedicated policy distribution and discussion platforms. Ensuring that these policies are communicated effectively helps to build a cohesive cybersecurity culture where everyone understands their role in protecting the organization's digital assets.

Recommendations

- **Institute a Cross-Functional Policy Review Board**: Establish a board comprising members from various departments, including IT, legal, HR, and operations, to oversee the policy review and update process. This diversity ensures that all cybersecurity aspects are considered and policies are aligned with the entire organization's needs and goals.
- **Leverage Automation for Efficiency**: Use automation tools to monitor for changes in the cybersecurity landscape, including legal requirements and emerging threats. These tools can alert the relevant teams when policy updates may be necessary, streamlining the review process.
- **Foster a Culture of Continuous Improvement**: Encourage feedback from employees at all levels on the cybersecurity policies and their implementation. This feedback can be invaluable in identifying areas for improvement and fostering a sense of ownership and engagement across the organization.
- **Implement Comprehensive Training Programs**: Develop training programs that are regularly updated to reflect changes in cybersecurity policies, technologies, and threats. This training should be mandatory for all employees and tailored to different organizational roles to ensure relevance and effectiveness.
- **Monitor Compliance and Enforce Policies**: Establish mechanisms for monitoring compliance with cybersecurity policies and procedures. This monitoring should be complemented by clear and consistent enforcement strategies to address violations, reinforcing the importance of adherence to cybersecurity practices.

NIST 800-53 Controls

(Refer to our appendix for a comprehensive description of the NIST 800-53 Controls, essential for organizations adhering to strict regulatory standards.)

- AC-01,AT-01,AU-01,CA-01,CM-01,CP-01,IA-01,IR-01,MA-01,MP-01,PE-01,PL-01,PM-01,PS-1, PT-01,RA-01,SA-01,SC-01,SI-01,SR-01

Simplified Security Controls (SSC)

Control 1: Scheduled Review and Update Process for Cybersecurity Policies

Control Objective: To ensure that cybersecurity policies remain effective and relevant in the face of changing requirements, threats, technologies, and the organizational mission through a structured review and update process.

Implementation Steps:

1.1 **Establish a Regular Review Schedule**: Set a regular schedule for reviewing all cybersecurity policies, at least annually or more frequently, as dictated by the pace of change in the cybersecurity landscape.

1.2 **Assemble Review Teams**: Form review teams composed of stakeholders from various departments such as IT, cybersecurity, legal, compliance, and business units to ensure a comprehensive evaluation of policies from multiple perspectives.

1.3 **Update and Approve Policies**: Revise policies based on review findings, emerging threats, legal and regulatory changes, and technological advancements. Submit updated policies for approval by senior leadership before dissemination.

Expected Outcome: Cybersecurity policies are regularly reviewed and updated to reflect the current cybersecurity environment, ensuring that the organization's defenses remain robust against new threats and aligned with legal requirements and business objectives.

Control 2: Incorporation of Stakeholder Feedback in Policy Updates

Control Objective: To leverage the insights and feedback of various stakeholders within and outside the organization in the cybersecurity policy review and update process, ensuring policies are comprehensive and practically applicable.

Implementation Steps:

2.1 **Solicit Stakeholder Feedback**: Implement mechanisms for regularly collecting feedback on cybersecurity policies from employees, suppliers, partners, and customers to identify areas for improvement.

2.2 **Review and Incorporate Feedback**: Analyze the feedback to identify common themes and areas requiring clarification or adjustment. Incorporate relevant feedback into policy updates to ensure policies are clear, practical, and effective.

2.3 **Feedback Acknowledgment**: Communicate to stakeholders how their feedback was considered and incorporated into the policies, reinforcing the value placed on their input.

Expected Outcome: Cybersecurity policies that are refined and improved through stakeholder feedback, enhancing their relevance, practicality, and effectiveness in managing cybersecurity risks across the organization.

Control 3: Enhanced Communication and Training for Updated Policies

Control Objective: To ensure that all organizational members know, understand, and adhere to updated cybersecurity policies through effective communication strategies and updated training programs.

Implementation Steps:

3.1 **Develop and Execute Communication Plans**: Create communication plans detailing how cybersecurity policy updates will be disseminated across the organization, using a variety of channels to ensure broad reach.

3.2 **Update Training Programs**: Revise existing cybersecurity training programs to reflect policy updates, ensuring all employees receive training on new policy requirements and changes.

3.3 **Monitor Understanding and Compliance**: Implement mechanisms to assess the effectiveness of communication and training efforts, ensuring employees understand and can comply with updated policies and adjust strategies as needed based on feedback.

Expected Outcome: An organization-wide understanding and adherence to updated cybersecurity policies, supported by comprehensive communication and training efforts, ensuring the workforce is informed and equipped to protect the organization against cybersecurity risks effectively.

9

Oversight (GV.OV)

> In cybersecurity, our vigilance is measured not by the defenses we build but by how swiftly and effectively we adapt them to meet the ever-changing threats, ensuring our resilience through constant evaluation and refinement.

In the dynamic realm of cybersecurity, the continuous evaluation and refinement of risk management strategies are paramount to safeguarding organizational assets and information. This iterative process involves systematically reviewing strategy outcomes, leveraging metrics and benchmarks to gauge performance, and analyzing data to assess the effectiveness of current cybersecurity measures. A collaborative approach, incorporating insights from a broad spectrum of stakeholders, is essential for identifying gaps in the strategy and uncovering opportunities for enhancement. Feedback loops are crucial in informing strategic adjustments and ensuring decisions that align with the organization's evolving needs and threat landscape. As changes are implemented, their impact on the cybersecurity posture is closely monitored, focusing on maintaining an agile, responsive strategy that meets organizational requirements and anticipates future cybersecurity challenges.

GV.OV-01: Cybersecurity Risk Management Strategy Outcomes Are Reviewed to Inform and Adjust Strategy and Direction

Understanding and refining a cybersecurity strategy is an iterative process that requires a comprehensive review mechanism to ensure its effectiveness over time. The cybersecurity strategy outcomes review process is critical for organizations aiming to maintain a robust defense against evolving threats. This process begins with clearly defining what constitutes successful outcomes, incorporating qualitative and quantitative measures. Metrics and benchmarks play a pivotal role in evaluating the performance of the cybersecurity strategy. Establishing these parameters upfront provides a clear framework for assessment and ensures alignment with organizational objectives.

Evaluating the performance of a cybersecurity strategy involves collecting and analyzing relevant data. This data encompasses a wide range of information, from incident response times and recovery costs to the effectiveness of security controls and employee cybersecurity awareness levels. Analyzing this data helps understand the strategy's impact on the organization's security posture. It is important to employ historical data analysis for trend identification and real-time monitoring to capture the current state of cybersecurity defenses.

Incorporating diverse perspectives is essential to a comprehensive review of cybersecurity strategy outcomes. Involving stakeholders from different departments and levels within the

A Comprehensive Guide to the NIST Cybersecurity Framework 2.0: Strategies, Implementation, and Best Practice, First Edition. Jason Edwards.
© 2025 John Wiley & Sons Ltd. Published 2025 by John Wiley & Sons Ltd.

organization ensures a multifaceted view of the strategy's effectiveness. This collaborative approach facilitates the identification of gaps and opportunities that may not be evident from a singular viewpoint. Stakeholder involvement also fosters a culture of security awareness and responsibility across the organization.

Identifying gaps and opportunities for strategy enhancement is crucial for maintaining an adaptive and resilient cybersecurity posture. This involves pinpointing areas of weakness and recognizing emerging threats and technologies that could impact the strategy's effectiveness. Continuous improvement is a key principle in cybersecurity management, and the strategy must evolve in response to new information and changing threat landscapes.

Feedback loops are instrumental in the strategic adjustment process. They enable the continuous flow of information from the outcome review process into the strategy development and implementation phases. This iterative process ensures the strategy remains aligned with organizational goals and the external threat environment. Feedback loops facilitate agile decision-making, allowing for timely adjustments to the cybersecurity strategy.

Decision-making processes for strategy direction changes require a structured approach to ensure adjustments are made based on sound analysis and strategic alignment. This involves evaluating the potential impact of proposed changes, considering both the benefits and risks. Decision-making should be supported by comprehensive data analysis and stakeholder input, ensuring that changes are both proactive and informed.

Communicating strategy adjustments across the organization is critical for ensuring alignment and compliance. Clear and concise communication helps to mitigate confusion and aligns the organization toward common cybersecurity goals. This communication should include the rationale behind adjustments, expected outcomes, and any changes in roles or responsibilities. Effective communication fosters organizational resilience by ensuring that all team members are informed and engaged in the cybersecurity strategy.

Implementing changes to the cybersecurity strategy and monitoring their impact is the final step in the review and adjustment process. This involves the technical aspects of deploying new controls or technologies and managing the change within the organization. Monitoring the impact of these changes is essential for validating their effectiveness and identifying any unintended consequences. Continuous monitoring and evaluation support the dynamic nature of cybersecurity, enabling organizations to adapt and thrive in the face of cyber threats.

Recommendations

- **Establish Clear Review Cycles**: Define and adhere to regular intervals for reviewing the cybersecurity strategy to ensure timely adjustments in response to new threats and organizational changes.
- **Leverage Advanced Analytics**: Utilize data analytics tools to process and analyze security data, enabling more accurate assessments of strategy effectiveness and identifying trends that may indicate potential vulnerabilities.
- **Enhance Stakeholder Engagement**: Develop mechanisms for more effective stakeholder involvement, ensuring a broader perspective in the review process and fostering a culture of cybersecurity across the organization.
- **Implement a Continuous Improvement Program**: Adopt a formal continuous improvement program for cybersecurity, focusing on regular updates to the strategy based on feedback loops and performance analysis.

- **Strengthen Communication Channels**: Improve the mechanisms for communicating strategy changes, ensuring clarity and understanding across all levels of the organization to facilitate smoother implementation and alignment.

NIST 800-53 Controls

(Refer to our appendix for a comprehensive description of the NIST 800-53 Controls, essential for organizations adhering to strict regulatory standards.)

- AC-01,AT-01,AU-01,CA-01,CM-01,CP-01,IA-01,IR-01,MA-01,MP-01,PE-01,PL-01,PM-01,PS-01, PT-01,RA-01,SA-01,SC-01,SI-01,SR-01,PM-09,PM-18,PM-30,PM-31,RA-07

Simplified Security Controls (SSC)

Control 1: Implementation of a Strategic Cybersecurity Review Committee

Control Objective: To establish a dedicated committee responsible for overseeing the periodic review and adjustment of the organization's cybersecurity risk management strategy, ensuring alignment with changing cybersecurity landscapes and business objectives.

Implementation Steps:

1.1 **Committee Formation**: Form a strategic cybersecurity review committee comprising members from various departments, including IT security, compliance, operations, and executive leadership. This diversity ensures that multiple perspectives are considered in the review process.

1.2 **Define Review Schedule and Criteria**: The committee should establish a regular schedule for reviewing the cybersecurity strategy (e.g., quarterly or bi-annually) and define clear criteria for evaluation based on established metrics and benchmarks.

1.3 **Recommend and Oversee Adjustments**: Based on the review findings, the committee recommends strategic adjustments and oversees their implementation, ensuring that changes are effectively communicated and monitored for impact.

Expected Outcome: Establishing a strategic cybersecurity review committee ensures a structured and continuous review process for the cybersecurity strategy. This approach enables the organization to remain agile, adjusting its cybersecurity direction to new threats, technological advancements, and business requirements, enhancing its overall security posture.

Control 2: Development of a Cybersecurity Strategy Dashboard

Control Objective: To create a comprehensive dashboard that aggregates and visualizes key metrics and benchmarks related to the performance of the cybersecurity strategy, facilitating real-time monitoring and informed decision-making.

Implementation Steps:

2.1 **Identify Key Metrics**: Identify critical metrics that accurately reflect the effectiveness of the cybersecurity strategy. These could include incident response times, the number of detected threats, system uptime, and compliance rates.

2.2 **Implement Dashboard Technology**: Utilize a suitable dashboard technology that can integrate data from various sources, providing a unified view of cybersecurity performance. Ensure the dashboard is customizable and scalable to adapt to changing requirements.

2.3 **Train and Deploy**: Train relevant personnel on how to use and interpret the dashboard. Deploy the dashboard across the organization, ensuring key decision-makers can access real-time data on cybersecurity strategy performance.

Expected Outcome: The cybersecurity strategy dashboard provides a dynamic, real-time view of how well the cybersecurity measures perform against established benchmarks. This visibility supports prompt identification of issues and informed decision-making, leading to quicker strategic adjustments and enhanced cybersecurity resilience.

Control 3: Formalizing Feedback Mechanisms for Continuous Strategy Improvement

Control Objective: To formalize mechanisms for collecting, analyzing, and integrating feedback from all organizational levels and external stakeholders into the cybersecurity strategy, ensuring it remains effective and relevant.

Implementation Steps:

3.1 **Establish Feedback Channels**: Set up structured channels for feedback collection, such as surveys, suggestion boxes, and regular stakeholder meetings, to gather insights from employees, customers, partners, and industry peers.

3.2 **Analyze and Prioritize Feedback**: Regularly review the feedback to identify common themes, pressing concerns, and innovative suggestions. Prioritize these inputs based on their potential impact on cybersecurity effectiveness and alignment with organizational goals.

3.3 **Incorporate Feedback into Strategy**: Update the cybersecurity strategy to reflect critical feedback, ensuring adjustments address identified gaps and leverage opportunities for enhancement. Document the rationale for changes and communicate these adjustments to all stakeholders.

Expected Outcome: Formalizing feedback mechanisms ensures that the cybersecurity strategy is continuously refined based on a wide range of insights and observations. This approach promotes a proactive, adaptive strategy better equipped to respond to emerging threats, evolving technologies, and changing business landscapes, ultimately leading to a stronger cybersecurity posture.

GV.OV-02: The Cybersecurity Risk Management Strategy Is Reviewed and Adjusted to Ensure Coverage of Organizational Requirements and Risks

A cybersecurity risk management strategy's meticulous review and adjustment are pivotal for ensuring it comprehensively covers an organization's unique requirements and risks. This iterative process entails understanding the multifaceted nature of cybersecurity threats, the evolving technological landscape, and the organization's strategic objectives. An effective strategy hinges on its initial design and its ability to adapt and evolve in response to new information and changing circumstances.

The introduction to ensuring comprehensive strategy coverage emphasizes the necessity of a dynamic approach to cybersecurity risk management. A static strategy quickly becomes obsolete due to rapidly advancing threats and technological innovations. Organizations must adopt a proactive stance, continuously scanning the horizon for emerging risks and assessing the adequacy of their existing strategies. This ongoing assessment ensures that the strategy remains aligned with the organization's risk appetite and capacity for risk management.

The framework for regularly reviewing the cybersecurity risk management strategy establishes a structured approach to periodic evaluations. This framework should incorporate scheduled reviews and triggers for ad-hoc assessments, such as significant changes in the threat landscape or the business environment. Regular reviews facilitate the identification of gaps in the strategy and opportunities for enhancement, ensuring the strategy's effectiveness over time.

Alignment checks between strategy, organizational objectives, and requirements are critical to ensuring that the cybersecurity strategy supports the organization's broader goals. This involves a detailed analysis of the strategy's objectives, the risks it addresses, and the controls it implements to ensure they harmonize with the organization's mission, values, and strategic objectives. Discrepancies between the strategy and organizational goals necessitate adjustments to ensure coherence and support in achieving organizational objectives.

Risk assessment updates to reflect the evolving threat landscape are fundamental to maintaining a relevant and effective cybersecurity strategy. As cyber threats continually evolve, so must the organization's understanding of its risk exposure. This involves monitoring for new types of cyberattacks and reassessing the likelihood and impact of known risks in light of recent developments. Such updates ensure that the strategy remains focused on the most pertinent threats and that resources are allocated efficiently.

Adjusting the strategy to address new technologies and processes acknowledges the rapid pace of technological advancement and its impact on cybersecurity. The adoption of new technologies can introduce new vulnerabilities, necessitating adjustments to the cybersecurity strategy. This process involves evaluating the security implications of new technologies, updating risk assessments, and implementing appropriate controls to mitigate new risks.

Engaging cross-functional teams for holistic strategy review ensures that the cybersecurity strategy benefits from diverse perspectives within the organization. Collaboration across departments allows for a more comprehensive understanding of risks, controls, and the implications of strategy adjustments. This interdisciplinary approach enhances the strategy's effectiveness by incorporating insights from various functional areas, including IT, operations, finance, and human resources.

Updating documentation and communication following strategy adjustments is essential for maintaining organizational alignment and ensuring all stakeholders are informed of changes to the cybersecurity strategy. This includes updating policy documents, procedures, and guidelines to reflect the revised plan and communicating changes to employees, partners, and other relevant parties. Effective communication ensures everyone understands their roles and responsibilities in supporting the cybersecurity strategy.

Training and awareness initiatives to reflect strategy updates are crucial for ensuring that all organizational members are equipped to contribute to the strategy's success. This involves developing and delivering training programs that address the specific changes to the cybersecurity strategy and broader cybersecurity awareness efforts. By fostering a culture of cybersecurity awareness, organizations can enhance their resilience to cyber threats.

Reviewing strategy efficacy in meeting organizational cybersecurity needs involves evaluating the performance of the cybersecurity strategy against predefined metrics and objectives. This assessment helps to determine whether the strategy effectively mitigates risks and supports the organization's cybersecurity goals. Insights gained from this review inform further adjustments to the strategy, closing the loop on the iterative process of strategy review and adjustment.

Recommendations

- **Establish a Continuous Review Process**: Implement a structured framework for regularly reviewing the cybersecurity strategy, incorporating periodic assessments and triggers for immediate review based on emerging threats or significant changes in the organizational or technological landscape.

- **Align with Organizational Objectives**: Regularly assess the alignment between the cybersecurity strategy and the organization's overall objectives and risk management capacity, adjusting the strategy as necessary to ensure coherence and support achieving broader goals.
- **Stay Informed of Evolving Threats**: Invest in intelligence gathering and threat analysis capabilities to continuously update the organization's risk assessments and adjust the cybersecurity strategy accordingly.
- **Foster Interdisciplinary Collaboration**: Engage cross-functional teams in reviewing and adjusting the cybersecurity strategy to benefit from diverse perspectives and expertise, enhancing the strategy's comprehensiveness and effectiveness.
- **Prioritize Communication and Training**: Update organizational documentation, policies, and training programs following adjustments to the cybersecurity strategy, ensuring that all stakeholders are informed and equipped to fulfill their roles in supporting the strategy.

NIST 800-53 Controls

(Refer to our appendix for a comprehensive description of the NIST 800-53 Controls, essential for organizations adhering to strict regulatory standards.)

- PM-09,PM-19,PM-30,PM-31,RA-07,SR-06

Simplified Security Controls (SSC)

Control 1: Periodic Cybersecurity Framework Review Control

Control Objective: To ensure the cybersecurity risk management strategy remains effective and aligned with organizational objectives and the evolving threat landscape, periodic reviews of the cybersecurity framework are essential. This control aims to identify gaps, assess the adequacy of current strategies, and make necessary adjustments to address new threats and technologies.

Implementation Steps:

1.1 **Schedule Regular Reviews**: Establish a regular schedule for reviewing the cybersecurity framework, such as bi-annually or annually, after any significant organizational changes or major incidents.

1.2 **Conduct Comprehensive Assessments**: Each review should include assessments of the alignment between the cybersecurity strategy and organizational goals, the current threat landscape, technological advancements, and regulatory requirements.

1.3 **Document and Communicate Changes**: Update the cybersecurity framework based on the review findings. Document and communicate all changes effectively across the organization, ensuring that all stakeholders are aware of updates and their roles in implementation.

Expected Outcome: The organization maintains an up-to-date cybersecurity strategy aligned with organizational goals and the current threat environment and effectively communicates across the organization, enhancing the overall cybersecurity posture.

Control 2: Cross-Functional Cybersecurity Review Team

Control Objective: A cross-functional team should be established to foster a holistic approach to reviewing and updating the cybersecurity risk management strategy. This team will bring diverse perspectives and expertise to the review process, ensuring that all aspects of the organization's operations and risk profile are considered.

Implementation Steps:

2.1 **Form the Team**: Assemble a team comprising members from IT, cybersecurity, operations, legal, HR, and other relevant departments. Ensure the team includes individuals with decision-making authority.

2.2 **Define Review Process**: The team should develop a clear process for reviewing the cybersecurity strategy, including criteria for evaluating alignment with organizational objectives, assessing new threats, and integrating new technologies.

2.3 **Implement and Monitor Adjustments**: Following each review, the team is responsible for overseeing the implementation of recommended adjustments to the strategy. They should also monitor the effectiveness of these changes over time.

Expected Outcome: Forming a cross-functional team ensures a comprehensive review of the cybersecurity strategy, leading to well-rounded adjustments that account for the full spectrum of organizational risks and objectives. This approach promotes a stronger, more agile cybersecurity posture.

Control 3: Cybersecurity Awareness and Training Program Enhancement

Control Objective: Continuous improvement of cybersecurity awareness and training programs is crucial to ensure they reflect cybersecurity risk management strategy updates. This control aims to enhance organizational resilience to cyber threats by ensuring all members are informed and prepared to act in accordance with the latest cybersecurity practices.

Implementation Steps:

3.1 **Assess Training Needs**: Regularly evaluate the organization's training needs, especially following updates to the cybersecurity strategy, to identify new areas of focus.

3.2 **Update Training Programs**: Revise existing training materials and develop new content to cover updated strategy elements, emerging threats, and new technologies and processes.

3.3 **Conduct Regular Training Sessions**: Deliver training sessions to all employees regularly, ensuring they understand their roles and responsibilities in upholding the cybersecurity strategy. Include scenario-based training and simulations to enhance engagement and retention.

Expected Outcome: The organization benefits from a workforce that is well-informed about the current cybersecurity strategy and prepared to contribute to its implementation. This ongoing commitment to cybersecurity education fosters a culture of security awareness and resilience against cyber threats.

GV.OV-03: Organizational Cybersecurity Risk Management Performance Is Evaluated and Reviewed for Adjustments Needed

Understanding the intricacies of organizational cybersecurity risk management performance evaluation is imperative for maintaining a robust security posture. This process begins with introducing methods to assess the effectiveness of existing cybersecurity measures. Assessing cybersecurity risk management performance is not merely about compliance checks; it delves deeper into how effectively an organization can identify, mitigate, and manage cybersecurity risks. The first step in this comprehensive evaluation involves establishing a baseline of current security practices and their outcomes, setting the stage for continuous improvement.

Developing key performance indicators (KPIs) tailored for cybersecurity effectiveness is pivotal in evaluating cybersecurity risk management performance. These KPIs are vital metrics that offer tangible evidence of how well an organization's cybersecurity initiatives are performing. They should be closely aligned with the organization's broader business objectives and cybersecurity goals, ensuring that they provide a clear picture of operational efficiency and security measures' effectiveness. By defining such KPIs, organizations can quantitatively measure their cybersecurity posture and make informed decisions based on concrete data.

Creating a performance evaluation framework is another critical component of this process. This framework outlines the methodology for assessing the performance of cybersecurity initiatives against the set KPIs. It encompasses various aspects, including the frequency of evaluations, assessment techniques, and criteria for judging performance levels. This structured approach ensures that evaluations are consistent, objective, and capable of identifying areas for improvement. The framework should be flexible enough to adapt to evolving cybersecurity threats and changes in organizational objectives.

The collection and analysis of performance data are integral to this evaluation process. Data gathered from various sources, including security incident reports, audit logs, and compliance assessments, provides insights into the effectiveness of cybersecurity measures. Analyzing this data helps identify trends, pinpoint vulnerabilities, and understand the impact of implemented security controls. This evidence-based approach facilitates informed decision-making and strategic planning for enhancing cybersecurity measures.

Incorporating stakeholder feedback into performance reviews is essential for a holistic evaluation. Stakeholders, including employees, customers, and partners, often provide valuable perspectives on the effectiveness of cybersecurity practices. Their feedback can highlight practical challenges, user experience issues, and areas where security measures may be too restrictive or not stringent enough. Engaging with stakeholders improves the evaluation process and fosters a culture of security awareness and collaboration across the organization.

Identifying performance shortfalls and successes is a crucial outcome of the evaluation process. Recognizing areas where cybersecurity measures fall short enables organizations to pinpoint and address vulnerabilities proactively. Similarly, acknowledging successes highlights effective strategies and practices that can be leveraged further or replicated in other areas. This balanced view encourages a proactive stance on cybersecurity, focusing on continuous improvement and strategic adjustments to enhance security posture.

Strategic planning for performance improvement initiatives is the next step, driven by insights from the evaluation process. This involves setting priorities, allocating resources, and defining timelines for implementing enhancements to cybersecurity measures. The planning process should be iterative, allowing for adjustments as new threats emerge and organizational goals evolve. Effective strategic planning ensures that cybersecurity initiatives remain aligned with the organization's objectives and the dynamic landscape of cyber threats.

Finally, communicating performance evaluation results and planned adjustments to all relevant stakeholders is paramount. Transparent communication fosters trust and demonstrates the organization's commitment to cybersecurity. It also encourages stakeholder engagement and support for implementing changes. Effective communication strategies might include regular reports, meetings, and updates through internal channels, ensuring that stakeholders are informed and involved in the cybersecurity enhancement process.

Recommendations

- **Establish a Continuous Improvement Cycle**: Implement a structured cycle for continuous evaluation and improvement of cybersecurity risk management performance, ensuring the process is iterative and responsive to new information and evolving threats.

- **Leverage Technology for Data Analysis**: Utilize advanced analytical tools and technologies to process and analyze performance data, enabling more accurate and faster identification of trends, vulnerabilities, and opportunities for improvement.
- **Enhance Stakeholder Engagement**: Develop a comprehensive stakeholder engagement plan that includes regular updates, feedback sessions, and collaborative reviews of cybersecurity practices to ensure all perspectives are considered in performance evaluations.
- **Prioritize Based on Risk**: Focus performance improvement initiatives on areas of highest risk to the organization, ensuring that resources are allocated efficiently to mitigate significant threats effectively.
- **Foster a Culture of Security Awareness**: Encourage a proactive security culture throughout the organization by providing training, resources, and support for all employees to understand their role in maintaining cybersecurity and contributing to performance improvement.

NIST 800-53 Controls

(Refer to our appendix for a comprehensive description of the NIST 800-53 Controls, essential for organizations adhering to strict regulatory standards.)

- PM-04,PM-06,RA-07,SR-06

Simplified Security Controls (SSC)

Control 1: Establishment of Comprehensive Cybersecurity KPIs

Control Objective: To develop and establish a comprehensive set of KPIs that accurately reflect the organization's cybersecurity effectiveness, aligned with its specific goals and risk landscape.

Implementation Steps:

1.1 **Identify Critical Security Domains**: Identify the critical domains of your cybersecurity program that require monitoring and evaluation, such as incident response times, patch management efficiency, and user awareness levels.

1.2 **Develop Specific KPIs**: For each identified domain, develop specific KPIs that are measurable, achievable, relevant, and time-bound (SMART). These KPIs should be capable of providing insights into the performance of cybersecurity initiatives and their impact on the overall security posture.

1.3 **Review and Approve KPIs**: Present the proposed KPIs to cybersecurity leadership for review and approval. Ensure these KPIs are documented with clear definitions, measurement methods, and targets.

Expected Outcome: Establishing comprehensive cybersecurity KPIs results in an actionable framework for measuring the effectiveness and efficiency of cybersecurity initiatives. Organizations can expect to understand their cybersecurity posture better, enabling informed decision-making and strategic improvements.

Control 2: Cybersecurity Performance Evaluation Framework

Control Objective: To create a structured framework that outlines the methodologies, frequencies, and responsibilities for evaluating the performance of cybersecurity initiatives against established KPIs.

Implementation Steps:

2.1 **Framework Development**: Develop a framework that details the processes for conducting performance evaluations. This should include methodologies for data collection, analysis techniques, evaluation schedules, and criteria for assessing KPI achievements.

2.2 **Assign Evaluation Responsibilities**: Clearly assign organizational roles and responsibilities for performance evaluations. Ensure that these roles have the authority and resources to conduct thorough assessments.

2.3 **Pilot and Refine the Framework**: Pilot the performance evaluation framework with a select set of KPIs. Gather feedback on the process and make necessary adjustments to ensure its effectiveness and efficiency.

Expected Outcome: Implementing a cybersecurity performance evaluation framework enables regular, systematic assessments of cybersecurity measures. Organizations can anticipate enhanced visibility into the performance of their security initiatives, facilitating continuous improvement and adjustment to cyber threats.

Control 3: Integration of Stakeholder Feedback into Cybersecurity Enhancements

Control Objective: To systematically gather, analyze, and integrate stakeholder feedback to improve cybersecurity risk management performance continuously.

Implementation Steps:

3.1 **Establish Feedback Channels**: Create multiple channels for stakeholders to provide feedback on cybersecurity practices and their perceived effectiveness. These channels could include surveys, interviews, and feedback forms during security training sessions.

3.2 **Analyze Feedback for Actionable Insights**: Regularly analyze the feedback collected to identify common themes, concerns, and suggestions for improvement. Prioritize the feedback based on its potential impact on cybersecurity effectiveness and risk management.

3.3 **Implement Feedback into Security Improvements**: Develop a plan for integrating the most critical feedback into cybersecurity enhancements. This plan should detail the adjustments needed, resources required, and timelines for implementation.

Expected Outcome: By integrating stakeholder feedback into cybersecurity enhancements, organizations can expect to achieve a more responsive and adaptive cybersecurity risk management strategy. This approach improves the effectiveness of security measures and fosters a culture of security awareness and collaboration across the organization.

10

Cybersecurity Supply Chain Risk Management (GV.SC)

> Establishing a C-SCRM program is akin to weaving a safety net beneath our digital endeavors. It requires every hand on deck, ensuring our strategies are as cohesive as they are comprehensive. This collective vigilance is our beacon against the shadows of supply chain vulnerabilities.

Establishing a comprehensive Cybersecurity Supply Chain Risk Management (C-SCRM) program is crucial for organizations navigating the complex web of modern supply chains. This approach necessitates a collaborative effort among stakeholders to define clear C-SCRM strategies and objectives, ensuring alignment with the organization's broader cybersecurity goals. Developing and documenting robust C-SCRM policies, alongside establishing effective implementation processes, are foundational steps in safeguarding against supply chain vulnerabilities. Integrating C-SCRM into the organizational risk management framework enhances the resilience and security posture, necessitating regular reviews, updates, and monitoring to adapt to the evolving cybersecurity landscape. Through continuous improvement and stakeholder engagement, organizations can foster a culture of security awareness and readiness, maintaining vigilance over their supply chain's integrity and security.

GV.SC-01: Establishing a Cybersecurity Supply Chain Risk Management Program

A C-SCRM program is crucial in today's interconnected digital ecosystem. The foundation of a robust C-SCRM program lies in its strategy, objectives, policies, and processes, which must be established and agreed upon by organizational stakeholders. This comprehensive approach ensures that the organization can effectively manage and mitigate the risks associated with its supply chain, safeguarding critical infrastructure and sensitive data against potential threats.

Establishing a comprehensive C-SCRM program begins with an in-depth understanding of the organization's supply chain and its cybersecurity risks. Collaboration with stakeholders across various departments is essential in defining a C-SCRM strategy that aligns with the organization's overall risk management goals. This collaborative effort involves identifying key assets and processes vulnerable to supply chain disruptions and formulating objectives to protect them. It is about creating a shared vision that guides the development and implementation of the C-SCRM program.

Developing and documenting C-SCRM policies are critical steps in operationalizing the strategy. These policies should outline the roles and responsibilities of all stakeholders, specify the

A Comprehensive Guide to the NIST Cybersecurity Framework 2.0: Strategies, Implementation, and Best Practice, First Edition. Jason Edwards.
© 2025 John Wiley & Sons Ltd. Published 2025 by John Wiley & Sons Ltd.

procedures for assessing and mitigating risks, and establish guidelines for monitoring supply chain security. Effective policies are clear, actionable, and adaptable to the evolving cybersecurity landscape. They serve as a foundation for the organization's C-SCRM practices and set the standards for compliance and governance.

For C-SCRM to be effective, it is necessary to establish processes that facilitate its implementation. These processes should include mechanisms for continuously identifying and assessing risks, implementing protective measures, and responding to supply chain incidents. By establishing structured processes, organizations can ensure a proactive and systematic approach to managing supply chain risks, enhancing their resilience against cyber threats.

Integration of C-SCRM into the organizational risk management framework is another critical component. This integration ensures that supply chain risks are considered part of the overall risk landscape and that C-SCRM initiatives are supported by the organization's broader risk management efforts. It facilitates a unified approach to risk management, where C-SCRM is not siloed but is an integral part of the organization's risk management practices.

Gaining stakeholder agreement on C-SCRM strategy and policy implementation is essential for the program's success. It requires clear communication of the program's value and alignment with the organization's strategic objectives. Stakeholder buy-in is crucial for securing the necessary resources and support for the C-SCRM program and fostering a culture of security awareness across the organization.

Training and awareness programs are vital in supporting C-SCRM policies. They equip employees with the knowledge and skills to identify and mitigate supply chain risks. Through regular training, employees can stay informed about the latest cyber threats and best practices for supply chain security. Awareness programs help build a security-conscious culture where employees understand their role in protecting the organization's supply chain.

The effectiveness of a C-SCRM program relies on its ability to adapt to changing risks and business needs. This requires regular review and update of the C-SCRM program and strategies. Monitoring and reporting mechanisms play a crucial role in this process, providing insights into the program's performance and areas for improvement. Continuous monitoring allows for the early detection of supply chain threats. At the same time, regular reviews ensure that the C-SCRM program remains aligned with the organization's objectives and the evolving cyber threat landscape.

Recommendations

- **Develop a Cross-Departmental C-SCRM Team**: Establish a dedicated team comprising members from different departments to ensure a holistic supply chain risk management approach. This promotes the integration of diverse perspectives and expertise in developing and implementing the C-SCRM program.
- **Leverage Technology for Risk Assessment and Monitoring**: Implement advanced technologies and tools for continuous risk assessment and monitoring of the supply chain. Technologies such as AI and blockchain can enhance the visibility and security of supply chain processes.
- **Foster Strong Partnerships with Suppliers**: Build strong relationships with suppliers and ensure they adhere to your organization's cybersecurity standards. Regular audits and compliance checks can help maintain a secure supply chain.
- **Implement Incident Response Plans for Supply Chain Risks**: Develop and test incident response plans specifically designed for addressing supply chain disruptions. This ensures preparedness and a swift response to mitigate the impact of supply chain incidents.

- **Encourage a Continuous Learning and Improvement Culture**: Promote ongoing education and training on supply chain risks and cybersecurity practices. Encourage feedback and suggestions from stakeholders to continuously improve the C-SCRM program.

NIST 800-53 Controls

(Refer to our appendix for a comprehensive description of the NIST 800-53 Controls, essential for organizations adhering to strict regulatory standards.)

- PM-30,SR-02,SR-03

Simplified Security Controls (SSC)

Control 1: Development of a Comprehensive C-SCRM Policy Framework

Control Objective: To create a robust framework of policies that guide the organization's approach to managing cybersecurity risks within its supply chain, ensuring that all aspects of C-SCRM are clearly defined, communicated, and implemented consistently across the organization.

Implementation Steps:

1.1 **Policy Development**: Draft comprehensive C-SCRM policies that cover identifying, assessing, and mitigating supply chain risks. This includes defining roles and responsibilities, setting risk management standards, and establishing response protocols.

1.2 **Stakeholder Review and Approval**: Present the drafted policies to key organizational stakeholders for review, input, and approval. Ensure that feedback from various departments is incorporated to address all aspects of supply chain security.

1.3 **Policy Dissemination and Training**: Once approved, disseminate the policies throughout the organization and conduct mandatory training sessions to ensure all employees understand their roles and responsibilities within the C-SCRM framework.

Expected Outcome: Developing and implementing a comprehensive C-SCRM policy framework will ensure a clear understanding and consistent application of supply chain risk management practices across the organization. This will enhance the organization's ability to identify, assess, and mitigate cybersecurity risks in its supply chain, improving its overall security posture.

Control 2: Integration of C-SCRM into the Organizational Risk Management Framework

Control Objective: To embed C-SCRM processes seamlessly into the existing organizational risk management framework, ensuring that supply chain risks are considered part of the organization's broader risk management efforts.

Implementation Steps:

2.1 **Cross-Functional Collaboration**: Facilitate collaboration between the C-SCRM and broader risk management teams to align strategies and share insights on managing supply chain risks.

2.2 **Incorporate C-SCRM into Risk Assessments**: Ensure that C-SCRM considerations are integrated into the organization's regular risk assessment processes, including identifying and evaluating supply chain-related risks.

2.3 **Update Risk Management Policies and Procedures**: Revise the organization's risk management policies and procedures to explicitly include supply chain risk management considerations, ensuring these risks are addressed systematically.

Expected Outcome: Integrating C-SCRM into the organizational risk management framework ensures supply chain risks are managed rigorously, just as other organizational risks are. This alignment facilitates a more holistic approach to risk management, enhancing the organization's ability to identify, assess, and mitigate risks across all areas of operation.

Control 3: Establishment of Continuous Monitoring and Reporting Mechanisms

Control Objective: To implement ongoing monitoring and reporting mechanisms for C-SCRM effectiveness, enabling the timely identification of issues and facilitating continuous improvement in supply chain security practices.

Implementation Steps:

3.1 **Develop Monitoring Criteria**: Define clear criteria and key performance indicators (KPIs) for monitoring the effectiveness of C-SCRM activities, including compliance with policies, incident response times, and the resolution of identified risks.

3.2 **Implement Monitoring Tools**: Deploy monitoring tools and technologies that enable real-time tracking of supply chain activities and the automated detection of security anomalies or risks.

3.3 **Regular Reporting and Review**: Establish a regular reporting schedule for C-SCRM performance, including creating detailed reports for management to review. Use these reviews as opportunities for feedback and the continuous refinement of C-SCRM strategies.

Expected Outcome: Establishing continuous monitoring and reporting mechanisms for C-SCRM ensures the organization has a clear and ongoing understanding of its supply chain security posture. This enables proactive management of supply chain risks, facilitates timely responses to issues, and supports the continuing enhancement of supply chain security measures.

GV.SC-02: Cybersecurity Roles and Responsibilities Within the Supply Chain

Defining cybersecurity roles and responsibilities within the supply chain is fundamental to securing an organization's assets and information. This process begins with a thorough supply chain analysis to understand the interdependencies and identify the key actors involved. Establishing clear roles and responsibilities is critical to ensuring that each party understands their duties in maintaining cybersecurity. This clarity supports the creation of a cohesive security posture that protects against threats and vulnerabilities across the supply chain.

Mapping roles and responsibilities across the supply chain is an intricate process requiring detailed planning and collaboration. Organizations must identify the specific cybersecurity tasks associated with each role and ensure that these tasks align with the individuals' skills and capabilities. This mapping also involves understanding the flow of information and resources through the supply chain to identify critical points where specific cybersecurity roles are essential. Effective mapping ensures that all parties are aware of their responsibilities and how their actions impact the overall security of the supply chain.

Communication strategies for disseminating role expectations are vital for the success of a cybersecurity program. Organizations must develop clear and concise communication plans that outline how information about roles and responsibilities will be shared with suppliers, customers, and partners. These strategies may include regular meetings, training sessions, and collaborative information-sharing platforms. Ensuring that all parties have access to the information they need to fulfill their cybersecurity responsibilities is crucial for the integrity of the supply chain.

Coordination mechanisms for cybersecurity activities across the supply chain are essential for promptly detecting and responding to threats. This coordination can be achieved by establishing joint cybersecurity task forces, shared threat intelligence platforms, and regular security audits. By working together, supply chain actors can develop a unified approach to cybersecurity that leverages each participant's strengths to enhance the overall security posture.

Establishing accountability and reporting structures within the supply chain is necessary to ensure that cybersecurity roles and responsibilities are taken seriously. Organizations should implement clear policies that define the consequences of failing to meet cybersecurity obligations. These structures must also include mechanisms for reporting cybersecurity incidents and vulnerabilities, ensuring that information is quickly communicated to the appropriate parties for action.

Training programs for suppliers, customers, and partners on their roles in maintaining cybersecurity are an integral part of a comprehensive security strategy. These programs should cover the basics of cybersecurity, the specific responsibilities of each role, and the actions to take in the event of a security breach. Training ensures that all supply chain actors have the knowledge and skills to contribute to security efforts.

Regular assessments to ensure clarity and compliance with defined roles are crucial for maintaining an effective cybersecurity posture. These assessments can include audits, reviews, and simulations designed to test the understanding and execution of roles and responsibilities. Through regular evaluations, organizations can identify areas of confusion or noncompliance and take corrective actions to reinforce the security of the supply chain.

Incident response coordination plans involving all supply chain actors are critical for managing and mitigating cybersecurity incidents. These plans should outline the steps to be taken in response to an incident, including communication protocols, roles in the response effort, and postincident analysis. Effective incident response requires the cooperation and coordination of all parties involved in the supply chain, ensuring a swift and organized response to cybersecurity threats.

Recommendations

- **Establish a Centralized Cybersecurity Coordination Hub**: Create a centralized platform or team responsible for coordinating cybersecurity efforts across the supply chain. This hub can facilitate communication, share threat intelligence, and streamline incident response activities.
- **Implement Standardized Security Protocols**: Develop and enforce standardized security protocols across the supply chain. These protocols should be based on industry best practices and tailored to address the specific risks of the supply chain.
- **Conduct Joint Security Exercises**: Organize regular security exercises involving all supply chain actors. These exercises can help test the effectiveness of communication and incident response plans, identify weaknesses, and improve overall security coordination.
- **Enhance Transparency and Trust**: Build a culture of transparency and trust among supply chain actors. Encourage sharing security practices and vulnerabilities to foster a collaborative approach to cybersecurity.
- **Leverage Blockchain for Secure Information Sharing**: Explore the use of blockchain technology to securely share information about cybersecurity roles, responsibilities, and incidents. Blockchain can provide a tamper-proof platform for transparent and efficient communication across the supply chain.

NIST 800-53 Controls

(Refer to our appendix for a comprehensive description of the NIST 800-53 Controls, essential for organizations adhering to strict regulatory standards.)

- SR-02,SR-03,SR-05

Simplified Security Controls (SSC)

Control 1: Comprehensive Role Mapping and Communication Plan

Control Objective: To clearly define and communicate all supply chain actors' cybersecurity roles and responsibilities, ensuring a unified understanding and approach to cybersecurity within the supply chain ecosystem.

Implementation Steps:

1.1 **Develop a Comprehensive Role Map**: Identify and document the specific cybersecurity roles and responsibilities of suppliers, customers, and partners, detailing how each contributes to the supply chain's overall security posture.

1.2 **Establish Communication Plans**: Create detailed communication plans that outline how roles and responsibilities will be disseminated across the supply chain, ensuring all parties have access to the information they need to fulfill their roles effectively.

1.3 **Regular Updates and Feedback Loops**: Implement a process for regularly updating the role map and communication plans based on feedback from supply chain actors and changes in the cybersecurity landscape.

Expected Outcome: Creating a comprehensive role mapping and communication plan ensures that all supply chain actors know their specific cybersecurity responsibilities. This clarity fosters a coordinated approach to managing cybersecurity risks, enhancing the overall resilience of the supply chain.

Control 2: Cybersecurity Accountability and Reporting Framework

Control Objective: To establish a structured framework that holds all supply chain actors accountable for their cybersecurity roles and responsibilities, with clear mechanisms for reporting cybersecurity incidents and performance.

Implementation Steps:

2.1 **Define Accountability Measures**: Outline specific accountability measures for cybersecurity roles, including performance metrics and compliance with established cybersecurity policies.

2.2 **Develop Reporting Mechanisms**: Implement standardized reporting mechanisms that enable suppliers, customers, and partners to report on cybersecurity incidents, vulnerabilities, and compliance issues.

2.3 **Review and Action Process**: Create a process for reviewing reports, taking corrective action where necessary, and providing feedback to involved parties to improve cybersecurity practices.

Expected Outcome: Implementing a cybersecurity accountability and reporting framework ensures supply chain actors are held accountable for their roles and responsibilities. This framework facilitates the timely identification and management of cybersecurity risks, promoting a proactive security posture across the supply chain.

Control 3: Integrated Incident Response Coordination Framework

Control Objective: To develop a cohesive framework for coordinating incident response activities among suppliers, customers, and partners, ensuring timely and effective management of cybersecurity incidents across the supply chain.

Implementation Steps:

3.1 **Establish Incident Response Protocols**: Draft clear incident response protocols that define the roles and responsibilities of all supply chain actors in the event of a cybersecurity incident.

3.2 **Coordinate Training and Simulation Exercises**: Conduct training sessions and simulation exercises with suppliers, customers, and partners to ensure everyone is prepared to act per the incident response protocols.

3.3 **Continuous Improvement Process**: Implement a process for continuously improving the incident response protocols based on lessons learned from exercises and actual incidents, incorporating feedback from all supply chain actors.

Expected Outcome: Creating an integrated incident response coordination framework enhances the supply chain's ability to respond to cybersecurity incidents swiftly and effectively. This coordinated approach minimizes the impact of incidents, ensures business continuity, and strengthens the overall security of the supply chain.

GV.SC-03: Integrating Cybersecurity Supply Chain Risk Management into Organizational Frameworks

Integrating C-SCRM into organizational risk frameworks represents a strategic alignment of cybersecurity and enterprise risk management processes. This approach ensures a comprehensive understanding of the threats and vulnerabilities within the supply chain and their potential impact on the organization. The initial step in this integration involves acknowledging the interconnected nature of cybersecurity and enterprise risks, enhancing the organization's ability to anticipate, identify, and mitigate potential threats more effectively.

Identifying intersections between C-SCRM and enterprise risk management is critical for creating a unified risk management strategy. This involves mapping out where supply chain vulnerabilities could intersect with broader organizational risks. By doing so, organizations can prioritize risk management efforts more effectively, ensuring that resources are allocated to the most significant potential impact areas. This alignment also facilitates a more strategic approach to risk mitigation, leveraging insights from across the organization to address complex challenges.

Processes for incorporating C-SCRM into ongoing risk assessments are essential for maintaining an up-to-date understanding of the risk landscape. These processes should include regular reviews of supply chain operations, assessments of suppliers' cybersecurity practices, and analyses of the potential impact of supply chain disruptions on organizational operations. By embedding C-SCRM into regular risk assessment activities, organizations can ensure that supply chain risks are continuously identified, evaluated, and addressed promptly.

The alignment of C-SCRM objectives with broader cybersecurity strategies ensures that efforts to manage supply chain risks are consistent with the organization's security goals. This requires clear communication and coordination between teams responsible for cybersecurity and those overseeing supply chain management. Through such alignment, organizations can develop cohesive strategies that address immediate and long-term security challenges, ensuring that supply chain considerations are an integral part of cybersecurity planning.

Developing joint risk mitigation and improvement initiatives is a collaborative effort that leverages expertise from across the organization. Organizations can devise comprehensive strategies that address multiple aspects of supply chain risk by involving stakeholders from various departments, including IT, procurement, and operations. These initiatives often include implementing more robust security controls with suppliers, enhancing incident response capabilities, and improving the resilience of supply chain operations against cyberattacks.

Cross-functional team involvement in C-SCRM integration efforts is crucial for fostering a culture of collaboration and shared responsibility for cybersecurity. These teams can provide diverse perspectives and insights, contributing to a more holistic understanding of the risks and opportunities within the supply chain. Additionally, cross-functional collaboration helps in breaking down silos, ensuring that risk management efforts are coordinated and aligned with organizational objectives.

Utilization of risk assessment tools and methodologies in C-SCRM is instrumental in quantifying and managing supply chain risks. These tools can range from qualitative assessments, such as supplier surveys and interviews, to quantitative methods, like risk scoring models. By leveraging these tools, organizations can better understand their supply chain vulnerabilities, enabling more informed decision-making and prioritization of risk mitigation efforts.

Continuous monitoring and adjustment of C-SCRM strategies are necessary to adapt to the evolving risk landscape. This entails regularly reviewing the effectiveness of implemented controls, staying informed about emerging threats, and adjusting strategy as required. Continuous improvement processes ensure that C-SCRM efforts remain relevant and practical, even as the organization's supply chain and the broader cybersecurity environment change.

Recommendations

- **Establish a Dedicated C-SCRM Function**: Create a dedicated team or function within the organization to integrate C-SCRM into overall risk management efforts. This team should have clear responsibilities and the authority to coordinate activities across different departments.
- **Leverage Advanced Analytics and AI**: Utilize advanced analytics and artificial intelligence tools to enhance the identification, assessment, and monitoring of supply chain risks. These technologies can provide deeper insights into complex data sets, helping organizations anticipate and mitigate potential threats more effectively.
- **Foster Stronger Supplier Relationships**: Develop closer, more collaborative relationships with key suppliers to improve transparency and cooperation in managing cybersecurity risks. Regular engagement and sharing of best practices can help raise the overall security posture of the supply chain.
- **Implement Comprehensive Training Programs**: Roll out extensive training and awareness programs focused on C-SCRM for all relevant stakeholders, including employees, suppliers, and partners. These programs should cover the latest cybersecurity threats, risk management practices, and the importance of a secure supply chain.
- **Adopt a Continuous Improvement Mindset**: Encourage a culture of continuous improvement in C-SCRM practices. Regularly solicit stakeholder feedback, conduct lessons learned reviews after incidents, and stay abreast of new technologies and methodologies to enhance the organization's risk management capabilities.

NIST 800-53 Controls

(Refer to our appendix for a comprehensive description of the NIST 800-53 Controls, essential for organizations adhering to strict regulatory standards.)

- AC-01,AT-01,AU-01,CA-01,CM-01,CP-01,IA-01,IR-01,MA-01,MP-01,PE-01,PL-01,PM-01,PS-01,PT-01,RA-01,SA-01,SC-01,SI-01,SR-01,PM-09,PM-18,PM-30,PM-31,SR-02,SR-03,RA-03,RA-07

Simplified Security Controls (SSC)

Control 1: Cross-Functional Cybersecurity Risk Integration Team

Control Objective: To ensure effective integration of C-SCRM with the organization's broader risk management and cybersecurity strategies by forming a cross-functional team.

Implementation Steps:

1.1 **Team Formation**: Establish a cross-functional team comprising members from cybersecurity, supply chain management, risk management, and business units to foster a holistic approach to C-SCRM.

1.2 **Integration Workshops**: Conduct workshops to map out the intersections between C-SCRM and enterprise risk management, identifying shared objectives and areas for collaboration.

1.3 **Joint Strategy Development**: Collaboratively develop integrated risk management strategies that align C-SCRM objectives with the organization's broader cybersecurity and risk management goals.

Expected Outcome: Forming a cross-functional team dedicated to integrating C-SCRM into the broader risk management framework ensures a unified approach to risk across the organization. This holistic strategy enhances the organization's ability to identify, assess, and mitigate cybersecurity risks within the supply chain, aligning with overall business objectives and improving resilience.

Control 2: Advanced Risk Assessment and Monitoring Tools

Control Objective: To enhance the organization's capability to continuously assess and monitor supply chain risks by leveraging advanced risk assessment tools and methodologies.

Implementation Steps:

2.1 **Tool Selection**: Identify and implement advanced risk assessment tools to analyze complex supply chain networks and identify potential cybersecurity vulnerabilities and threats.

2.2 **Integration into Existing Processes**: Ensure these tools are integrated into the organization's ongoing risk assessment processes, allowing for real-time monitoring and analysis of supply chain risks.

2.3 **Training and User Adoption**: Provide comprehensive training to relevant staff on the operation and capabilities of these tools, ensuring widespread adoption and effective use across the organization.

Expected Outcome: By incorporating advanced risk assessment tools into C-SCRM processes, the organization can better understand supply chain risks, enabling proactive identification and mitigation strategies. This continuous monitoring capability ensures the organization can quickly adapt to emerging threats, maintaining a robust cybersecurity posture.

Control 3: Comprehensive C-SCRM Training and Awareness Program

Control Objective: To bolster the effectiveness of integrated risk management efforts by developing and implementing a comprehensive training and awareness program focused on C-SCRM principles and practices.

Implementation Steps:

3.1 **Program Development**: Create a training and awareness program that covers key C-SCRM concepts, including risk assessment, mitigation strategies, and the importance of supply chain security to the organization's overall risk posture.

3.2 **Targeted Training Sessions**: Deliver training sessions tailored to different organizational roles, ensuring that staff at all levels understand their responsibilities and how they contribute to supply chain security.
3.3 **Ongoing Awareness Campaigns**: Implement ongoing awareness campaigns using newsletters, intranet updates, and workshops to keep C-SCRM at the forefront of all employees' and stakeholders' minds.

Expected Outcome: The development and implementation of a comprehensive C-SCRM training and awareness program ensure that all organization members know the importance of supply chain security and their role in maintaining it. This heightened awareness fosters a culture of security across the organization, enhancing the collective ability to effectively manage and mitigate supply chain risks.

GV.SC-04: Prioritizing Suppliers by Criticality in Cybersecurity Supply Chain Risk Management

Identifying and prioritizing suppliers by criticality are pivotal steps in bolstering an organization's C-SCRM strategy. Recognizing the varying levels of risk and dependency each supplier introduces to the operational ecosystem is the first step toward creating a resilient and secure supply chain. This initial phase involves a detailed analysis of the supply chain to distinguish suppliers based on the criticality of their goods or services to the organization's operations and cybersecurity posture.

Developing criteria for assessing supplier criticality requires a multifaceted approach that considers several dimensions of risk and impact. Factors such as the uniqueness of the supplier's products or services, the difficulty of replacing them, their integration level within your systems, and the potential impact of a disruption on your operations are all critical considerations. Developing these criteria is not a one-size-fits-all process; it must be tailored to reflect the organization's specific needs, vulnerabilities, and risk appetite.

Creating a comprehensive inventory of suppliers, complete with criticality ratings, enables organizations to visualize their supply chain landscape and identify potential points of failure. This inventory is a foundational element for C-SCRM, supporting the strategic allocation of resources toward suppliers that pose the highest risk to cybersecurity and operational continuity. It facilitates a more structured supply chain risk management approach, allowing for targeted risk mitigation strategies.

The dynamic nature of cyber threats and business operations necessitates regular reviews and updates of supplier criticality. As market conditions shift, new vulnerabilities emerge, and business priorities evolve, the criticality of suppliers can change. Organizations must establish processes for periodically reassessing their suppliers, ensuring that the criticality ratings accurately reflect the current risk landscape and business needs.

Integrating supplier criticality into risk management and decision-making processes ensures that supply chain risks are considered in strategic planning, operational decisions, and risk mitigation efforts. This integration fosters a holistic view of risk that encompasses both direct and indirect sources of cyber threats. By understanding the criticality of suppliers, organizations can make informed decisions that enhance supply chain resilience and cybersecurity.

Communicating expectations and requirements to critical suppliers is essential for establishing a mutual understanding of the cybersecurity standards and practices that must be upheld. This communication should include clear guidelines on security requirements, incident reporting protocols, and compliance expectations. Organizations can work collaboratively with their suppliers to fortify the supply chain against cyber threats by setting these standards.

Implementing enhanced controls and monitoring is imperative for those suppliers identified as high-criticality. These controls may include more stringent security audits, regular vulnerability assessments, and establishing secure communication channels. Enhanced monitoring of these suppliers helps in the early detection of potential security breaches, enabling timely and effective responses to mitigate risks.

Tracking supplier performance concerning their criticality involves monitoring compliance with agreed-upon cybersecurity standards and their overall impact on the organization's risk posture. This ongoing assessment helps identify suppliers that consistently meet or exceed expectations and those that present increasing risks. Performance tracking is crucial for maintaining an effective and secure supply chain, facilitating continuous improvement in C-SCRM practices.

Recommendations

- **Develop a Scalable Framework for Supplier Assessment**: Establish a flexible and scalable framework to accommodate changes in supplier criticality, business needs, and the threat landscape. This framework should allow for easy updating and reassessment of suppliers.
- **Leverage Technology for Supplier Risk Management**: Utilize advanced technology solutions, such as AI and machine learning, to continuously monitor and assess supplier risks. These technologies can provide real-time insights and alerts, enhancing the ability to respond to emerging threats swiftly.
- **Establish Strong Collaboration Channels with Suppliers**: Create formal mechanisms for ongoing collaboration with suppliers, such as joint cybersecurity task forces or information-sharing platforms. These channels can facilitate the exchange of best practices, threat intelligence, and timely communication of security incidents.
- **Conduct Regular Cybersecurity Audits and Drills with Suppliers**: Organize periodic cybersecurity audits and response drills involving critical suppliers. These activities can help assess the effectiveness of security measures and the preparedness of both parties to handle cyber incidents.
- **Incorporate Supplier Risk Management into Business Continuity Planning**: Integrate supplier risk management strategies into the organization's broader business continuity and disaster recovery plans. This ensures that the organization can maintain operations and recover quickly in the event of a significant disruption in the supply chain.

NIST 800-53 Controls

(Refer to our appendix for a comprehensive description of the NIST 800-53 Controls, essential for organizations adhering to strict regulatory standards.)

- RA-09,SA-09,SR-06

Simplified Security Controls (SSC)

Control 1: Supplier Criticality Assessment Framework

Control Objective: To systematically assess and prioritize suppliers based on their criticality to the organization's operations and cybersecurity posture, ensuring focused risk management efforts where they are most needed.

Implementation Steps:

1.1 **Develop Assessment Criteria**: Create a set of criteria for evaluating supplier criticality, including the supplier's access to sensitive data, the uniqueness of their services, and the potential impact of their compromise on the organization.

1.2 **Conduct Criticality Assessments**: Apply the assessment criteria to each supplier to determine their criticality level to the organization's operations and cybersecurity.

1.3 **Categorize and Prioritize Suppliers**: Categorize suppliers into tiers based on their assessed criticality, prioritizing them for further risk management actions and resource allocation.

Expected Outcome: Establishing a supplier criticality assessment framework enables the organization to identify and prioritize high-criticality suppliers. This prioritization ensures that risk management efforts and resources are focused on areas of most significant potential impact, enhancing the organization's cybersecurity resilience.

Control 2: Enhanced Security Measures for High-Criticality Suppliers

Control Objective: To implement more robust cybersecurity controls and ongoing monitoring for suppliers identified as high criticality, reducing the risk of cybersecurity incidents that could impact the organization.

Implementation Steps:

2.1 **Define Enhanced Security Requirements**: Specify additional cybersecurity requirements for high-criticality suppliers, including more substantial encryption standards, frequent security audits, and advanced incident response capabilities.

2.2 **Communicate and Implement Requirements**: Communicate the enhanced security requirements to high-criticality suppliers and work collaboratively to ensure their implementation.

2.3 **Establish Continuous Monitoring**: Set up continuous monitoring mechanisms to oversee the compliance of high-criticality suppliers with enhanced security requirements, ensuring any deviations are quickly identified and addressed.

Expected Outcome: By implementing enhanced security measures and continuous monitoring for high-criticality suppliers, the organization can significantly reduce the risk of cybersecurity incidents originating from or affecting these key supply chain elements. This control strengthens the overall security of the supply chain and the organization's resilience against cyber threats.

Control 3: Dynamic Supplier Criticality Management Process

Control Objective: To establish a dynamic process for regularly reviewing and adjusting the criticality ratings of suppliers as business needs and the threat landscape evolve, ensuring that supplier management remains aligned with current risks.

Implementation Steps:

3.1 **Schedule Regular Criticality Reviews**: Set a regular schedule for reviewing the criticality of suppliers, taking into account changes in the organization's operations, the threat environment, and suppliers' practices.

3.2 **Adjust Criticality Ratings as Needed**: Based on the review, adjust suppliers' criticality ratings to reflect any changes in their importance or risk profile to the organization.

3.3 **Update Risk Management Practices Accordingly**: Modify risk management practices and controls applied to suppliers based on their updated criticality ratings, ensuring that resources and efforts are appropriately aligned with the current risk landscape.

Expected Outcome: Implementing a dynamic supplier criticality management process ensures the organization's approach to managing supply chain cybersecurity risks remains flexible and responsive to changes. This adaptability ensures that cybersecurity efforts are always aligned with the most current understanding of risks, enhancing the effectiveness of the organization's overall cybersecurity strategy.

GV.SC-05: Establishing Cybersecurity Requirements in Supply Chain Contracts

In the current digital age, ensuring the cybersecurity of supply chains is paramount. This necessity has led organizations to establish, prioritize, and integrate cybersecurity requirements into contracts and agreements with suppliers and other relevant third parties. The process begins with a clear understanding of the cybersecurity risks inherent within the supply chain and extends to developing comprehensive strategies to mitigate these risks effectively. Establishing robust cybersecurity requirements forms the backbone of a secure and resilient supply chain, safeguarding sensitive information and critical infrastructure from potential cyber threats.

Developing and prioritizing cybersecurity requirements are contingent upon a thorough risk assessment. This assessment identifies the specific vulnerabilities and threats within the supply chain, allowing organizations to tailor their cybersecurity requirements accordingly. Prioritization is based on the potential impact of identified risks, ensuring that resources are allocated to address the most critical vulnerabilities first. This strategic approach ensures that cybersecurity measures are practical and efficient, providing maximum protection where needed.

Integrating cybersecurity requirements into legal and contractual frameworks is a critical step in formalizing the expectations and responsibilities of all parties involved. Contracts serve as binding agreements and tools for clearly communicating cybersecurity expectations to suppliers and third parties. These frameworks should detail the specific cybersecurity measures required, the standards to be adhered to, and the responsibilities of each party in maintaining the security of the supply chain.

Effective communication of cybersecurity requirements to suppliers and third parties is essential for ensuring compliance and fostering a collaborative approach to supply chain security. Organizations must convey the importance of these requirements and the rationale behind them, encouraging suppliers to adopt similar cybersecurity standards and practices. This collaborative effort enhances the overall security posture of the supply chain, creating a unified front against cyber threats.

Mechanisms for verifying compliance with cybersecurity requirements are critical for ensuring that suppliers and third parties adhere to agreed-upon standards. These mechanisms can include regular audits, security assessments, and reviews of compliance documentation. Verifying compliance helps identify gaps or weaknesses in suppliers' cybersecurity practices, allowing for timely remediation and continuous improvement of the supply chain's security.

Contractual provisions for ongoing cybersecurity compliance monitoring are necessary to maintain a dynamic and responsive approach to supply chain security. These provisions should outline the frequency and scope of compliance reviews and the processes for reporting and addressing any identified issues. Ongoing monitoring ensures that cybersecurity measures remain effective over time and can adapt to evolving threats and vulnerabilities.

Processes for updating cybersecurity requirements in response to emerging threats are crucial for maintaining the relevance and effectiveness of contractual agreements. The cyber threat landscape constantly evolves, necessitating regular updates to cybersecurity requirements to reflect new threats and vulnerabilities. Organizations must establish agile processes for revising contracts and agreements, ensuring they can quickly respond to changes in the threat environment.

Training for procurement and contract management teams on cybersecurity requirements is essential for ensuring that these teams are equipped to negotiate and manage contracts that effectively address supply chain risks. Training should cover the basics of cybersecurity, the specific risks associated with the supply chain, and the best practices for integrating cybersecurity requirements into contracts. Well-informed teams can more effectively advocate for robust cybersecurity measures, enhancing the security of the supply chain.

Recommendations

- **Adopt a Standardized Set of Cybersecurity Requirements**: Develop a standardized set of cybersecurity requirements that can be customized and integrated into contracts with suppliers and third parties. This approach ensures consistency and comprehensiveness in addressing cybersecurity risks across the supply chain.
- **Implement a Supplier Certification Program**: Establish a certification program for suppliers that meet or exceed your organization's cybersecurity requirements. This program can incentivize suppliers to adopt higher cybersecurity standards and provide a mechanism for recognizing and promoting compliant suppliers.
- **Leverage Third-Party Risk Management Solutions**: Utilize third-party risk management solutions to streamline the process of assessing and monitoring the cybersecurity posture of suppliers. These solutions can offer valuable insights and automation capabilities, reducing the administrative burden on your organization.
- **Include Cybersecurity Performance Metrics in Contracts**: Integrate specific cybersecurity performance metrics into contracts to quantify compliance and effectiveness. These metrics can serve as benchmarks for evaluating the cybersecurity practices of suppliers and facilitating continuous improvement.
- **Foster Open Communication and Collaboration**: Encourage open communication and collaboration with suppliers regarding cybersecurity challenges and best practices. This collaborative approach can lead to more effective mitigation strategies and a stronger overall cybersecurity posture for the supply chain.

NIST 800-53 Controls

(Refer to our appendix for a comprehensive description of the NIST 800-53 Controls, essential for organizations adhering to strict regulatory standards.)

- SA-04,SA-09,SR-03,SR-05,SR-06,SR-10

Simplified Security Controls (SSC)

Control 1: Cybersecurity Requirement Integration in Supplier Contracts

Control Objective: To ensure all suppliers and third parties comply with the organization's cybersecurity standards by integrating specific cybersecurity requirements into contracts and agreements.

Implementation Steps:

1.1 **Define Cybersecurity Standards**: Develop clear and comprehensive cybersecurity standards that suppliers must adhere to based on the organization's risk assessment and industry best practices.

1.2 **Contractual Integration**: Incorporate these cybersecurity standards into all new and renewed contracts, specifying the requirements, responsibilities, and expectations for compliance.

1.3 **Legal Review**: Have legal teams review contractual clauses to ensure they are enforceable and align with local and international regulations.

Expected Outcome: Integrating cybersecurity requirements into contracts ensures suppliers are legally obligated to comply with the organization's cybersecurity standards. This control measure significantly reduces the risk of security breaches originating from the supply chain, enhancing the organization's overall security posture.

Control 2: Compliance Verification Mechanism

Control Objective: To establish a systematic approach for verifying supplier compliance with the organization's cybersecurity requirements, ensuring that standards are met consistently.

Implementation Steps:

2.1 **Develop Verification Processes**: Create a detailed process for periodically assessing supplier compliance, which may include audits, self-assessments, and third-party certifications.

2.2 **Implement Monitoring Tools**: Utilize technology solutions for continuous monitoring of key cybersecurity metrics and indicators of compromise that could signal a breach of contract terms.

2.3 **Remediation and Penalties**: Define clear procedures for addressing noncompliance, including remediation steps and potential penalties, to ensure suppliers correct deficiencies promptly.

Expected Outcome: Implementing a compliance verification mechanism assures the organization that suppliers are meeting cybersecurity requirements. This systematic monitoring and enforcement facilitate quick detection and remediation of compliance issues, maintaining the integrity of the supply chain's security.

Control 3: Dynamic Cybersecurity Clauses in Contracts

Control Objective: To ensure supplier contracts remain responsive to the evolving cybersecurity landscape by including provisions for updating cybersecurity requirements in response to new threats.

Implementation Steps:

3.1 **Clause Development**: Draft dynamic cybersecurity clauses that adjust security requirements based on emerging threats, technological advancements, and changes in regulatory environments.

3.2 **Communication Plan**: Establish a formal communication plan for notifying suppliers of changes to cybersecurity requirements, ensuring they are informed and prepared to implement necessary updates.

3.3 **Review and Update Schedule**: Set a regular schedule for reviewing the cybersecurity landscape and assessing the need for updates to contractual cybersecurity requirements.

Expected Outcome: Including dynamic cybersecurity clauses in supplier contracts ensures that cybersecurity requirements can be adapted to address new and emerging threats. This flexibility enhances the organization's resilience against cybersecurity risks by ensuring that the supply chain's defense mechanisms evolve with the threat landscape.

GV.SC-06: Enhancing Cybersecurity Through Diligent Supplier Selection and Management

In cybersecurity, due diligence in selecting and managing suppliers or third-party relationships is not just a preliminary step; it is a continuous, essential process for minimizing risks and safeguarding assets. This chapter delves into the systematic approach organizations should adopt to ensure that their suppliers and third parties uphold the same high standards of cybersecurity that they do. This multistep process begins with an introduction to due diligence, emphasizing its significance in C-SCRM.

Establishing comprehensive due diligence processes for new and existing suppliers is pivotal. These processes must be meticulously designed to evaluate the cybersecurity posture of suppliers, assessing their practices, policies, and compliance with relevant standards. The goal is to create a framework that scrutinizes new suppliers and continuously evaluates existing ones, ensuring they evolve in tandem with the organization's cybersecurity requirements.

Criteria for cybersecurity risk assessment during the due diligence process are central to identifying potential vulnerabilities and threats that a supplier might introduce into the organization. These criteria should encompass a range of factors, including the supplier's access to sensitive data, their cybersecurity practices, and their history of security incidents. This risk assessment is a critical component of the due diligence process, guiding organizations in making informed decisions about which suppliers to engage with.

Integration of due diligence findings into supplier selection and contracting ensures that cybersecurity considerations are front and center when formalizing relationships. Contracts should reflect the outcomes of the due diligence process, incorporating requirements and obligations that safeguard the organization's cybersecurity interests. This integration is key to creating a transparent, accountable relationship between the organization and its suppliers.

Processes for regular reassessment of supplier risks and performance are necessary to respond to the dynamic nature of cybersecurity threats. Organizations must establish ongoing evaluation mechanisms, leveraging internal assessments and external audits. This continuous reassessment helps identify emerging risks and ensure suppliers adapt their practices to mitigate them effectively.

Risk mitigation strategies for identified risks during the due diligence are crucial for preventing potential breaches. Organizations should work closely with suppliers to develop and implement these strategies, from enhancing data encryption to improving incident response protocols. Effective risk mitigation requires a proactive approach, anticipating vulnerabilities before exploitation.

Documentation and record-keeping of due diligence activities serve multiple purposes. They provide a historical record of the evaluation process, support compliance efforts, and offer insights into decision-making. Adequate documentation ensures that due diligence activities are transparent and that lessons learned can be applied to future evaluations.

Training for teams involved in supplier evaluation and management is fundamental to the success of the due diligence process. These teams should be equipped with the knowledge and skills needed to assess cybersecurity risks accurately and understand supplier relationships' legal and regulatory implications. Training enhances the effectiveness of the due diligence process, enabling teams to identify and mitigate risks more effectively.

Collaboration with suppliers to address identified risks underscores the notion that cybersecurity is a shared responsibility. Open dialogue and partnership are vital in developing robust security measures that benefit the organization and its suppliers. This collaborative approach fosters a culture of security that extends throughout the supply chain.

Recommendations

- **Establish Clear Cybersecurity Expectations**: Before engaging with suppliers, clearly define the cybersecurity standards and practices expected of them. This clarity helps ensure that suppliers know your security requirements from the outset.
- **Leverage Cybersecurity Frameworks and Standards**: Utilize recognized cybersecurity frameworks and standards, such as the NIST Cybersecurity Framework, to evaluate suppliers' cybersecurity practices. This provides a common language and expectations for cybersecurity within the supply chain.
- **Conduct Comprehensive Risk Assessments**: Perform thorough risk assessments of potential and existing suppliers, focusing on their ability to protect sensitive data and their resilience against cyber threats. These assessments should inform all stages of the supplier relationship, from selection to ongoing management.
- **Incorporate Cybersecurity Clauses in Contracts**: Ensure contracts with suppliers include specific cybersecurity clauses that outline requirements, responsibilities, and penalties for non-compliance. These clauses are crucial for legally binding suppliers to maintain adequate cybersecurity practices.
- **Foster Continuous Improvement**: Encourage suppliers to improve their cybersecurity measures continuously. Offer guidance and support to help them align with best practices and address any identified gaps. This ongoing improvement strengthens the organization's and its suppliers' overall cybersecurity posture.

NIST 800-53 Controls

(Refer to our appendix for a comprehensive description of the NIST 800-53 Controls, essential for organizations adhering to strict regulatory standards.)

- SA-04,SA-09,SR-05,SR-06,PM-1

Simplified Security Controls (SSC)

Control 1: Comprehensive Supplier Due Diligence Program

Control Objective: To systematically assess potential and existing suppliers' cybersecurity risks before formalizing relationships, ensuring suppliers adhere to the organization's cybersecurity standards.

Implementation Steps:

1.1 **Develop Due Diligence Criteria**: Establish detailed criteria for cybersecurity risk assessment, focusing on suppliers' security policies, incident response capabilities, compliance with relevant regulations, and security culture.

1.2 **Conduct Due Diligence Assessments**: Perform thorough cybersecurity risk assessments for all potential suppliers and regularly reassess existing suppliers to ensure ongoing compliance with the organization's security requirements.

1.3 **Documentation and Review Process**: Document all findings from the due diligence assessments and establish a review process involving key stakeholders to make informed decisions regarding supplier selection and management.

Expected Outcome: Implementing a comprehensive supplier due diligence program ensures that the organization engages with suppliers meeting its cybersecurity standards, significantly reducing the risk of security breaches from the supply chain.

Control 2: Integration of Cybersecurity Clauses in Supplier Contracts

Control Objective: To legally bind suppliers to comply with the organization's cybersecurity requirements by integrating specific cybersecurity clauses into contracts and agreements.

Implementation Steps:

2.1 **Define Cybersecurity Requirements**: Clearly outline the cybersecurity requirements that suppliers must meet, including data protection measures, incident reporting protocols, and compliance with relevant cybersecurity frameworks.

2.2 **Incorporate Requirements into Contracts**: Work with legal and procurement teams to ensure cybersecurity requirements are incorporated into all supplier contracts, including penalties for noncompliance.

2.3 **Continuous Contractual Compliance Monitoring**: Establish a system for continuously monitoring suppliers' compliance with the cybersecurity clauses in their contracts, including regular audits and assessments.

Expected Outcome: Including specific cybersecurity clauses in supplier contracts ensures that suppliers are legally obligated to adhere to the organization's cybersecurity requirements, providing a mechanism for enforcement and recourse in the event of noncompliance.

Control 3: Supplier Collaboration and Risk Mitigation Program

Control Objective: To proactively work with suppliers to identify and mitigate cybersecurity risks through a collaborative risk management program.

Implementation Steps:

3.1 **Establish Communication Channels**: Set up dedicated channels for ongoing communication with suppliers about cybersecurity risks and best practices.

3.2 **Joint Risk Mitigation Planning**: Collaborate with suppliers to develop and implement mitigation plans for identified cybersecurity risks, leveraging both parties' expertise and resources.

3.3 **Regular Performance and Risk Review Meetings**: Schedule regular meetings with suppliers to review the effectiveness of implemented risk mitigation strategies and discuss any new or evolving cybersecurity risks.

Expected Outcome: A supplier collaboration and risk mitigation program foster a partnership approach to managing cybersecurity risks, ensuring the organization and its suppliers are aligned to maintain a secure supply chain. This collaborative effort enhances the overall cybersecurity resilience of the supply chain and enables more effective risk mitigation strategies.

GV.SC-07: Mastering Supplier Risk Management in the Cybersecurity Landscape

Managing risks posed by suppliers, their products, and services, as well as other third parties, is critical to an organization's cybersecurity posture. A comprehensive approach to risk management for suppliers and third parties not only identifies and records these risks but also prioritizes, assesses, responds to, and monitors them throughout the duration of the relationship. This process ensures that organizations can maintain a strong defense against potential vulnerabilities their supply chain introduces.

Developing a framework for identifying and recording supplier risks is the foundation of effective supplier risk management. This framework should encompass the various dimensions of risk, including operational, financial, reputational, and, most importantly, cybersecurity risks. It should provide a structured method for capturing risk-related information, ensuring that all potential threats are documented and available for analysis.

Prioritizing risks based on their impact and likelihood is essential for allocating resources effectively. Not all risks carry the same weight; thus, understanding which risks pose the greatest threat to organizational objectives is crucial. This prioritization guides the organization in focusing its efforts on mitigating the most significant vulnerabilities first, optimizing the use of limited cybersecurity resources.

Continuous assessment of supplier-related risks acknowledges the dynamic nature of the threat landscape. As suppliers evolve, so too do the risks associated with them. Regular evaluations ensure that new risks are identified promptly and that existing risk assessments remain accurate and relevant. This ongoing process supports a proactive stance in the face of emerging cybersecurity threats.

Implementing response plans for key supplier risks is about preparing to act when a risk materializes. These plans should outline the steps to mitigate a risk's impact, specifying roles, responsibilities, and timelines. Effective response plans are tailored to the specific nuances of supplier risks, ensuring that the organization can react swiftly and decisively to protect its interests.

Ongoing monitoring of supplier risks and mitigation measures is vital for understanding the effectiveness of the organization's risk management strategies. This continuous oversight helps in detecting when risk levels change or when mitigation measures fail to perform as expected. Monitoring provides the feedback necessary to adjust strategies and tactics in real time, enhancing the overall resilience of the supply chain.

Communication and reporting structures for supplier risk management facilitate transparency and accountability within the organization and with its suppliers. Clear communication channels ensure that all stakeholders are informed of potential risks and the steps being taken to manage them. Reporting structures, meanwhile, provide a formal mechanism for escalating issues and sharing insights about the risk landscape with decision-makers.

Supplier collaboration in risk identification and mitigation efforts recognizes suppliers' role in managing supply chain risks. Organizations and their suppliers can develop more effective strategies for addressing mutual vulnerabilities by working together. Collaboration leads to a more integrated approach to cybersecurity, where both parties contribute to a stronger defense against threats.

Reviewing and updating supplier risk profile processes ensures that the organization's understanding of supplier risks remains current. As the business environment and threat landscape evolve, so too must the organization's assessment of its suppliers. Regular reviews allow for the timely identification of changes in supplier risk profiles, enabling the organization to adapt its risk management strategies accordingly.

Recommendations

- **Leverage Technology for Enhanced Risk Monitoring**: Implement advanced monitoring tools and technologies to gain real-time insights into supplier risk. Automated risk analysis platforms can help identify emerging threats more quickly, enabling proactive risk management.
- **Foster a Culture of Risk Awareness Among Suppliers**: Encourage suppliers to adopt a risk-aware culture by providing them with training and resources on cybersecurity best practices. A well-informed supplier is a valuable ally in the fight against cyber threats.
- **Establish Clear Risk Reporting Guidelines**: Develop and communicate clear guidelines for reporting cybersecurity incidents and vulnerabilities. Timely reporting by suppliers can significantly reduce the potential impact of security breaches.

- **Incorporate Risk Management into Supplier Performance Metrics**: Integrate supplier risk management performance into overall supplier evaluation criteria. Suppliers that demonstrate strong risk management practices should be recognized and potentially rewarded.
- **Conduct Joint Risk Management Exercises**: Organize regular exercises with suppliers to simulate cybersecurity incidents and test response plans. These exercises can strengthen the organization's and its suppliers' preparedness for real-world cyber threats.

NIST 800-53 Controls

(Refer to our appendix for a comprehensive description of the NIST 800-53 Controls, essential for organizations adhering to strict regulatory standards.)

- RA-09,SA-04,SA-09,SR-03,SR-06

Simplified Security Controls (SSC)

Control 1: Supplier Risk Identification and Documentation System

Control Objective: To ensure all risks associated with suppliers, their products, and services are systematically identified, documented, and maintained in a centralized repository throughout the relationship.

Implementation Steps:

1.1 **Develop a Risk Identification Framework**: Create a comprehensive framework for identifying potential risks associated with suppliers, including security vulnerabilities, compliance issues, and operational risks.

1.2 **Implement a Centralized Documentation System**: Utilize a centralized platform or database to document and store all identified risks, ensuring easy access and visibility for relevant organizational stakeholders.

1.3 **Continuous Risk Identification Process**: Establish a continuous process for updating the documentation system with new risks as they are identified through ongoing assessments, audits, and monitoring activities.

Expected Outcome: Implementing a supplier risk identification and documentation system ensures that all risks are systematically captured and accessible to relevant parties. This central repository supports effective risk management by providing a comprehensive overview of supplier-related risks, facilitating prioritization and response planning.

Control 2: Dynamic Supplier Risk Assessment and Prioritization Tool

Control Objective: To dynamically assess and prioritize supplier risks based on their impact on the organization and the likelihood of occurrence, ensuring focused risk management efforts on the most critical areas.

Implementation Steps:

2.1 **Tool Development**: Develop or implement an existing tool capable of dynamically assessing and prioritizing risks based on predefined criteria, such as the impact on the organization and the likelihood of occurrence.

2.2 **Integration with Documentation System**: Integrate this tool with the centralized risk documentation system to ensure that all identified risks are assessed and prioritized in real time.

2.3 **Regular Review and Update of Assessment Criteria**: Review and update the risk assessment criteria to reflect changes in the organization's risk appetite, business objectives, and the external threat landscape.

Expected Outcome: The deployment of a dynamic risk assessment and prioritization tool enables the organization to identify high-priority risks associated with suppliers quickly. This facilitates timely and focused risk management actions on areas with the most significant potential impact, enhancing the organization's overall risk posture.

Control 3: Collaborative Risk Management Framework with Suppliers

Control Objective: To foster collaboration with suppliers in the identification, assessment, and mitigation of risks, ensuring a shared approach to managing security challenges.

Implementation Steps:

3.1 **Establish Communication Channels**: Set up dedicated communication channels for risk management discussions with suppliers, including regular meetings, secure messaging platforms, and shared documentation tools.

3.2 **Joint Risk Management Workshops**: Organize workshops with suppliers to collaboratively identify and assess risks, develop mitigation strategies, and establish joint response plans.

3.3 **Implement Shared Risk Monitoring Mechanisms**: Deploy monitoring mechanisms that allow the organization and its suppliers to track the effectiveness of implemented mitigation measures and identify new risks.

Expected Outcome: Implementing a collaborative risk management framework with suppliers ensures that risk management efforts are aligned and leverages the expertise and resources of both parties. This collaborative approach enhances the effectiveness of risk identification, assessment, and mitigation efforts, leading to more robust security measures across the supply chain.

GV.SC-08: Collaborative Incident Management with Suppliers

Integrating suppliers and third parties into an organization's cybersecurity incident management processes is crucial for ensuring a coordinated response to security incidents. This integration enhances the organization's ability to respond effectively and ensures that all parties involved are prepared to act swiftly to mitigate the impact of incidents. This collaborative approach to incident management is becoming increasingly necessary as the interdependencies between organizations and their suppliers grow deeper in the digital age.

Identifying key suppliers for inclusion in incident response planning is the first step in building a robust incident management framework. Organizations must assess which suppliers have access to their networks, systems, or data and determine the potential impact of their involvement in or response to a cybersecurity incident. Prioritizing these suppliers ensures that efforts are focused on those whose participation is critical to the organization's incident response capabilities.

Developing joint incident response protocols with suppliers is essential for establishing clear guidelines and expectations for managing incidents. These protocols should outline the steps to be taken by the organization and its suppliers in the event of an incident, including initial response, investigation, containment, and recovery activities. Joint protocols ensure that all parties are aligned and can act quickly and efficiently to address threats.

Training and exercises to test supplier inclusion in incident response are vital for preparedness. Regular training sessions and simulated exercises help to familiarize suppliers with the organization's incident response procedures and test the effectiveness of joint protocols. These activities also identify areas for improvement and help build a strong working relationship between the organization and its suppliers.

Establishing communication channels for incident reporting and coordination is critical for effective incident management. These channels should facilitate the timely exchange of information between the organization and its suppliers, enabling swift action to mitigate the impact of incidents. Clear communication protocols ensure that relevant information is shared with the right parties at the right time, enhancing the incident response.

Defining the roles and responsibilities of suppliers in incident recovery activities is a key aspect of collaborative incident management. Suppliers should understand their role in the organization's recovery efforts, including the restoration of services and support for investigations. Clear expectations help ensure recovery activities are conducted smoothly and efficiently, minimizing the downtime and impact on the organization.

Evaluation and feedback mechanisms for incident response activities involving suppliers are essential for continuous improvement. After an incident, organizations should review the effectiveness of their response, including the role played by suppliers, and identify lessons learned. This evaluation process should involve feedback from all parties, leading to refinements in joint protocols and training programs.

Contractual obligations related to incident response and notification play a significant role in formalizing the expectations for suppliers. Contracts should specify the requirements for incident reporting, cooperation during investigations, and assistance in recovery efforts. These contractual obligations ensure that suppliers are legally bound to participate in incident management activities, providing a foundation for collaboration.

Improving incident response processes based on lessons learned is a continuous enhancement cycle. Organizations should use the insights from incident evaluations and feedback to refine their incident management protocols, training programs, and communication strategies. This iterative process ensures that the organization and its suppliers are better prepared to handle future incidents.

Recommendations

- **Foster a Culture of Transparency and Collaboration**: Encourage open communication and collaboration regarding cybersecurity incident management with suppliers. A culture of transparency fosters trust and ensures suppliers are more willing and prepared to engage in joint incident response efforts.
- **Incorporate Cybersecurity Incident Management into Supplier Selection**: Consider a supplier's ability and willingness to participate in incident management activities as a criterion during the supplier selection. Suppliers with solid incident response capabilities should be given preference.
- **Regularly Update and Review Incident Management Protocols**: Joint incident response protocols with suppliers should be reviewed and updated regularly to reflect changes in the threat landscape, business processes, or supplier relationships. This ensures that the protocols remain relevant and effective.
- **Leverage Technology for Efficient Incident Coordination**: Utilize incident management and communication tools that facilitate efficient coordination between the organization and its suppliers. Technology can play a crucial role in streamlining the exchange of information and coordinating response activities.
- **Build Redundancy into Supplier Networks**: Develop relationships with multiple suppliers for critical services or materials, when possible, to ensure that the organization has alternatives

in the event of a supplier-related incident. This redundancy can enhance the organization's resilience and reduce the potential impact of incidents.

NIST 800-53 Controls

(Refer to our appendix for a comprehensive description of the NIST 800-53 Controls, essential for organizations adhering to strict regulatory standards.)

- SA-04,SA-09,SR-02,SR-03,SR-08,CP-01,IR-01

Simplified Security Controls (SSC)

Control 1: Joint Incident Response Planning with Key Suppliers

Control Objective: To ensure a coordinated and efficient response to cybersecurity incidents that involve or affect key suppliers by developing joint incident response plans.

Implementation Steps:

1.1 **Identify Key Suppliers**: Determine which suppliers are critical to the organization's operations and would require a coordinated incident response effort.

1.2 **Develop Joint Incident Response Plans**: Collaboratively develop incident response plans with these key suppliers, outlining roles, responsibilities, communication protocols, and specific actions to be taken in the event of an incident.

1.3 **Formalize Plans through Agreements**: Solidify these plans by including them in contracts or service level agreements (SLAs) with the suppliers, ensuring both parties are legally committed to the response strategy.

Expected Outcome: Establishing joint incident response plans with key suppliers ensures that the organization and its suppliers are prepared to act swiftly and effectively in a cybersecurity incident. This preparation minimizes the impact of incidents on operations and strengthens the overall security posture.

Control 2: Supplier Incident Response Training and Simulation Exercises

Control Objective: To validate the effectiveness of the joint incident response plans and enhance the organization's and its suppliers' readiness through regular training and simulation exercises.

Implementation Steps:

2.1 **Develop Training Programs**: Create comprehensive training programs for internal teams and key suppliers covering joint incident response protocols, communication procedures, and role-specific actions.

2.2 **Conduct Simulation Exercises**: Regularly schedule simulation exercises that mimic potential cybersecurity incidents involving key suppliers to test the effectiveness of the joint response plans.

2.3 **Review and Revise Plans**: Use the insights gained from training and exercises to review and update the incident response plans as necessary, addressing any weaknesses or areas for improvement.

Expected Outcome: Regular training and simulation exercises with suppliers ensure that all parties are familiar with their roles and responsibilities in the event of an incident. This preparedness allows for a more effective and cohesive response to cybersecurity incidents, reducing potential damage and recovery time.

Control 3: Continuous Communication and Feedback Mechanism for Incident Management

Control Objective: To establish ongoing communication and feedback mechanisms that support the continuous improvement of incident response activities involving suppliers.

Implementation Steps:

3.1 **Set Up Communication Channels**: Implement secure and reliable communication channels dedicated to incident management, allowing for the swift exchange of information between the organization and its suppliers during an incident.

3.2 **Create Feedback Loops**: After an incident response, conduct debriefing sessions with all involved parties to gather feedback on the effectiveness of the response, communication, and coordination.

3.3 **Implement Improvement Processes**: Based on the feedback collected, systematically review and refine incident response plans, training programs, and communication protocols to address identified shortcomings and enhance future responses.

Expected Outcome: Creating continuous communication and feedback mechanisms ensures that lessons learned from incident responses are captured and used to strengthen future incident management efforts. This ongoing improvement process enhances the organization's and its suppliers' resilience against cybersecurity threats.

GV.SC-09: Fortifying Cybersecurity Through Strategic Supply Chain Security Integration

Integrating supply chain security practices into cybersecurity and enterprise risk management programs is pivotal in fortifying an organization's defense against cyber threats. This integration emphasizes the importance of supply chain security in the broader context of organizational security. It ensures a unified approach to managing risks associated with suppliers and third-party vendors. Organizations can achieve a more comprehensive and resilient security posture by treating supply chain security as an integral component of the cybersecurity strategy.

Aligning supply chain security with overall cybersecurity objectives necessitates a strategic approach considering the entire ecosystem of suppliers and partners. This alignment ensures that supply chain security measures support the broader goals of protecting sensitive data, maintaining IT systems' integrity, and ensuring operations' continuity. It requires a concerted effort to synchronize policies, controls, and incident response strategies across all facets of the organization and its extended network.

Adopting a lifecycle approach to supply chain security within the organization involves addressing security considerations at every stage of the product and service lifecycle, from development and acquisition to deployment and decommissioning. This holistic perspective ensures that security is not an afterthought but a fundamental aspect of all lifecycle phases. It enables organizations to anticipate and mitigate risks before they materialize, ensuring that supply chain security evolves with products and services.

Performance monitoring techniques for supply chain security practices are critical for assessing the effectiveness of security measures and identifying areas for improvement. These techniques may include regular security assessments, audits, and using KPIs to quantify security outcomes. Monitoring provides insights for making informed decisions about enhancing supply chain security and responding to emerging threats.

Risk assessment and management processes throughout the product and service lifecycle are essential for identifying, analyzing, and mitigating risks associated with supply chain relationships.

These processes should be integrated into the organization's overall risk management framework, ensuring supply chain risks are considered alongside operational and strategic risks. Effective risk management requires ongoing vigilance and adaptation to the changing threat landscape.

Collaboration with suppliers on security practices and performance metrics fosters a partnership approach to supply chain security. By working closely with suppliers, organizations can ensure that security measures are aligned and that suppliers understand the organization's security expectations. Collaborative efforts may include sharing best practices, conducting joint security assessments, and developing shared response strategies for incidents that affect the supply chain.

Continuous improvement processes for supply chain security are necessary to adapt to new challenges and technologies. These processes should be based on regular reviews of security practices, lessons learned from security incidents, and feedback from suppliers and other stakeholders. Continuous improvement ensures that supply chain security practices remain effective and responsive to evolving threats and business needs.

Reporting and communication strategies for supply chain security performance are crucial in keeping stakeholders informed and engaged. Effective communication ensures that decision-makers, employees, and suppliers know security issues, understand their roles in mitigating risks, and are committed to maintaining a secure supply chain. Reporting mechanisms should provide clear, actionable information that supports strategic decision-making and fosters a culture of security awareness.

Recommendations

- **Implement a Centralized Supply Chain Risk Management Function**: Establish a dedicated team or function responsible for overseeing supply chain security, facilitating coordination, and ensuring that security practices are integrated across the organization and with suppliers.
- **Leverage Advanced Technologies for Risk Monitoring**: Utilize technology solutions, such as blockchain and machine learning, to enhance visibility into the supply chain and automate monitoring security practices and risks.
- **Establish Comprehensive Supplier Security Assessments**: Conduct thorough security assessments of potential and existing suppliers, focusing on their security practices, compliance with industry standards, and ability to respond to incidents.
- **Develop Incident Response Plans that Include Suppliers**: Create incident response plans that clearly define the roles and responsibilities of suppliers, ensuring coordinated action in the event of a security breach or other incidents affecting the supply chain.
- **Promote Transparency and Information Sharing**: Encourage an open exchange of information about security threats, vulnerabilities, and incidents among suppliers and partners. This collaboration can enhance the collective ability to detect, respond to, and recover from security incidents.

NIST 800-53 Controls

(Refer to our appendix for a comprehensive description of the NIST 800-53 Controls, essential for organizations adhering to strict regulatory standards.)

- PM-09,PM-19,PM-28,PM-30,PM-31,RA-03,RA-07,SA-04,SA-09,SR-02,SR-03,SR-05,SR-06

Simplified Security Controls (SSC)

Control 1: Lifecycle Security Management for Supply Chain

Control Objective: To embed supply chain security considerations into every stage of the technology product and service lifecycle, ensuring that security practices evolve with products from inception through retirement.

Implementation Steps:

1.1 **Define Lifecycle Security Criteria**: Establish security criteria and checkpoints for each phase of the product and service lifecycle, from design and development to deployment, maintenance, and decommissioning.

1.2 **Integrate Criteria into Development Processes**: Integrate these lifecycle security criteria into the standard operating procedures for product development and service provisioning, ensuring that all teams adhere to these requirements.

1.3 **Regular Review and Update**: Schedule periodic reviews of the lifecycle security criteria to update them based on emerging threats, technological advancements, and feedback from lifecycle assessments.

Expected Outcome: Implementing a lifecycle approach to supply chain security ensures that products and services are designed, developed, maintained, and retired with security as a central consideration. This approach minimizes vulnerabilities throughout the product or service lifecycle and enhances the overall security posture of the supply chain.

Control 2: Supplier Security Performance Monitoring System

Control Objective: To systematically monitor and evaluate the security performance of suppliers against agreed-upon metrics and standards, ensuring continuous adherence to security best practices.

Implementation Steps:

2.1 **Develop Security Performance Metrics**: Collaborating with suppliers, develop clear, measurable security performance metrics that align with the organization's cybersecurity objectives.

2.2 **Implement Monitoring Tools and Processes**: Utilize monitoring tools and establish processes for regularly assessing supplier performance against the defined metrics, including regular security audits and assessments.

2.3 **Feedback and Remediation Process**: Create mechanisms for providing suppliers with performance feedback and require remediation plans for addressing any identified security deficiencies.

Expected Outcome: A supplier security performance monitoring system provides continuous oversight of supplier adherence to best practices, enabling early detection of compliance gaps or security weaknesses. This control proactively manages supply chain security risks and strengthens supplier security practices over time.

Control 3: Continuous Improvement Program for Supply Chain Security

Control Objective: To establish a formal program to continuously improve supply chain security practices, ensuring they remain effective and responsive to the evolving threat landscape.

Implementation Steps:

3.1 **Establish a Continuous Improvement Framework**: Set up a framework for the continuous review and enhancement of supply chain security practices, incorporating feedback loops, incident analysis, and emerging threat intelligence.

3.2 **Conduct Regular Security Practice Reviews**: Schedule regular reviews of existing supply chain security practices, assessing their effectiveness and identifying areas for improvement.

3.3 **Foster Innovation and Adaptation**: Encourage innovation in security practices and adapt the organization's supply chain security strategy based on review findings, best practices, and technological advancements.

Expected Outcome: Implementing a continuous improvement program for supply chain security ensures that security practices are regularly scrutinized and enhanced. This ongoing refinement process helps the organization avoid potential security challenges, fostering a culture of resilience and adaptability within the supply chain.

GV.SC-10: Navigating Cybersecurity After the Conclusion of Supplier Partnerships

In the complex tapestry of modern cybersecurity strategies, managing the conclusion of partnerships and service agreements with suppliers stands as a pivotal, yet often overlooked, component. This chapter delves into the essential considerations and actions required to ensure C-SCRM plans remain robust, even after the dissolution of partnerships. The aim is to safeguard sensitive information, maintain the integrity of cybersecurity defenses, and ensure business continuity.

Developing exit strategies and handover procedures for cybersecurity information is a crucial initial step in postpartnership considerations. These strategies are designed to ensure a smooth transition of responsibilities and the secure transfer of any shared information. Effective handover procedures safeguard against data loss and unauthorized access, ensuring that all cybersecurity-related information remains controlled and protected throughout the transition process.

Incorporating data retention, destruction, and transfer provisions into termination clauses is essential for managing the lifecycle of sensitive information postpartnership. These provisions should clearly define how data should be handled, ensuring compliance with legal and regulatory requirements while protecting the organization's interests. Whether data is to be returned, destroyed, or retained under specific conditions, clear guidelines help prevent potential data breaches and ensure the secure handling of information.

Managing access rights and credentials upon contract conclusion is critical in maintaining cybersecurity postures. Revoking access rights and collecting or deactivating credentials prevent former partners from accessing the organization's systems and information. This process is vital for preventing unauthorized access and safeguarding against potential cyber threats from former suppliers or partners.

Processes for reviewing and analyzing supplier-related incidents postpartnership are necessary for understanding and mitigating any lingering risks. These reviews can provide valuable insights into potential vulnerabilities or threats associated with past partnerships, informing future risk management strategies. Learning from these incidents helps organizations enhance their cybersecurity practices and prevent similar issues in the future.

Continuity planning for critical services postsupplier relationship ensures that the organization can maintain operations despite the conclusion of crucial partnerships. Identifying alternative suppliers or developing in-house capabilities can mitigate the risk of disruptions. Effective continuity planning is integral to preserving operational resilience and safeguarding against potential vulnerabilities during transition.

Ensuring legal and regulatory compliance in the termination of supplier contracts is paramount. Organizations must navigate the conclusion of partnerships with due diligence, adhering to contractual obligations and regulatory requirements. This adherence protects against legal disputes and compliance issues, safeguarding the organization's reputation and financial standing.

Extracting lessons learned and transferring knowledge following the conclusion of the partnership is invaluable for continuous improvement. Documenting experiences, challenges, and successes throughout the partnership provides a knowledge base that can inform future supplier relationships and cybersecurity strategies. This practice fosters organizational learning and resilience, enhancing the overall effectiveness of C-SCRM efforts.

Strategies for maintaining organizational resilience postpartnership focus on ensuring that the organization remains secure, compliant, and operationally robust despite the end of supplier relationships. This involves a holistic approach to cybersecurity, considering the implications of partnership conclusions on the organization's risk profile and implementing measures to address these risks proactively.

Recommendations

- **Implement Robust Offboarding Processes**: Develop comprehensive offboarding processes for ending supplier relationships, covering data handling, access revocation, and asset retrieval to mitigate security risks.
- **Establish Clear Data Handling Guidelines**: Specify data retention, destruction, and transfer protocols in supplier agreements to ensure sensitive information is handled securely postpartnership.
- **Conduct Thorough Access Review and Revocation**: Regularly review and promptly revoke access rights and credentials for former suppliers to prevent unauthorized access to systems and data.
- **Engage in Postpartnership Reviews**: Analyze incidents and performance issues related to past suppliers to identify lessons learned and inform future risk management strategies.
- **Prioritize Continuity and Resilience Planning**: Develop strategies for ensuring the continuity of critical services postpartnership, including identifying alternative suppliers and enhancing in-house capabilities.
- **Ensure Compliance and Legal Due Diligence**: Adhere to legal and regulatory requirements when terminating supplier contracts to avoid potential legal disputes and compliance issues.
- **Foster a Culture of Continuous Improvement**: Use the conclusion of partnerships as an opportunity for learning and improvement, applying lessons learned to strengthen future supplier relationships and cybersecurity practices.
- **Adopt a Proactive Stance on Cybersecurity Postpartnership**: Maintain vigilance and implement proactive security measures to manage potential risks arising from the conclusion of supplier relationships, ensuring the organization's resilience and security posture are uncompromised.

NIST 800-53 Controls

(Refer to our appendix for a comprehensive description of the s 800-53 Controls, essential for organizations adhering to strict regulatory standards.)

- PM-31,RA-03,RA-05,RA-07,SA-04,SA-09,SR-02,SR-03,SR-05,SR-06

Simplified Security Controls (SSC)

Control 1: Secure Handover and Exit Protocol for Supplier Transition

Control Objective: To ensure the secure and orderly transition of services, data, and access controls when a supplier relationship concludes, safeguarding confidential information and maintaining cybersecurity integrity.

Implementation Steps:

1.1 **Develop Exit Protocols**: Create detailed exit protocols outlining the steps for secure data transfer, access revocation, and the handover of services and materials upon termination of a supplier relationship.

1.2 **Implement Data Handling Agreements**: Integrate data retention, destruction, and transfer clauses into all supplier contracts to ensure data is handled securely during and after the termination of the agreement.

1.3 **Conduct Exit Reviews**: Perform comprehensive exit reviews with departing suppliers to ensure all contractual obligations are met and the transition poses no security risks.

Expected Outcome: Establishing a secure handover and exit protocol ensures that the conclusion of supplier relationships does not expose the organization to data breaches or other security vulnerabilities. This control helps maintain the organization's cybersecurity posture and protects sensitive information during supplier transitions.

Control 2: Postpartnership Incident Analysis and Lessons Learned Process

Control Objective: To analyze cybersecurity incidents involving suppliers postpartnership to identify lessons learned and apply these insights to strengthen future supplier risk management strategies.

Implementation Steps:

2.1 **Incident Analysis Framework**: Set up a framework for conducting postpartnership incident analyses that examine supplier-related cybersecurity incidents' root causes, impact, and response effectiveness.

2.2 **Lessons Learned Workshops**: Organize workshops with relevant internal teams to discuss the findings of incident analyses and identify actionable lessons learned.

2.3 **Update Risk Management Practices**: Integrate the lessons learned into existing risk management policies and practices to mitigate similar risks in future supplier relationships.

Expected Outcome: Implementing a postpartnership incident analysis and lessons-learned process enables the organization to continuously improve its supplier risk management practices. By understanding past incidents, the organization can adapt its strategies to better manage and mitigate similar risks in the future, enhancing overall supply chain security.

Control 3: Continuity Planning for Critical Supplier Dependencies

Control Objective: To develop and implement continuity plans that address the potential loss of critical services or data due to the conclusion of supplier relationships, ensuring uninterrupted business operations.

Implementation Steps:

3.1 **Identify Critical Dependencies**: Analyze the supply chain to identify and document critical dependencies on suppliers, focusing on those whose loss could significantly impact business operations.

3.2 **Develop Continuity Plans**: For each critical dependency, develop specific continuity plans that detail alternative solutions, such as secondary suppliers or in-house capabilities, to maintain operations if a supplier relationship ends.

3.3 **Regular Review and Testing**: Regularly review and test continuity plans to ensure they are effective and updated to reflect changes in the business environment and supply chain dynamics.

Expected Outcome: Establishing continuity plans for critical supplier dependencies ensures the organization can quickly adapt and maintain operations during a supplier transition. This proactive approach minimizes the potential disruption to business activities and reinforces the organization's resilience to changes within the supply chain.

11

Asset Management (ID.AM)

> In the digital age, the strength of our cybersecurity and operational efficiency lies in the precision with which we manage every asset, turning meticulous oversight into our most formidable shield against disruption.

In cybersecurity and operational efficiency, the meticulous management of an organization's assets—from hardware, software, services, and data to the detailed mapping of network communications—stands paramount. This comprehensive approach involves the initial inventory and categorization of these assets. It integrates these processes with broader IT and cybersecurity frameworks, ensuring that every phase of an asset's lifecycle, from acquisition to disposal, is carefully monitored and managed. Key to this management is the regular auditing and updating of inventories to reflect changes, the strategic allocation of security resources based on asset criticality, and the development of robust contingency plans for service disruptions. Moreover, the role of documentation, compliance, and leveraging automated tools cannot be overstated, as they provide the backbone for effective asset and risk management. Through these practices, organizations can enhance their cybersecurity posture and ensure operational resilience and compliance with regulatory standards, making asset management a cornerstone of modern organizational strategy.

Author's Note: ID.AM-06 was deprecated from the CSF 2 and is not included in this chapter.

ID.AM-01: Inventories of Hardware Managed by the Organization Are Maintained

Managing hardware assets efficiently is a foundational aspect of cybersecurity, ensuring that every piece of hardware within an organization's network is accounted for, secured, and properly utilized. The initiation of a robust hardware inventory management process marks the beginning of a comprehensive cybersecurity strategy, as it identifies what assets exist, where they are located, and the role they play in the organization's operations. This process starts with compiling a complete and accurate inventory, which includes every hardware component from servers to end-user devices.

The initial compilation of the hardware inventory is a meticulous process that requires a systematic approach. Organizations should deploy automated tools to discover hardware assets on their network, supplemented by manual checks to ensure no asset goes unnoticed. This step is crucial for

A Comprehensive Guide to the NIST Cybersecurity Framework 2.0: Strategies, Implementation, and Best Practice, First Edition. Jason Edwards.
© 2025 John Wiley & Sons Ltd. Published 2025 by John Wiley & Sons Ltd.

establishing a baseline of all hardware that needs to be managed, monitored, and protected. This inventory must be updated to reflect new acquisitions and phasing out decommissioned assets.

Categorization and tagging of hardware assets follow the compilation process, serving multiple purposes. It aids in the management by allowing for the quick identification of hardware types, their locations, and their users. This systematization facilitates easier tracking, maintenance, and security of these assets. Tagging can include physical labels and digital tags in inventory management systems, providing vital information at a glance, such as asset status, sensitivity level, and compliance requirements.

Regular audits and update procedures are essential to maintaining the integrity of the hardware inventory. These audits should be scheduled at regular intervals and after any significant change to the IT environment, such as introducing new hardware or decommissioning old equipment. Audits help identify discrepancies between the inventory records and the actual hardware assets, allowing for corrective actions to ensure the inventory's accuracy and reliability.

Integrating hardware inventory management with broader IT asset management (ITAM) systems enhances visibility and control over an organization's assets. This integration supports optimizing asset utilization, lifecycle management, and strategic planning for IT infrastructure upgrades. Additionally, it reinforces security measures by aligning hardware inventory management with cybersecurity protocols, ensuring that all hardware is appropriately secured and monitored.

Security considerations are paramount in hardware inventory management. Each asset should be evaluated for its security impact, with appropriate controls implemented based on its risk profile. This includes ensuring that hardware is configured securely, maintained with up-to-date firmware and software, and monitored for any signs of unauthorized access or compromise. The security of hardware assets is a continuous process that evolves with the changing threat landscape.

The role of hardware inventory in incident response and recovery is significant. In a security incident, having a detailed and accurate inventory allows response teams to quickly identify affected assets, assess the impact, and prioritize their response efforts. This inventory also supports recovery activities by providing information on hardware configurations, enabling faster restoration of services.

Disposal and decommissioning processes for end-of-life hardware are critical components of inventory management. Proper procedures ensure that all data is securely erased and that hardware is disposed of in an environmentally responsible manner. Decommissioning processes must be documented and reflected in the hardware inventory to maintain accuracy.

Recommendations

- **Implement Automated Discovery Tools:** Utilize automated discovery tools to maintain a real-time inventory of hardware assets, complemented by manual audits to catch discrepancies.
- **Develop a Categorization System:** Establish a clear system for categorizing and tagging hardware assets to simplify management, security, and compliance processes.
- **Schedule Regular Audits:** Conduct regular audits of the hardware inventory to ensure its accuracy and completeness, adjusting the inventory to reflect the current state of hardware assets.
- **Integrate with ITAM Systems:** Leverage integration with ITAM systems to enhance visibility, security, and management of hardware assets across their lifecycle.
- **Enforce Secure Decommissioning:** Adhere to secure data erasure and environmentally responsible disposal practices for end-of-life hardware, ensuring that decommissioning steps are documented in the inventory management system.

NIST 800-53 Controls

(Refer to our appendix for a comprehensive description of the NIST 800-53 Controls, essential for organizations adhering to strict regulatory standards.)

- CM-08,PM-05

Simplified Security Controls (SSC)

This section outlines our simplified security controls (SSC), tailored for environments beyond the scope of NIST compliance. These controls offer streamlined, effective security solutions suitable for any organization seeking to enhance its cybersecurity defenses with clear, actionable measures.

Control 1: Automated Hardware Inventory Scanning and Management System

Control Objective: To ensure accurate and up-to-date inventories of all hardware assets the organization manages, facilitating effective asset management, security, and compliance through automated tools.

Implementation Steps:

1.1 **System Selection and Deployment:** Identify and deploy an automated hardware inventory management system to discover, catalog, and monitor hardware assets across the organization's network. This system should support integration with existing ITAM systems.

1.2 **Configuration and Customization:** Configure the automated system to recognize and categorize the organization's hardware assets according to predefined categories. Customize the system to generate alerts for unregistered devices and report hardware inventory changes.

1.3 **Continuous Monitoring and Updating:** Set the system to continuously monitor the network for new devices, changes, or removals of existing hardware, automatically updating the inventory in real time.

Expected Outcome: Implementing an automated hardware inventory scanning and management system results in a constantly updated, accurate inventory of all hardware assets. This automation reduces manual errors, enhances the organization's ability to respond to security incidents involving hardware assets, and supports compliance with regulatory requirements related to asset management.

Control 2: Comprehensive Hardware Asset Tagging Protocol

Control Objective: To establish a uniform and comprehensive protocol for the physical and digital tagging of all hardware assets, enhancing the traceability, security, and management of these assets throughout their lifecycle.

Implementation Steps:

2.1 **Develop Tagging Standards:** Create a standardized set of guidelines for physical and digital tagging of hardware assets. This should include unique identifiers, classification levels, and ownership information.

2.2 **Implement Tagging at Acquisition:** Integrate the tagging protocol into the procurement process, ensuring every new hardware asset is tagged according to the established standards before deployment.

2.3 **Regular Tagging Audits:** Schedule and conduct regular audits to ensure compliance with the tagging protocol and to update or replace tags as necessary, especially for assets changing classification, ownership, or location.

Expected Outcome: Establishing a comprehensive hardware asset tagging protocol improves asset traceability and management. This system enhances the organization's ability to identify and respond to incidents quickly, manage asset lifecycles effectively, and ensure compliance with internal and external asset management standards.

Control 3: Hardware Disposal and Decommissioning Policy

Control Objective: To securely and responsibly dispose of or decommission end-of-life hardware assets, ensuring that sensitive data is irrecoverably erased and disposal practices comply with environmental regulations.

Implementation Steps:

3.1 **Policy Development:** Develop and document a comprehensive policy for the disposal and decommissioning of hardware assets. The policy should outline procedures for data wiping, physical destruction, and recycling or disposal in compliance with environmental standards.

3.2 **Secure Data Erasure:** Implement standardized processes for the secure erasure of all data from hardware assets before disposal or decommissioning. This may include using certified data destruction software or physical destruction methods.

3.3 **Verification and Documentation:** Establish procedures for verifying the thoroughness of data erasure and the safe decommissioning of hardware. Document each asset's disposal or decommissioning process, maintaining records for compliance and audit purposes.

Expected Outcome: Implementing a hardware disposal and decommissioning policy ensures that end-of-life hardware assets are disposed of securely and responsibly. This policy protects sensitive information from potential data breaches and supports the organization's commitment to environmental sustainability.

ID.AM-02: Inventories of Software, Services, and Systems Managed by the Organization Are Maintained

Effective software, services, and systems inventory management is critical to an organization's cybersecurity posture. A comprehensive inventory provides a foundation for securing the digital infrastructure against potential threats. This process begins with a thorough introduction to software and services inventory management, emphasizing the necessity for organizations to clearly understand all software applications and services operating within their environments. This understanding is crucial for identifying vulnerabilities, managing patches, and ensuring only authorized software is used.

Strategies for comprehensive software inventory collection involve a mix of automated tools and manual processes to ensure no software or service goes undocumented. Organizations must employ techniques that allow continuous monitoring and updating of the software inventory. This might include deploying software discovery tools that scan networks for installed applications, cloud services, and unauthorized software. The goal is to achieve complete visibility into the software landscape, which is essential for effectively managing cybersecurity risks.

Classification and documentation of software and services form the backbone of inventory management. Organizations can prioritize their security efforts by categorizing software assets based on their function, sensitivity, and criticality to the business operations. Proper documentation, including version numbers, deployment environments, and dependencies, aids in vulnerability management and compliance checks. This structured approach streamlines management processes and enhances the organization's ability to respond to incidents swiftly.

Implementing automated tools for inventory management transforms the daunting task of tracking and managing software assets into a manageable process. These tools can dynamically detect and inventory software across the enterprise, including remote and cloud environments.

Automation ensures that the inventory remains current and accurate, reflecting real-time changes and reducing the manual effort required for inventory updates.

Integrating software inventories with cybersecurity risk management embeds security into the lifecycle of software assets. By understanding the inventory landscape, organizations can assess the risk profile of their software assets, identify vulnerabilities, and prioritize remediation based on the asset's criticality. This integration is pivotal for building a resilient cybersecurity framework that proactively addresses potential threats before they can be exploited.

Lifecycle management of software assets addresses the need to manage software from procurement through retirement. Effective lifecycle management ensures that software is kept up to date, licenses are handled efficiently, and end-of-life software is retired securely. This process reduces the risk of security vulnerabilities associated with outdated software and helps maintain a clean and optimized software environment.

Licensing compliance and optimization in software inventory management are key to avoiding legal and financial risks. Organizations must have mechanisms to ensure that software licenses are up-to-date and compliant with the terms of use. This includes monitoring over-licensed software to reduce unnecessary expenditures and under-licensed software to avoid compliance issues. Effective license management contributes to overall cybersecurity by ensuring all software is legally acquired and properly registered.

Periodic review and rationalization of software assets ensure the inventory remains relevant and aligned with business needs. This involves regularly assessing software assets' utilization and value, identifying redundancies, and discontinuing unnecessary applications. Rationalization helps reduce the attack surface by eliminating unused or obsolete software and focusing security resources on protecting critical assets.

Recommendations

- **Leverage Automated Discovery Tools:** Utilize advanced automated discovery tools for real-time monitoring and management of software assets. This ensures that the inventory is always current and reduces manual oversight.
- **Establish a Classification System:** Develop a robust classification system for software and services based on their importance and role in the organization. This aids in prioritizing security measures and managing resources efficiently.
- **Integrate with Risk Management Processes:** Seamlessly integrate software inventories with the organization's risk management framework to proactively identify and mitigate potential vulnerabilities.
- **Conduct Regular Audits:** Schedule regular audits and reviews of the software inventory to ensure accuracy, compliance with licensing agreements, and alignment with the organization's cybersecurity strategy.
- **Optimize Software Licensing:** Implement a comprehensive license management program to ensure compliance and optimize software spending. This includes regularly reviewing software usage to adjust licenses accordingly and avoid financial or legal penalties.

NIST 800-53 Controls

(Refer to our appendix for a comprehensive description of the NIST 800-53 Controls, essential for organizations adhering to strict regulatory standards.)

- AC-20,CM-08,PM-05,SA-05,SA-09

Simplified Security Controls (SSC)

This section outlines our SSC, tailored for environments beyond the scope of NIST compliance. These controls offer streamlined, effective security solutions suitable for any organization seeking to enhance its cybersecurity defenses with clear, actionable measures.

Control 1: Centralized Software and Services Inventory System

Control Objective: To develop and maintain a centralized inventory system that comprehensively tracks all software and services used within the organization, ensuring accurate visibility and management across the enterprise.

Implementation Steps:

1.1 **System Selection and Implementation:** Select and deploy a centralized inventory management system that integrates various environments and platforms to track software and service usage and deployment across the organization.

1.2 **Continuous Discovery and Integration:** Configure the system to automatically discover new software installations and service subscriptions, integrating this information into the real-time centralized inventory.

1.3 **Regular Verification and Reconciliation:** Establish a routine for regularly verifying the inventory data against physical and digital assets and reconcile any discrepancies to ensure the inventory remains accurate and comprehensive.

Expected Outcome: Implementing a centralized software and services inventory system will give the organization a single source of truth for all software assets, facilitating better management, security risk assessment, and compliance. This system will enhance operational efficiency by enabling accurate software resource tracking, allocation, and optimization.

Control 2: Automated Compliance and License Management Tool

Control Objective: To automate the management of software licenses and ensure compliance with licensing agreements, minimizing the risk of legal issues and optimizing software expenditure.

Implementation Steps:

2.1 **Tool Selection and Deployment:** Choose and implement an automated license management tool to track and manage software licenses across multiple platforms and environments within the organization.

2.2 **Integration with Inventory System:** Integrate the license management tool with the existing software and services inventory system to leverage real-time data on software usage and deployment.

2.3 **Continuous License Monitoring and Optimization:** Set the tool to continuously monitor software usage against license entitlements, generating alerts for compliance risks or opportunities for license optimization and cost savings.

Expected Outcome: By automating license management, the organization will maintain continuous compliance with software licensing agreements, avoid legal and financial penalties, and optimize software spending by reallocating or discontinuing underutilized licenses.

Control 3: Periodic Software Rationalization Process

Control Objective: Regularly review and rationalize the software asset portfolio, eliminating redundant, obsolete, or underutilized software and services to reduce complexity and costs.

Implementation Steps:

3.1 **Establish Rationalization Criteria:** Develop criteria for evaluating software and services based on usage metrics, business value, security posture, and cost-effectiveness.

3.2 **Conduct Regular Software Audits:** Conduct regular audits of the software and services inventory against the rationalization criteria to identify candidates for removal or consolidation.

3.3 **Implement Rationalization Decisions:** Execute decisions to retire, consolidate, or replace identified software and services, ensuring that changes are communicated to affected stakeholders and that any data migration or decommissioning is securely managed.

Expected Outcome: The periodic software rationalization process will ensure the software asset portfolio remains aligned with business needs, optimized for cost and performance, and free from unnecessary complexity. This will enhance the organization's agility, reduce security risks associated with obsolete software, and result in cost savings from reduced software licensing and support requirements.

ID.AM-03: Representations of the Organization's Authorized Network Communication and Internal and External Network Data Flows Are Maintained

Understanding and managing an organization's complex network communications and data flows is crucial for maintaining a secure cyber environment. This understanding begins with developing detailed network communication and data flow mappings, foundational elements in cybersecurity. Such mappings provide clear visualizations of how data moves through an organization's internal and external networks, identifying potential vulnerabilities and points of unauthorized access.

Techniques for visualizing network topologies and data flows have evolved, offering cybersecurity professionals sophisticated tools to map out the intricacies of network communications. These visualizations can range from simple diagrams to complex, interactive models that provide real-time insights into network behavior. Utilizing these techniques helps comprehend the complete picture of network activities, aiding in detecting anomalies that could signify security threats.

The importance of accurate data flow diagrams in cybersecurity cannot be overstated. They are essential for identifying the paths data takes as it moves within and outside the organization, highlighting critical assets and potential security gaps. Accurate diagrams assist in planning security measures, ensuring that protective controls are placed at strategic points to safeguard sensitive information. They also facilitate the understanding of how changes to the network can impact security, guiding the implementation of necessary adjustments.

Tools and software for network mapping and management play a pivotal role in maintaining an up-to-date representation of an organization's network. These tools automate discovering network devices and mapping data flows, significantly reducing the manual effort required. They also offer capabilities for monitoring network performance and detecting irregularities, which can indicate cybersecurity incidents.

Regular updates to network maps are essential to reflect network changes and additions. As organizations expand and evolve, new devices are added, and existing configurations are modified, all of which can alter data flow paths and introduce new vulnerabilities. Maintaining current network diagrams ensures that cybersecurity measures remain effective and aligned with the organization's current state.

Identifying and managing unauthorized network communications is critical to network communication management. Unauthorized communications can indicate the presence of malware, data exfiltration activities, or other security threats. Effective network mapping and monitoring enable the early detection of such activities, allowing for swift mitigation actions to protect the organization's digital assets.

Integration of network maps with incident response planning enhances the organization's preparedness for cybersecurity incidents. Having detailed network representations available during an incident enables response teams to quickly understand the scope of an incident, identify affected systems, and implement targeted containment and remediation strategies. This integration is a key component of a robust cybersecurity framework, ensuring swift and effective responses.

Using network representations in risk assessment activities enables organizations to make informed decisions about where to allocate resources for maximum cybersecurity impact. Cybersecurity teams can identify high-risk areas and prioritize security controls by understanding the data flow and communication patterns. This strategic approach to risk management ensures that protective measures are both efficient and effective, safeguarding the organization against potential threats.

Recommendations

- **Adopt Advanced Mapping Tools:** Utilize sophisticated network mapping tools to generate accurate and detailed representations of network topologies and data flows. These tools should offer capabilities for real-time monitoring and alerting on unusual network activities.
- **Ensure Regular Diagram Updates:** Implement a routine for regularly updating network diagrams to reflect any changes in the network infrastructure, ensuring that the organization's cybersecurity posture is based on the most current information.
- **Integrate with Cybersecurity Practices:** Seamlessly integrate network representations into broader cybersecurity practices, including incident response planning and risk assessment activities. This integration enhances the organization's ability to detect, respond to, and mitigate cybersecurity threats.
- **Monitor for Unauthorized Communications:** Establish monitoring protocols to detect unauthorized network communications promptly. This includes setting up alerts for unexpected data flows, which could indicate a breach or other security incidents.
- **Conduct Comprehensive Risk Assessments:** Utilize network diagrams in risk assessment processes to identify and prioritize areas of the network that may be vulnerable to attack. This targeted approach enables the efficient allocation of resources to areas of greatest need, strengthening the organization's overall cybersecurity defenses.

NIST 800-53 Controls

(Refer to our appendix for a comprehensive description of the NIST 800-53 Controls, essential for organizations adhering to strict regulatory standards.)

- AC-04,CA-03,CA-09,PL-02,PL-08,PM-07

Simplified Security Controls (SSC)

This section outlines our SSC, tailored for environments beyond the scope of NIST compliance. These controls offer streamlined, effective security solutions suitable for any organization seeking to enhance its cybersecurity defenses with clear, actionable measures.

Control 1: Deployment of Advanced Network Mapping and Visualization Tools

Control Objective: To employ advanced tools for the dynamic visualization of network topologies and data flows, enabling real-time detection of anomalies and a comprehensive understanding of network behavior.

Implementation Steps:

1.1 **Tool Selection and Deployment:** Identify and deploy network mapping and visualization tools that support dynamic, real-time representation of the network's structure and data flows.

1.2 **Integration with Existing Systems:** Ensure the selected tools are integrated with existing network management and security systems to enhance network visibility and monitoring capabilities.

1.3 **Training for Cybersecurity Teams:** Provide specialized training for cybersecurity teams on utilizing these tools effectively for network monitoring, anomaly detection, and security incident response.

Expected Outcome: Implementing advanced network mapping and visualization tools provides cybersecurity professionals with deep insights into network activities, enhancing their ability to identify and respond to security threats quickly. This real-time visibility into network behavior and data flows significantly improves the organization's security posture.

Control 2: Continuous Update Protocol for Network Maps

Control Objective: To maintain updated network diagrams that accurately reflect current network configurations and data flows, ensuring that cybersecurity measures align appropriately with the network's operational state.

Implementation Steps:

2.1 **Establish Regular Update Schedules:** Implement a protocol for regularly reviewing and updating network diagrams, ensuring they accurately represent the current network topology and data flows.

2.2 **Automated Discovery and Documentation:** Utilize automated network discovery tools to detect changes in the network configuration and automatically update the network diagrams accordingly.

2.3 **Review and Verification Process:** Conduct periodic reviews involving IT and cybersecurity teams to verify network diagrams' accuracy and make necessary manual adjustments.

Expected Outcome: Establishing a continuous update protocol for network maps ensures that all network diagrams are current and accurately reflect the organization's evolving network configuration. This ongoing accuracy supports effective security planning, risk management, and incident response activities, reducing the risk of security breaches.

Control 3: Integration of Network Mapping with Cybersecurity Risk Management

Control Objective: To integrate network mapping and visualization efforts with the organization's cybersecurity risk management processes, enhancing identifying and mitigating network-based risks.

Implementation Steps:

3.1 **Cross-Functional Collaboration:** Collaborate network management, cybersecurity, and risk management teams to align network mapping efforts with risk assessment and mitigation strategies.

3.2 **Risk-Based Analysis of Data Flows:** Utilize network diagrams to conduct risk-based data flow analysis, identify potential vulnerabilities, and assess the impact of possible security breaches.

3.3 **Implementation of Strategic Security Controls:** Based on the analysis, strategically implement security controls at critical points within the network to mitigate identified risks and protect sensitive data flows.

Expected Outcome: Integrating network mapping with cybersecurity risk management processes ensures that network-based risks are systematically identified, assessed, and mitigated. This approach enhances the security of network data flows and aligns network operations with the organization's overall risk management strategy, leading to a more resilient cybersecurity posture.

ID.AM-04: Inventories of Services Provided by Suppliers Are Maintained

Managing inventories of services provided by suppliers is a crucial aspect of an organization's cybersecurity framework, ensuring that all external services are accounted for, assessed for risks, and managed efficiently. This management begins with a thorough understanding of suppliers' scope and variety of services, encompassing everything from cloud-based solutions to IT support services. Recognizing these services' integral role in daily operations underscores the importance of maintaining a comprehensive inventory.

Establishing criteria for including supplier services in the inventory is the first step toward effective management. These criteria should consider factors such as the criticality of the service to the organization's operations, the sensitivity of data handled by the supplier, and the connectivity of the supplier's services to the organization's network. This selective approach ensures that the inventory focuses on those services that pose the most significant potential risk to the organization's cybersecurity posture.

Documentation and categorization of supplier services form the foundation of the inventory management process. Each service should be documented with details such as the scope of the service, the data it accesses or stores, and the security measures in place to protect that data. Categorizing these services based on their function, data sensitivity, and risk level aids in effectively prioritizing cybersecurity efforts and resources.

Assessment of supplier services for cybersecurity risks is a critical component of inventory management. This assessment should evaluate the supplier's security policies, incident response capabilities, and compliance with relevant cybersecurity standards. It involves a detailed review of how the supplier manages data security, responds to security incidents, and ensures the continuity of service in the face of cyber threats.

Integration of the supplier services inventory with vendor management processes enhances the organization's ability to manage supplier-related risks. This integration ensures cybersecurity considerations are embedded in the vendor selection and management lifecycle, from initial selection to ongoing monitoring and evaluation. It facilitates a holistic approach to managing supplier risks, incorporating cybersecurity assessments into broader vendor performance reviews.

Regular reviews and updates to the supplier services inventory are essential for keeping it relevant and effective. The dynamic nature of the cybersecurity landscape and the organization's use of supplier services means that the inventory must be regularly reviewed and updated to reflect new services, changes in existing services, and the discontinuation of services. This ongoing maintenance is critical for ensuring that the inventory remains a useful tool for managing cybersecurity risks associated with supplier services.

Communication and coordination with suppliers about service changes significantly maintain an accurate and up-to-date inventory. Establishing clear lines of communication ensures that the organization is promptly informed about any changes to the services provided, including updates, upgrades, and any changes to security practices. This proactive engagement helps manage risks more effectively and ensures that both parties are aligned in their approach to cybersecurity.

Contingency planning for supplier service disruptions is integral to managing the cybersecurity risks associated with external services. This planning involves identifying potential points of failure, assessing the impact of service disruptions on the organization, and developing strategies for maintaining operations during such disruptions. It ensures that the organization is prepared to respond effectively to incidents affecting supplier services, minimizing the impact on operations and security.

Recommendations

- **Develop Comprehensive Inclusion Criteria:** Establish clear criteria for including supplier services in the inventory, focusing on the criticality and risk associated with each service. This ensures that the inventory prioritizes services significantly impacting the organization's cybersecurity posture.
- **Implement Rigorous Documentation Practices:** Maintain detailed documentation for each supplier service, including service descriptions, data sensitivity levels, and applicable security measures. This documentation provides a solid foundation for risk assessment and management.
- **Conduct Regular Risk Assessments:** Perform ongoing assessments of cybersecurity risks associated with supplier services. These assessments should evaluate the supplier's security measures, incident response capabilities, and compliance with industry standards.
- **Integrate with Vendor Management Processes:** Seamlessly integrate the supplier services inventory with broader vendor management processes. This integration ensures that cybersecurity considerations are a central aspect of the vendor lifecycle management, from selection to ongoing evaluation.
- **Establish Robust Communication Protocols:** Foster open and effective communication with suppliers regarding service changes. Regular updates and coordination help manage risks and maintain the accuracy of the supplier services inventory.

NIST 800-53 Controls

(Refer to our appendix for a comprehensive description of the NIST 800-53 Controls, essential for organizations adhering to strict regulatory standards.)

- AC-20,SA-09,SR-02

Simplified Security Controls (SSC)

This section outlines our SSC, tailored for environments beyond the scope of NIST compliance. These controls offer streamlined, effective security solutions suitable for any organization seeking to enhance its cybersecurity defenses with clear, actionable measures.

Control 1: Centralized Supplier Services Inventory Management System

Control Objective: To maintain a centralized and up-to-date inventory of all supplier services, facilitating effective management, risk assessment, and compliance verification of supplier-provided services.

Implementation Steps:

1.1 **Implement Inventory Management Software:** Deploy a centralized inventory management system that can catalog and track all supplier services, including details on service type, deployment, and associated cybersecurity risk assessments.

1.2 **Integration with Vendor Management Processes:** Ensure the inventory management system is fully integrated with the organization's vendor management processes, allowing for seamless tracking of supplier performance, compliance, and risk status.

1.3 **Regular Inventory Review and Update:** Establish a routine for regularly reviewing and updating the inventory to reflect new services, modifications to existing services, or termination of services, ensuring the inventory remains accurate and comprehensive.

Expected Outcome: The implementation of a centralized supplier services inventory management system provides the organization with a comprehensive view of its supplier services landscape. This system supports effective risk management, compliance verification, and strategic decision-making regarding supplier services.

Control 2: Supplier Communication and Change Coordination Framework

Control Objective: To establish structured communication and coordination mechanisms with suppliers, ensuring timely notification and joint management of service changes that could impact the organization's cybersecurity posture.

Implementation Steps:

2.1 **Formalize Communication Channels:** Establish formal communication channels with suppliers to notify service changes, including updates, vulnerabilities, and incidents.

2.2 **Develop Change Management Procedures:** Define clear procedures for managing changes to supplier services, including risk reassessment, impact analysis, and necessary updates to security controls and documentation.

2.3 **Conduct Joint Reviews of Service Changes:** Schedule regular meetings with suppliers to review any proposed service changes, assess potential impacts, and coordinate implementation strategies to maintain security standards.

Expected Outcome: Establishing a supplier communication and change coordination framework ensures that the organization is promptly informed and can effectively manage changes to supplier services. This proactive approach minimizes cybersecurity risks associated with service changes and supports continuous alignment with the organization's security requirements and standards.

ID.AM-05: Assets Are Prioritized Based on Classification, Criticality, Resources, and Impact on the Mission

Asset prioritization within an organization's cybersecurity strategy is an essential practice that ensures the most critical assets receive the highest level of protection. This process begins with a clear understanding of asset prioritization principles, which emphasize identifying and protecting assets most vital to an organization's mission and operations. By focusing on these principles, organizations can allocate their cybersecurity resources more effectively, ensuring that the most critical assets are safeguarded against potential threats.

The asset classification and criticality assessment framework is a structured approach that categorizes assets based on their importance to the organization. This categorization typically considers factors such as the sensitivity of the data, the asset's role in operational processes, and the potential impact of a security breach on the asset. A thorough criticality assessment helps determine which assets are indispensable to the organization's mission and, therefore, require more robust security measures.

Methodologies for assessing the asset impact on the organizational mission involve evaluating how each asset supports the organization's core functions and objectives. This assessment considers the consequences of asset compromise, including operational disruption, financial loss, and reputational damage. Understanding the relationship between assets and the organization's mission enables targeted protection strategies that minimize risk and ensure operational continuity.

Integrating asset prioritization with risk management processes ensures that cybersecurity efforts align with the organization's overall risk strategy. This integration involves identifying threats and vulnerabilities associated with high-priority assets and implementing controls tailored to mitigate these risks. By linking asset prioritization to risk management, organizations can ensure a cohesive approach to cybersecurity that effectively protects critical assets while managing overall risk exposure.

Allocation of security resources based on asset prioritization is a strategic approach that optimizes using limited cybersecurity resources. By concentrating efforts on the most critical assets, organizations can ensure that these assets are protected by the most effective and advanced security measures available. This approach enhances the security of high-priority assets and contributes to the overall resilience of the organization's cybersecurity posture.

Using prioritization in incident response and business continuity planning is crucial for minimizing the impact of security incidents on critical operations. By understanding which assets are most important, incident response teams can prioritize their efforts to restore essential functions quickly, minimizing downtime and reducing the impact on the organization. Similarly, business continuity planning focuses on maintaining or promptly restoring operations of high-priority assets to ensure the organization can continue to fulfill its mission under adverse conditions.

Regular review and adjustment of asset prioritization are necessary to adapt to changing organizational priorities, emerging threats, and evolving cybersecurity landscapes. This process ensures that asset prioritization remains aligned with the organization's mission, operational needs, and risk environment. Organizations can maintain a dynamic and responsive cybersecurity strategy that addresses the most pressing threats and vulnerabilities by periodically reassessing and adjusting priorities.

Communication of asset priorities within the organization ensures that all stakeholders understand the significance of protecting critical assets. This communication should clearly explain why certain assets are prioritized, the specific threats to these assets, and the role individuals play in safeguarding them. Additionally, training and awareness programs on asset prioritization can reinforce the significance of cybersecurity measures and encourage a culture of security across the organization.

Recommendations

- **Develop a Dynamic Framework for Asset Classification:** Implement a flexible asset classification framework that can be adjusted as organizational priorities and the threat landscape evolve. This framework should facilitate the regular reassessment of asset criticality and ensure that cybersecurity resources are allocated efficiently.
- **Integrate Asset Prioritization with Enterprise Risk Management:** Seamlessly integrate asset prioritization into the broader enterprise risk management strategy. This alignment ensures that cybersecurity risks are considered in the context of overall organizational risk, promoting a holistic approach to risk mitigation.

- **Focus on Training and Awareness:** Invest in training and awareness programs highlighting the importance of asset prioritization and every employee's role in protecting critical assets. This education can enhance the effectiveness of cybersecurity measures by fostering a culture of security awareness throughout the organization.
- **Leverage Advanced Analytics for Asset Impact Assessment:** Utilize advanced analytics and threat intelligence to assess the impact of assets on the organizational mission continuously. These tools can provide insights into emerging threats and vulnerabilities, allowing for proactive adjustments to asset prioritization and protection strategies.
- **Foster Open Communication and Collaboration:** Encourage open communication and collaboration across departments to ensure a unified understanding of asset priorities and the collective responsibility for protecting these assets. Cross-functional collaboration can lead to more effectively identifying critical assets and developing comprehensive strategies to protect them.

NIST 800-53 Controls

(Refer to our appendix for a comprehensive description of the NIST 800-53 Controls, essential for organizations adhering to strict regulatory standards.)

- RA-03,RA-09,RA-02

Simplified Security Controls (SSC)

This section outlines our SSC, tailored for environments beyond the scope of NIST compliance. These controls offer streamlined, effective security solutions suitable for any organization seeking to enhance its cybersecurity defenses with clear, actionable measures.

Control 1: Dynamic Asset Classification and Prioritization System

Control Objective: To develop and maintain a dynamic system for the classification and prioritization of assets based on their criticality, sensitivity, and impact on the organizational mission, ensuring effective allocation of security resources.

Implementation Steps:

1.1 **Define Classification Criteria:** Establish clear criteria for classifying assets based on factors such as sensitivity, regulatory requirements, value to the organization, and potential impact on the mission if compromised.

1.2 **Implement Prioritization Mechanism:** Develop a mechanism within the asset management system to assign priority levels to assets based on their classification, considering the likelihood and impact of potential threats.

1.3 **Integrate with Risk Management Processes:** Ensure the asset prioritization system fully integrates with the organization's risk management processes. This allows for the continuous reassessment and reprioritization of assets as risk profiles and business objectives evolve.

Expected Outcome: Establishing a dynamic asset classification and prioritization system ensures that security resources are allocated efficiently, focusing on protecting the most critical assets. This approach enhances the organization's resilience by ensuring high-value assets are adequately secured, and risk exposure is minimized.

Control 2: Continuous Improvement Program for Asset Management

Control Objective: To establish a continuous improvement program for asset management that regularly reviews, assesses, and refines asset prioritization, classification, and security practices based on evolving threats, business objectives, and technological advancements.

Implementation Steps:

2.1 **Establish a Review Committee:** Form a cross-functional committee responsible for overseeing the continuous improvement of asset management practices, including regular reviews of asset prioritization and classification criteria.

2.2 **Incorporate Feedback and Lessons Learned:** Implement mechanisms for gathering feedback from incident responses, audits, and stakeholder input, using this information to identify areas for improvement in asset management practices.

2.3 **Schedule Regular Updates and Training:** Set a regular schedule for updating asset management policies and procedures based on committee recommendations and provide ongoing training to relevant staff on any changes and best practices.

Expected Outcome: Creating a continuous improvement program for asset management ensures that asset prioritization, classification, and protection measures are always aligned with the current threat landscape, business objectives, and technological capabilities. This program promotes an adaptive and proactive approach to asset management, enhancing the organization's overall security posture and operational efficiency.

ID.AM-07: Inventories of Data and Corresponding Metadata for Designated Data Types Are Maintained

Complemented by meticulous metadata management, the creation and maintenance of data inventories represent critical facets of modern cybersecurity and data governance strategies. This process begins with an in-depth introduction to data inventory and metadata management, underscoring their significance in the broader context of information security. Effective management of these assets enables organizations to safeguard sensitive information and optimize their data utilization, ensuring compliance with regulatory requirements and enhancing operational efficiencies.

Processes for identifying and categorizing critical data types lay the groundwork for robust data governance. These processes involve a comprehensive analysis of the data landscape within an organization, pinpointing data types that are crucial for business operations, sensitive in nature, or subject to regulatory requirements. Categorization facilitates a structured approach to data protection, tailoring security measures to the sensitivity and importance of the data, thereby optimizing resource allocation and mitigating risks more effectively.

Documentation practices for data and metadata are paramount in clearly understanding the data's origin, nature, and purpose. This documentation should include detailed descriptions of data types, their sources, usage patterns, and any processing they undergo. Such practices not only aid in maintaining the integrity and usability of data but also play a vital role in compliance efforts, particularly in environments regulated by strict data protection standards.

Integrating data inventories with data governance frameworks ensures that data management practices align with organizational policies and objectives. This integration facilitates a cohesive data usage, protection, and quality management strategy, embedding data inventories within the larger context of enterprise data governance. It underscores the necessity of aligning data inventory management with organizational goals, ensuring that data assets are leveraged responsibly and effectively.

Security and privacy considerations are central to managing data inventories. These considerations involve implementing controls to protect data from unauthorized access, breaches, and leaks. Privacy measures are equally critical, ensuring that personal and sensitive data are handled in compliance with legal and ethical standards. This dual focus on security and privacy underscores the complex landscape in which data inventory management operates, navigating the challenges of protecting sensitive information while maintaining its utility.

Regular audits and updates of data inventories are essential for ensuring their accuracy and relevance. The dynamic nature of data, characterized by constant changes, additions, and deletions, necessitates ongoing maintenance of data inventories. These audits help identify discrepancies, outdated information, and opportunities for optimization, ensuring that the data inventory remains a reliable resource for decision-making, compliance, and risk management.

The role of metadata in data management and security is multifaceted, enhancing the discoverability, management, and protection of data assets. Metadata provides contextual information about data, such as creation dates, ownership, and classification levels, playing a crucial role in enforcing access controls, managing data lifecycles, and facilitating audits. Its significance extends beyond mere data description, serving as a pivotal element in the broader data security and governance frameworks.

Data lifecycle management, from creation to disposal, is integral to effective data inventory management. This comprehensive approach ensures that data is managed appropriately at every stage, from initial creation and active use to archiving and eventual disposal. Effective lifecycle management includes implementing policies for data retention, archival, and destruction, ensuring that data is handled securely, efficiently, and in compliance with legal and regulatory requirements.

Recommendations

- **Establish Clear Categorization Criteria:** Develop and implement clear criteria for categorizing data, focusing on sensitivity, regulatory requirements, and business importance. This structured approach enhances the effectiveness of data protection and governance strategies.
- **Implement Comprehensive Documentation Practices:** Adopt thorough documentation practices for data and metadata, detailing their characteristics, usage, and processing. This documentation supports compliance efforts and ensures data integrity and usability.
- **Integrate Inventories with Governance Frameworks:** Seamlessly integrate data inventories with organizational data governance frameworks. This alignment ensures data management practices are consistent with enterprise policies and objectives.
- **Prioritize Security and Privacy:** Emphasize security and privacy in managing data inventories, implementing robust controls to protect sensitive information, and complying with privacy regulations.
- **Conduct Regular Audits and Updates:** Schedule and execute regular audits of data inventories, updating them to reflect changes in the data landscape. This ongoing maintenance ensures the accuracy and relevance of the inventory.
- **Leverage Metadata for Enhanced Management and Security:** Utilize metadata to improve data discoverability, management, and security. Employ metadata to enforce access controls, support data lifecycle management, and facilitate audits.
- **Adopt a Holistic Approach to Data Lifecycle Management:** Implement policies and practices that address every stage of the data lifecycle, from creation to disposal. This approach ensures secure, efficient, and compliant data management.
- **Utilize Data Inventories in Compliance and Reporting Efforts:** Leverage data inventories to support compliance with regulatory requirements and facilitate reporting. The detailed insight provided by inventories can streamline compliance processes and enhance transparency.

NIST 800-53 Controls

(Refer to our appendix for a comprehensive description of the NIST 800-53 Controls, essential for organizations adhering to strict regulatory standards.)

- CM-12,CM-13,SI-12

Simplified Security Controls (SSC)

This section outlines our SSC, tailored for environments beyond the scope of NIST compliance. These controls offer streamlined, effective security solutions suitable for any organization seeking to enhance its cybersecurity defenses with clear, actionable measures.

Control 1: Automated Data Discovery and Classification Tool

Control Objective: To automate the identification, classification, and documentation of critical data types across the organization's digital environment, ensuring that data inventories are accurate, comprehensive, and current.

Implementation Steps:

1.1 **Tool Selection and Deployment:** Choose and deploy an automated data discovery and classification tool to scan the organization's systems and repositories to identify and categorize data based on predefined criteria and sensitivity levels.

1.2 **Define Classification Criteria:** Collaborate with data governance and cybersecurity teams to define clear criteria for data classification, ensuring that all critical data types are accurately identified and categorized according to their sensitivity and importance to the organization.

1.3 **Continuous Scanning and Updating:** Configure the tool to continuously scan for new or modified data, automatically updating the data inventories to reflect changes and ensuring ongoing accuracy and completeness of the data inventory.

Expected Outcome: Deploying an automated data discovery and classification tool streamlines the process of maintaining accurate and comprehensive data inventories. This control ensures that critical data is promptly identified, classified, and protected in accordance with its value and sensitivity, enhancing data security and governance.

Control 2: Integration of Data Inventories with Privacy and Security Frameworks

Control Objective: To integrate data inventories with the organization's privacy and security frameworks, ensuring that all critical data is managed and protected in line with regulatory requirements and best practices.

Implementation Steps:

2.1 **Establish Integration Processes:** Develop processes for integrating data inventories with existing data governance, privacy, and cybersecurity frameworks, ensuring data handling practices comply with applicable regulations and standards.

2.2 **Map Data Flows to Privacy Requirements:** Utilize the data inventories to map how data moves within and outside the organization, aligning data handling practices with privacy requirements and identifying areas where additional controls are needed.

2.3 **Implement and Monitor Control Measures:** Based on the integration and mapping efforts, implement necessary control measures to protect critical data and regularly monitor these controls for effectiveness, making adjustments as needed.

Expected Outcome: Integrating data inventories with privacy and security frameworks ensures that critical data is consistently managed and protected throughout its lifecycle. This control enhances compliance with data protection regulations, reduces the risk of data breaches, and strengthens the organization's overall data governance posture.

Control 3: Regular Audits and Reviews of Data Inventories

Control Objective: Conduct regular audits and reviews of data inventories, ensuring they accurately reflect the current data landscape and comply with data governance policies and regulatory requirements.

Implementation Steps:

3.1 **Schedule Regular Audits:** Establish a regular schedule for auditing data inventories, including checks for completeness, accuracy, and alignment with data governance policies.

3.2 **Conduct Compliance and Accuracy Reviews:** Perform detailed reviews of data inventories to verify compliance with regulatory requirements and internal data governance policies and to ensure that data classifications remain accurate.

3.3 **Update and Refine Inventories Based on Findings:** Based on audit and review findings, make necessary updates and refinements to data inventories, adjusting classifications and documentation to reflect changes in the data environment or regulatory landscape.

Expected Outcome: Regular audits and reviews of data inventories ensure they remain accurate, comprehensive, and compliant over time. This control enables the organization to swiftly adapt to changes in the data landscape or regulatory requirements, maintaining effective data governance and enhancing data security and privacy.

ID.AM-08: Systems, Hardware, Software, Services, and Data Are Managed Throughout Their Life Cycles

The comprehensive management of IT assets throughout their life cycles is a cornerstone of effective cybersecurity and operational efficiency. This management encompasses a broad spectrum of assets, including systems, hardware, software, services, and data, from initial acquisition to final disposal. The introduction to lifecycle management of IT assets sets the stage for understanding the complexities and necessities of tracking, securing, and optimizing these resources over time. A strategic approach to lifecycle management enhances security and maximizes the value derived from each asset.

A robust framework for the lifecycle management of various asset types is essential for ensuring that each phase of an asset's life is handled with due diligence. This framework outlines processes for acquisition, deployment, operation, maintenance, and disposal, tailored to the unique requirements and risks associated with different types of IT assets. Such a framework serves as a guide for organizations to systematically manage assets that align with business goals and security requirements.

Integrating security considerations throughout the asset lifecycle is crucial for safeguarding information and systems against cyber threats. From the outset, security must be a primary consideration in selecting and acquiring IT assets, followed by secure deployment practices, regular maintenance, including patching and updates, and safe disposal methods that ensure sensitive data is irrecoverably erased. This comprehensive security approach prevents vulnerabilities at each lifecycle stage from being exploited.

Processes for the acquisition, deployment, maintenance, and disposal of IT assets are elaborated within the framework, emphasizing the importance of security at each phase. Acquisition processes should include assessments of vendor security practices and product vulnerabilities. Deployment should follow secure configuration guidelines. Maintenance processes must prioritize timely updates and patches. Disposal practices should ensure the secure deletion or destruction of data and hardware, preventing unauthorized access to residual information.

Automated tools for lifecycle management streamline the monitoring and maintenance of IT assets, enhancing the organization's ability to respond to vulnerabilities and changes in the asset environment. These tools can automate inventory management, monitor the health and performance of assets, and facilitate the efficient deployment of updates and patches, thereby reducing the risk of security breaches.

Incorporating change management processes in lifecycle management ensures that modifications to IT assets, whether they are updates, upgrades, or decommissioning, are conducted in a controlled and secure manner. Change management processes help assess the security implications of changes, plan for potential disruptions, and communicate changes across the organization, thereby maintaining the integrity and security of IT assets.

Documentation and record-keeping for each stage of the asset lifecycle are indispensable for effective management and compliance. Detailed records support the tracking of asset performance, the auditability of security practices, and the verification of compliance with regulations and standards. This documentation serves as a vital resource for understanding the history and status of an asset, facilitating decision-making and incident response.

Assessing and mitigating risks at different lifecycle stages involves continuous evaluation of the threats and vulnerabilities associated with IT assets. Risk assessments inform the implementation of appropriate security controls and mitigation strategies tailored to the asset's current phase in its lifecycle. This ongoing risk management process ensures that assets are protected against evolving cyber threats throughout their operational life.

Recommendations

- **Develop Comprehensive Lifecycle Frameworks:** Implement detailed frameworks for the lifecycle management of all IT asset types, incorporating security considerations at every stage.
- **Leverage Automation:** Utilize automated tools to streamline lifecycle management processes, from inventory management to patch deployment, enhancing efficiency and security.
- **Integrate Change Management:** Embed change management processes into lifecycle management to ensure that all changes are assessed for security implications and managed effectively.
- **Maintain Rigorous Documentation:** Keep thorough records for each asset through its lifecycle, supporting effective management, compliance, and security incident response.
- **Conduct Regular Risk Assessments:** Perform continuous risk assessments of IT assets, adjusting security controls and mitigation strategies as assets progress through their lifecycles and as the threat landscape evolves.
- **Promote Training and Awareness:** Foster an organizational culture that values lifecycle management best practices through ongoing training and awareness initiatives, empowering employees to contribute to the security and efficiency of ITAM.

NIST 800-53 Controls

(Refer to our appendix for a comprehensive description of the NIST 800-53 Controls, essential for organizations adhering to strict regulatory standards.)

- CM-09,CM-13,MA-02,MA-06,PL-02,PM-22,PM-23,SA-03,SA-04,SA-08,SA-22,SI-12,SI-18, SR-05,SR-12

Simplified Security Controls (SSC)

This section outlines our SSC, tailored for environments beyond the scope of NIST compliance. These controls offer streamlined, effective security solutions suitable for any organization seeking to enhance its cybersecurity defenses with clear, actionable measures.

Control 1: Secure Asset Acquisition and Deployment Protocol

Control Objective: To establish secure processes for acquiring and deploying IT assets, ensuring they meet the organization's security standards from the outset and throughout their operational life.

Implementation Steps:

1.1 **Develop Security Criteria for Acquisition:** Formulate comprehensive security criteria for acquiring IT assets, including requirements for vendor security assessments, product security features, and compliance with relevant standards.

1.2 **Secure Deployment Procedures:** Implement standardized, secure deployment procedures that include secure configuration settings, initial security testing, and integration into existing security monitoring tools before the asset is implemented.

1.3 **Vendor Security Collaboration:** Work closely with vendors to understand the security measures of their products and services, ensuring these measures align with the organization's security requirements before purchase and deployment.

Expected Outcome: Implementing a secure asset acquisition and deployment protocol ensures that all IT assets are safe by design and remain secure throughout their deployment. This control minimizes vulnerabilities in newly acquired and deployed assets, reducing the organization's security risk.

Control 2: Automated Lifecycle Management and Monitoring System

Control Objective: To utilize automated tools for continuously managing and monitoring IT assets throughout their lifecycle, ensuring assets remain compliant with security policies from acquisition to disposal.

Implementation Steps:

2.1 **System Selection and Integration:** Select and implement an automated lifecycle management system to track and manage IT assets throughout their entire lifecycle, from acquisition through disposal, and integrate this system with existing security tools.

2.2 **Continuous Monitoring for Compliance:** Configure the system to monitor IT assets for security compliance, automatically flagging assets that deviate from established security configurations or require updates and patches.

2.3 **Lifecycle Event Tracking:** Ensure the system tracks significant lifecycle events for each asset, including deployment, maintenance activities, security incidents, and decommissioning, facilitating comprehensive oversight.

Expected Outcome: Deploying an automated lifecycle management and monitoring system provides continuous visibility and control over the security posture of IT assets throughout their lifecycle. This system ensures assets comply with security policies, facilitates timely updates, and supports effective risk management.

Control 3: Comprehensive Lifecycle Documentation and Risk Management

Control Objective: To maintain thorough documentation and perform regular risk assessments for IT assets at each stage of their lifecycle, ensuring that risks are identified, managed, and mitigated proactively.

Implementation Steps:

3.1 **Document Lifecycle Activities:** Implement procedures for the comprehensive documentation of lifecycle activities for each IT asset, including acquisition details, deployment configurations, maintenance records, and disposal methods.

3.2 **Conduct Lifecycle Risk Assessments:** Regularly assess IT assets for new and evolving risks at each stage of their lifecycle, utilizing detailed documentation to inform the assessment process.

3.3 Mitigation Strategy Implementation: Based on risk assessment outcomes, develop and implement mitigation strategies tailored to the specific risks identified at different lifecycle stages, updating documentation to reflect these strategies and their outcomes.

Expected Outcome: Comprehensive lifecycle documentation and regular risk assessments ensure that IT assets are continuously evaluated for potential risks, with effective mitigation strategies implemented in a timely manner. This control enhances the organization's ability to proactively manage and reduce risks associated with IT assets throughout their lifecycle, contributing to overall cybersecurity resilience.

12

Risk Assessment (ID.RA)

> Securing our digital infrastructure demands more than vigilance; it requires a foundational commitment to meticulously validating every component, from hardware to human resources, ensuring our resilience in the face of relentless threats.

In cybersecurity, identifying, validating, and managing vulnerabilities, alongside the strategic acquisition of hardware, software, and supplier services, form the bedrock of a robust security posture. Businesses can significantly mitigate potential threats through meticulously designed processes for assessing the authenticity and integrity of IT assets before their integration into organizational frameworks. This is complemented by establishing formal vulnerability disclosure programs (VDPs) and integrating cyber threat intelligence (CTI) into security strategies, ensuring a proactive stance against emerging threats. Moreover, the critical assessment of suppliers before acquisition safeguards the supply chain—a frequently overlooked yet vulnerable aspect of cybersecurity. Regular reviews, continuous monitoring, and the adaptation of risk assessment and response strategies in line with evolving threats and organizational objectives underscore the dynamic nature of effective cybersecurity risk management. These concerted efforts, underpinned by comprehensive training and a culture of continuous improvement, empower organizations to navigate the complexities of the digital landscape securely.

ID.RA-01: Vulnerabilities in Assets Are Identified, Validated, and Recorded

Understanding and managing vulnerabilities within organizational assets is critical to robust cybersecurity. Initiating a vulnerability management process marks the first step toward safeguarding systems against potential threats. This process encompasses a systematic approach to identifying, validating, and recording vulnerabilities, providing an organization with the knowledge necessary to mitigate risks effectively. Organizations must recognize the significance of these processes as they lay the groundwork for a proactive security posture, enabling the detection and resolution of vulnerabilities before malicious actors can exploit them.

The methodology for identifying vulnerabilities in organizational assets involves a combination of automated tools and manual testing techniques. Computerized tools like vulnerability scanners are crucial in continuously monitoring systems for known vulnerabilities. However, these tools must be

A Comprehensive Guide to the NIST Cybersecurity Framework 2.0: Strategies, Implementation, and Best Practice, First Edition. Jason Edwards.
© 2025 John Wiley & Sons Ltd. Published 2025 by John Wiley & Sons Ltd.

complemented by manual testing methods, such as penetration testing, to uncover vulnerabilities that automated tools might miss. This blended approach ensures a comprehensive assessment of organizational assets, highlighting potential security weaknesses that must be addressed.

Once vulnerabilities are identified, the next step involves the validation of these findings. Validation techniques are crucial to confirm the existence of vulnerabilities and to assess their true impact on the organization's security posture. This process often involves replicating the conditions under which a vulnerability could be exploited in a controlled environment to verify its authenticity. Such validation not only confirms the presence of vulnerabilities but also aids in prioritizing remediation efforts based on the potential impact of each vulnerability.

Documenting and recording identified vulnerabilities is an essential practice that facilitates efficient tracking and management of cybersecurity risks. This documentation should include detailed information about the vulnerability, such as its nature, severity, affected assets, and potential mitigation strategies. Effective recording practices ensure that all stakeholders have access to up-to-date information, enabling informed decision-making in vulnerability management and risk mitigation.

Integrating vulnerability identification processes with patch management systems is a best practice that enhances an organization's ability to respond to identified vulnerabilities swiftly. This integration enables organizations to streamline the deployment of patches or updates that address vulnerabilities, reducing the window of opportunity for attackers. By automating the flow of information between vulnerability identification and patch management systems, organizations can ensure that vulnerabilities are addressed promptly and efficiently.

Automated tools for continuous vulnerability scanning are a cornerstone of an effective vulnerability management program. These tools enable organizations to be vigilant about their digital assets constantly, identifying new vulnerabilities as they emerge. Continuous scanning provides the advantage of timely detection, allowing swift action to mitigate risks associated with newly discovered vulnerabilities.

Prioritizing vulnerabilities based on their severity and potential impact on the organization is critical to the vulnerability management process. This prioritization helps allocate resources effectively, focusing remediation efforts on vulnerabilities that pose the most significant risk to the organization. Factors such as the ease of exploitation, potential impact on confidentiality, integrity, and availability, and the presence of active exploits in the wild are considered during the prioritization process.

Training staff to identify and report vulnerabilities is essential in enhancing an organization's cybersecurity posture. Educating employees about the importance of vigilance and their role in the organization's security strategy empowers them to act as the first line of defense. Regular training sessions should cover identifying suspicious activities and the proper channels for reporting potential vulnerabilities, reinforcing the collective responsibility toward cybersecurity.

Recommendations

- **Implement a Blended Approach**: Utilize automated tools and manual testing methods to assess vulnerabilities comprehensively. This dual approach ensures the identification of a more comprehensive range of vulnerabilities, enhancing the organization's security posture.
- **Establish a Continuous Monitoring Strategy**: Deploy automated vulnerability scanning tools for continuously monitoring organizational assets. This strategy enables the timely detection and remediation of new vulnerabilities, reducing the organization's exposure to cyber threats.
- **Integrate Vulnerability Management with Patch Management**: Streamline vulnerability remediation by integrating vulnerability management practices with patch management

systems. This integration ensures that vulnerabilities are swiftly addressed through the timely deployment of patches.

- **Prioritize Vulnerabilities**: Develop a system for prioritizing vulnerabilities based on their severity and the potential impact on the organization. This prioritization assists in focusing remediation efforts on the most critical vulnerabilities, optimizing the use of resources.
- **Foster a Culture of Security Awareness**: Conduct regular training sessions for staff on identifying and reporting vulnerabilities. Creating a culture of security awareness among employees strengthens the organization's defense against cyber threats by involving everyone in the cybersecurity efforts.

NIST 800-53 Controls

(Refer to our appendix for a comprehensive description of the NIST 800-53 Controls, essential for organizations adhering to strict regulatory standards.)

- CA-02,CA-07,CA-08,RA-03,RA-05,SA-11,SA-15,SI-04,SI-05

Simplified Security Controls (SSC)

This section outlines our simplified security controls (SSC), tailored for environments beyond the scope of NIST compliance. These controls offer streamlined, effective security solutions suitable for any organization seeking to enhance its cybersecurity defenses with clear, actionable measures.

Control 1: Development of Comprehensive Vulnerability Management Program

Control Objective: To systematically identify, assess, prioritize, and mitigate vulnerabilities within the organization's IT infrastructure, thereby reducing the attack surface and enhancing the overall security posture.

Implementation Steps:

1.1 **Program Development**: Establish a comprehensive vulnerability management program that outlines procedures for regular vulnerability scanning, assessment, prioritization, remediation, and documentation.

1.2 **Integration with Existing Systems**: Integrate the vulnerability management program with existing patch management and incident response systems to streamline the remediation process and ensure timely response to identified vulnerabilities.

1.3 **Continuous Improvement**: Implement a continuous improvement process for the vulnerability management program, incorporating feedback from postremediation reviews, changes in organizational assets, and emerging threat intelligence.

Expected Outcome: Establishing a comprehensive vulnerability management program ensures that vulnerabilities are systematically identified, assessed, and mitigated. This approach significantly reduces the organization's exposure to cyber threats and enhances its ability to respond effectively to emerging vulnerabilities.

Control 2: Implementation of Automated Vulnerability Scanning Tools

Control Objective: To leverage automated tools for continuously and systematically scanning the organization's network and systems for known vulnerabilities, ensuring timely identification and response.

Implementation Steps:

2.1 **Tool Selection and Deployment**: Select and deploy automated vulnerability scanning tools that align with the organization's IT infrastructure and security requirements.

Ensure tools are configured to perform scans at optimal intervals without impacting system performance.

2.2 **Scan Analysis and Reporting**: Analyze scan results to identify actionable vulnerabilities and generate reports that detail findings, including vulnerability severity and recommended remediation steps.

2.3 **Integration with Remediation Processes**: Integrate scanning tool outputs with the organization's remediation processes, ensuring that vulnerabilities are promptly addressed based on their prioritization.

Expected Outcome: Using automated vulnerability scanning tools enables the organization to identify and address vulnerabilities proactively. This continuous scanning and integration with remediation processes enhance the organization's ability to maintain a strong security posture against evolving cyber threats.

Control 3: Staff Training and Awareness Programs on Vulnerability Management

Control Objective: To enhance the organization's security culture by equipping staff with the knowledge and skills to recognize and report vulnerabilities, contributing to the early detection and response to security threats.

Implementation Steps:

3.1 **Develop Training Content**: Create comprehensive training modules on vulnerability management processes, including the importance of vulnerability identification, reporting mechanisms, and the role of staff in the organization's security posture.

3.2 **Conduct Regular Training Sessions**: Schedule and conduct regular training sessions for all staff, ensuring participation across departments. Include practical exercises and simulations to reinforce learning outcomes.

3.3 **Continuous Awareness Campaigns**: Implement ongoing awareness campaigns using posters, emails, and intranet articles to keep security at the forefront of staff consciousness and encourage vigilant behavior.

Expected Outcome: Implementing staff training and awareness programs on vulnerability management fosters a proactive security culture within the organization. This cultural shift enhances the organization's resilience to cyber threats by promoting early detection and reporting of vulnerabilities, contributing significantly to the overall effectiveness of the vulnerability management program.

ID.RA-02: Cyber Threat Intelligence Is Received from Information Sharing Forums and Sources

Cybersecurity is constantly evolving, with new threats emerging at an alarming rate. In this context, CTI has become an indispensable resource for organizations aiming to stay one step ahead of potential threats. CTI encompasses actionable information about adversaries, their methods, motives, and capabilities. An introduction to CTI underscores its significance in enabling organizations to make informed decisions about their security posture and to strategize their defenses more effectively. The pivotal role of CTI lies in its ability to offer insights into potential cyber threats, facilitating a proactive rather than reactive approach to cybersecurity.

For organizations to leverage CTI effectively, establishing channels for receiving intelligence from trusted sources is crucial. These channels can range from formal partnerships with governmental agencies to subscriptions with private threat intelligence providers. The reliability and relevance of the intelligence received heavily depend on the credibility of the sources. Therefore, organizations must diligently select authoritative and pertinent sources for their specific industry or sector. This careful curation of sources ensures that the intelligence received is accurate and applicable to the organization's threat landscape.

Active participation in information-sharing forums and partnerships significantly enhances an organization's ability to gather diverse and insightful CTI. These forums, which can be industry-specific or cross-sectoral, facilitate the exchange of intelligence on emerging threats, vulnerabilities, and incidents. Participation in such forums is not merely about receiving intelligence; it also involves contributing insights from one's own experiences, thereby enriching the collective knowledge base. This reciprocal exchange of information fosters a sense of community and shared responsibility among participants, driving collective resilience against cyber threats.

Analyzing and validating the received intelligence is a critical process that ensures its relevance and accuracy before integration into the organization's security strategy. This process involves scrutinizing the information to verify its sources, assessing its credibility, and determining its applicability to the organization. Such rigorous analysis helps filter out noise and focus on intelligence that could have a tangible impact on the organization's security. The ability to distinguish between genuine threats and false alarms is key to allocating resources efficiently and enhancing the organization's defensive measures.

Integrating CTI into organizational security strategies is a complex but rewarding endeavor. It requires the alignment of intelligence insights with the organization's overall security objectives and risk management framework. This integration enables organizations to tailor their security measures based on the specific threats and vulnerabilities identified through CTI. By embedding intelligence insights into security policies, incident response plans, and proactive defense mechanisms, organizations can significantly improve their resilience against cyberattacks.

Mechanisms for sharing intelligence within the organization ensure that relevant insights reach all stakeholders, from top-level management to operational teams. Effective internal dissemination of CTI is crucial for fostering a unified security stance and ensuring coordinated action in response to threats. This can be achieved through regular briefings, integration of CTI into internal security platforms, and the establishment of cross-departmental teams dedicated to threat intelligence analysis and response.

The use of CTI for proactive threat detection and response marks a shift from traditional, reactive security measures to a more anticipatory approach to cybersecurity. Leveraging CTI allows organizations to identify potential threats and vulnerabilities before they are exploited, enabling preemptive measures to mitigate risks. This proactive stance is supported by advanced analytical tools and technologies that parse through vast amounts of intelligence to identify patterns and indicators of compromise (IoCs).

Training programs on utilizing CTI are essential for building staff's requisite skills and knowledge to interpret and act on intelligence insights effectively. These programs should cover various topics, including the sources of CTI, methods of analysis, and practical applications in security operations. By empowering employees with knowledge on leveraging CTI, organizations can cultivate a culture of awareness and vigilance that contributes significantly to their cybersecurity efforts.

Recommendations

- **Cultivate Trusted Intelligence Sources**: Prioritize establishing channels to receive CTI from sources with proven credibility and relevance to your organizational context. This strategic selection ensures that the intelligence is both actionable and applicable.

- **Engage in Information-Sharing Forums**: Participate in information-sharing forums and partnerships to enhance intelligence-gathering capabilities. This engagement facilitates access to a broader range of intelligence, enriching your understanding of the cyber threat landscape.
- **Implement Rigorous Analysis Processes**: Develop robust processes for analyzing and validating received CTI. This ensures that the intelligence integrated into your security strategy is accurate and high-quality, enhancing decision-making.
- **Integrate CTI Across the Organization**: Ensure the effective dissemination of CTI within your organization. Mechanisms for internal sharing play a crucial role in aligning security efforts and fostering a cohesive defense posture against cyber threats.
- **Invest in Training**: Establish comprehensive training programs on utilizing CTI for your staff. These programs should enhance their ability to effectively interpret and apply intelligence insights, bolstering your organization's proactive cybersecurity capabilities.

NIST 800-53 Controls

(Refer to our appendix for a comprehensive description of the NIST 800-53 Controls, essential for organizations adhering to strict regulatory standards.)

- SI-05,PM-15,PM-16

Simplified Security Controls (SSC)

This section outlines our SSC, tailored for environments beyond the scope of NIST compliance. These controls offer streamlined, effective security solutions suitable for any organization seeking to enhance its cybersecurity defenses with clear, actionable measures.

Control 1: Establishment of a Dedicated CTI Team

Control Objective: To create a specialized team responsible for the aggregation, analysis, and dissemination of CTI, ensuring that actionable insights are effectively integrated into the organization's security strategy.

Implementation Steps:

1.1 **Team Formation**: Form a dedicated CTI team comprising individuals with cybersecurity, intelligence analysis, and information technology expertise. This team will manage the CTI lifecycle, from collection to action.

1.2 **Define Operational Procedures**: Develop standard operating procedures for the CTI team, outlining how intelligence will be collected, analyzed, validated, and shared within the organization. Include protocols for prioritizing intelligence based on relevancy and urgency.

1.3 **Establish Communication Channels**: Set up secure and efficient communication channels for disseminating intelligence insights to relevant stakeholders, ensuring that the information is actionable and accessible to those who need it.

Expected Outcome: Establishing a dedicated CTI team enhances the organization's ability to collect, analyze, and act on CTI. This structured approach to intelligence management ensures that the organization can proactively identify and respond to threats, thereby improving its overall cybersecurity posture.

Control 2: Participation in Industry-Specific Information Sharing and Analysis Centers (ISACs)

Control Objective: To engage in collaborative threat intelligence sharing and analysis through participation in industry-specific ISACs, leveraging collective knowledge to enhance the organization's threat awareness and response capabilities.

Implementation Steps:

2.1 **Identify Relevant ISACs**: Determine which ISACs are most relevant to the organization's industry and cybersecurity needs, and initiate the membership process to join these forums.

2.2 **Active Participation**: Participate in ISAC meetings, threat briefing sessions, and collaborative projects. Encourage CTI team members to contribute to discussions and share insights from the organization's experiences.

2.3 **Integrate ISAC Insights**: Develop a process for integrating intelligence and best practices from ISAC participation into the organization's cybersecurity strategy, including updates to security policies, defenses, and incident response plans.

Expected Outcome: Participation in ISACs allows the organization to benefit from a broader perspective on cyber threats specific to its industry. This engagement fosters a collaborative approach to cybersecurity, enhancing the organization's ability to detect, prevent, and respond to threats through shared intelligence and collective defense strategies.

Control 3: Implementation of an Intelligence Analysis and Validation Platform

Control Objective: To deploy an advanced platform that supports the in-depth analysis and validation of CTI, ensuring that the organization acts on verified and relevant information.

Implementation Steps:

3.1 **Platform Selection**: Evaluate and select an intelligence analysis and validation platform that meets the organization's technical requirements and supports the integration of various intelligence feeds and internal data sources.

3.2 **Training and Configuration**: Train CTI team members on the effective use of the platform and configure the system to accurately filter, analyze, and prioritize intelligence based on predefined criteria relevant to the organization's threat landscape.

3.3 **Integration with Response Processes**: Ensure the platform is integrated with the organization's incident response and risk management processes, enabling automated alerts and swiftly reacting to verified threats.

Expected Outcome: Implementing an intelligence analysis and validation platform streamlines the processing of vast amounts of cyber threat information, enabling the organization to quickly identify and respond to verified threats. This control enhances the organization's operational efficiency and agility in countering cyber threats, significantly improving its defensive capabilities.

ID.RA-03: Internal and External Threats to the Organization Are Identified and Recorded

Proactively identifying and documenting internal and external threats are foundational elements in establishing a resilient cybersecurity posture. This process begins with a clear understanding of an organization's various threats, ranging from malicious insider activities to sophisticated external cyberattacks. An effective threat identification and documentation strategy not only aids in the early detection of potential security incidents but also supports the development of tailored mitigation and response strategies. The complexity and dynamism of the cyber threat landscape necessitate an ongoing commitment to this aspect of cybersecurity management.

A structured framework for classifying threats is essential to efficiently distinguish between internal and external sources of risk. Internal threats may include, but are not limited to negligent or malicious employee actions. In contrast, external threats can encompass a range of adversaries, including cybercriminals, nation-state actors, and competitors. This classification aids in allocating resources and developing specific controls designed to mitigate each type of threat. Furthermore, understanding the nature of these threats allows organizations to prioritize their cybersecurity efforts more effectively.

Techniques for identifying emerging threats are diverse and should be selected based on the organization's specific context and risk profile. These techniques often include using advanced cybersecurity tools, such as threat intelligence platforms, and methodologies, such as red teaming exercises and threat modeling. Organizations can detect potential security issues by adopting a proactive stance toward threat identification before they escalate into serious incidents. This proactive approach is critical in an environment where adversaries continually evolve their tactics, techniques, and procedures.

Recording and tracking identified threats in a central repository is vital for maintaining an organized and accessible record of potential risks. Such repositories enable the systematic analysis of threats over time, facilitating trend analysis and the identification of patterns that may indicate broader security issues. Additionally, these records serve as a crucial reference point for incident response teams, allowing for quicker and more informed decision-making in the face of active security incidents.

Integrating threat intelligence with risk assessment processes enhances the organization's ability to accurately evaluate its vulnerability to identified threats. This integration ensures that threat intelligence informs all stages of the risk management lifecycle, from identification and assessment to mitigation and monitoring. By aligning threat intelligence with risk assessment, organizations can ensure that their security measures are appropriate and proportional to the threats they face.

Collaboration with external entities for threat information exchange is an increasingly important component of a comprehensive threat identification strategy. Such collaborations can take many forms, including participation in sector-specific ISACs or partnerships with law enforcement agencies. These collaborative efforts enrich an organization's understanding of the threat landscape and facilitate the sharing of best practices for threat mitigation.

The development of IoCs for threat detection is a technique that allows organizations to identify specific signs of malicious activity within their networks. IoCs, such as unusual network traffic patterns or suspicious file modifications, serve as early warning signs that enable security teams to swiftly detect and respond to threats. The effectiveness of IoCs as a detection tool relies on their continuous update and refinement in response to evolving threat behaviors.

Continuous monitoring for identifying new threats is a critical capability that requires both technological and procedural elements. Technologies such as intrusion detection systems and security information and event management (SIEM) systems play a key role in this ongoing monitoring effort. However, the success of continuous monitoring also depends on establishing processes that ensure the practical analysis and escalation of detected anomalies. This dual focus on technology and process is essential for maintaining vigilance against emerging threats.

Recommendations

- **Adopt a Proactive Threat Identification Strategy**: Leverage advanced cybersecurity tools and methodologies to identify emerging threats proactively. This approach enables organizations to stay ahead of potential security issues.
- **Establish a Centralized Threat Repository**: Implement a system for recording and tracking identified threats. A centralized repository facilitates efficient threat analysis and supports quick decision-making in response to incidents.

- **Integrate Threat Intelligence with Risk Assessment**: Ensure that threat intelligence informs all aspects of the risk assessment process. This integration helps align security measures with the actual threats the organization faces.
- **Foster External Collaboration for Threat Information Exchange**: Engage in partnerships and information-sharing initiatives with other organizations and entities. Such collaboration enriches the organization's threat intelligence and strengthens collective cybersecurity resilience.
- **Implement Continuous Monitoring**: Develop and maintain capabilities for continuously monitoring the organization's networks and systems. Continuous monitoring, supported by both technology and process, is vital for the early detection of new threats.

NIST 800-53 Controls

(Refer to our appendix for a comprehensive description of the NIST 800-53 Controls, essential for organizations adhering to strict regulatory standards.)

- PM-12,PM-16,RA-03,SI-05

Simplified Security Controls (SSC)

This section outlines our SSC, tailored for environments beyond the scope of NIST compliance. These controls offer streamlined, effective security solutions suitable for any organization seeking to enhance its cybersecurity defenses with clear, actionable measures.

Control 1: Establishment of a Centralized Threat Intelligence Platform

Control Objective: To create a centralized repository for aggregating, analyzing, and disseminating information on identified internal and external threats, ensuring timely and informed decision-making across the organization.

Implementation Steps:

1.1 **Platform Selection and Implementation**: Evaluate and select a comprehensive threat intelligence platform that supports the integration of diverse intelligence feeds and facilitates the analysis of threat data.

1.2 **Customization and Integration**: Customize the platform to align with the organization's specific threat landscape and integrate it with existing security tools and infrastructure for seamless information sharing.

1.3 **Training and Awareness**: Conduct training sessions for security analysts and relevant stakeholders on effectively utilizing the platform for threat analysis and response.

Expected Outcome: Establishing a centralized threat intelligence platform enhances the organization's ability to collect, analyze, and act upon diverse threat intelligence, ensuring a proactive and coordinated response to emerging threats. This centralization facilitates the swift identification of threats and the dissemination of actionable intelligence across the organization.

Control 2: Development and Implementation of a Threat Classification Schema

Control Objective: To systematically categorize internal and external threats based on their nature, source, impact, and likelihood, facilitating prioritized and effective risk management responses.

Implementation Steps:

2.1 **Schema Development**: Develop a comprehensive threat classification schema that includes criteria for categorizing threats based on their characteristics and potential impact on the organization.

2.2 **Integration with Risk Management**: Integrate the threat classification schema with the organization's existing risk management processes, ensuring that threats are assessed and prioritized according to their classification.

2.3 **Continuous Update and Revision**: Establish a process for regularly reviewing and updating the classification schema to reflect the evolving threat landscape and the organization's changing risk profile.

Expected Outcome: Developing and implementing a threat classification schema enables the organization to prioritize threats effectively and allocate resources more efficiently. This systematic approach to threat classification supports enhanced situational awareness and improved risk management strategies.

Control 3: Continuous Threat Monitoring and Indicator of Compromise (IoC) Development

Control Objective: To implement continuous monitoring capabilities for identifying new threats and developing IoCs for early threat detection.

Implementation Steps:

3.1 **Monitoring Tool Deployment**: Deploy advanced monitoring tools and technologies capable of detecting anomalies and potential threats across the organization's digital infrastructure.

3.2 **IoC Development**: Develop and maintain a dynamic library of IoCs based on the latest threat intelligence, past incidents, and ongoing monitoring insights, enabling rapid detection of potential security breaches.

3.3 **Integration and Automation**: Integrate continuous monitoring tools and IoC libraries with the organization's SIEM systems, automating alert generation and preliminary response actions for potential threats.

Expected Outcome: Implementing continuous threat monitoring and developing IoCs enables the organization to swiftly detect and respond to emerging threats. This proactive stance enhances the organization's defensive capabilities, reducing the likelihood and impact of security breaches.

ID.RA-04: Potential Impacts and Likelihoods of Threats Exploiting Vulnerabilities Are Identified and Recorded

Understanding the potential impact and likelihood of threats exploiting vulnerabilities is pivotal to developing an effective risk management strategy in cybersecurity. This understanding begins with an introduction to impact and likelihood assessment in threat analysis, which provides a foundation for estimating the severity and probability of threats materializing into actual cyber incidents. Such assessments are crucial for informed decision-making and prioritizing cybersecurity efforts, ensuring that resources are allocated efficiently to address the most significant risks.

Assessing the impact of threat exploitation involves evaluating the potential consequences that a successful cyberattack could have on an organization. This evaluation considers various factors, including the sensitivity of the compromised data, the disruption to business operations, and the financial and reputational damage that could result. This assessment's methods often incorporate qualitative scales, such as low, medium, and high, or quantitative measures, such as estimated financial loss. This multifaceted approach enables organizations to gain a comprehensive understanding of the ramifications of potential security breaches.

Techniques for determining the likelihood of threat realization play an equally important role in threat analysis. These techniques involve analyzing various factors, such as known vulnerabilities, the capabilities and motivations of potential attackers, and the effectiveness of existing security controls. By estimating the probability of different threat scenarios, organizations can better focus their cybersecurity efforts on the most plausible risks, enhancing their overall security posture.

Documentation practices for impact and likelihood assessments are critical for maintaining a record of the risk analysis process and its outcomes. Proper documentation provides a historical account of threat assessments, supports accountability, and facilitates continuous improvement in the risk management process. This documentation should be detailed, including the rationale behind impact and likelihood ratings, and should be readily accessible to relevant stakeholders for review and decision-making.

Integrating impact and likelihood assessments with overall risk assessment is essential for a holistic understanding of cybersecurity risks. This integration combines the insights gained from impact and likelihood assessments with other components of risk analysis, such as vulnerability assessment and threat identification, to develop a comprehensive risk profile. Such an integrated approach ensures that all aspects of cybersecurity risk are considered when developing security strategies and policies.

Scenario analysis and modeling for threat assessment are valuable tools that allow organizations to simulate various threat scenarios and their potential impacts. This approach helps visualize the consequences of different types of cyberattacks and can aid in identifying vulnerabilities that might otherwise be overlooked. Organizations can better prepare for various security challenges by exploring multiple hypothetical scenarios.

Prioritization of threats based on impact and likelihood ratings is a logical next step in threat analysis. This prioritization enables organizations to focus their resources and efforts on mitigating the risks that pose the most significant potential impact and are most likely to occur. Such a targeted approach to cybersecurity enhances the efficiency and effectiveness of risk management strategies, ensuring that the most critical threats are addressed promptly.

Continuous review and update of impact and likelihood assessments are necessary to adapt to the ever-evolving cyber threat landscape. As new vulnerabilities are discovered and threat actors develop new tactics, it is imperative for organizations to regularly reassess their risk analyses to ensure they remain relevant and accurate. This ongoing review and update process is crucial for maintaining an up-to-date understanding of cybersecurity risks and ensuring that risk management strategies remain effective over time.

Recommendations

- **Enhance Documentation Practices**: Adopt comprehensive documentation practices for impact and likelihood assessments to ensure a detailed and accessible record of the risk analysis process. This facilitates informed decision-making and supports continuous improvement.
- **Utilize Scenario Analysis**: Employ scenario analysis and modeling to simulate potential threat scenarios and assess their impact. This approach aids in identifying vulnerabilities and preparing for a range of security challenges.
- **Integrate Assessments with Overall Risk Management**: Ensure that impact and likelihood assessments fully integrate with the organization's overall risk management framework. This holistic approach provides a comprehensive understanding of cybersecurity risks.
- **Prioritize Based on Ratings**: Implement a threat prioritization strategy based on impact and likelihood ratings. Focusing on the most significant risks enhances the efficiency of cybersecurity efforts.

- **Commit to Continuous Review**: Establish a routine for the continuous review and update of impact and likelihood assessments to keep pace with the dynamic cyber threat environment. Regular reassessment is vital for the relevance and effectiveness of risk management strategies.

NIST 800-53 Controls

(Refer to our appendix for a comprehensive description of the NIST 800-53 Controls, essential for organizations adhering to strict regulatory standards.)

- PM-09,PM-11,RA-02,RA-03,RA-08,RA-09

Simplified Security Controls (SSC)

This section outlines our SSC, tailored for environments beyond the scope of NIST compliance. These controls offer streamlined, effective security solutions suitable for any organization seeking to enhance its cybersecurity defenses with clear, actionable measures.

Control 1: Risk Impact and Likelihood Assessment Framework

Control Objective: To establish a standardized framework for assessing the potential impact and likelihood of threats exploiting vulnerabilities within the organization, enabling a structured approach to risk prioritization and management.

Implementation Steps:

1.1 **Framework Development**: Develop a comprehensive framework that outlines methods for assessing the impact and likelihood of threats exploiting vulnerabilities. This framework should include qualitative and quantitative assessment techniques tailored to the organization's context.

1.2 **Integration with Existing Risk Management Processes**: Seamlessly integrate the impact and likelihood assessment framework with the organization's existing risk management processes, ensuring that threat assessments contribute directly to overall risk prioritization and response strategies.

1.3 **Regular Training and Awareness**: Provide ongoing training for relevant personnel on applying the framework effectively. This includes understanding the assessment methodologies and incorporating assessment findings into risk management decisions.

Expected Outcome: Implementing a risk impact and likelihood assessment framework enables the organization to systematically evaluate and prioritize risks based on their potential impact and likelihood. This structured approach enhances the organization's ability to allocate resources efficiently and respond to the most critical threats, improving its overall security posture.

Control 2: Scenario Analysis and Threat Modeling Program

Control Objective: To utilize scenario analysis and threat modeling as proactive tools for identifying potential threats and assessing their impact and likelihood, facilitating a deeper understanding of the organization's threat landscape.

Implementation Steps:

2.1 **Program Establishment**: Establish a scenario analysis and threat modeling program with regular sessions focused on current and emerging threats. This program should involve cross-functional teams to cover various perspectives and expertise.

2.2 **Scenario Development and Analysis**: Develop realistic threat scenarios based on recent incidents, intelligence reports, and hypothetical situations. Analyze these scenarios to assess the potential impact and likelihood of threats materializing.

2.3 **Integration with Risk Management**: Integrate findings from scenario analysis and threat modeling into the broader risk management process, using these insights to inform risk assessments, mitigation strategies, and incident response planning.

Expected Outcome: Implementing a scenario analysis and threat modeling program provides the organization with a proactive mechanism for understanding potential threats in a structured and analytical manner. This control contributes to more effective risk management by enabling the organization to anticipate and prepare for potential threats before they occur.

Control 3: Continuous Impact and Likelihood Assessment Review Process

Control Objective: To ensure that the organization's approach to assessing the impact and likelihood of threats remains relevant and effective in the face of an evolving threat landscape through continuous review and updates.

Implementation Steps:

3.1 **Review Schedule Establishment**: Establish a regular review schedule for evaluating the organization's impact and likelihood assessment processes, ensuring these reviews occur at least annually or in response to significant changes in the threat landscape.

3.2 **Feedback Mechanism Implementation**: Implement a feedback mechanism that allows stakeholders to provide input on the effectiveness and applicability of the current impact and likelihood assessment methodologies.

3.3 **Process Updates**: Based on the review findings and stakeholder feedback, make necessary adjustments to the assessment methodologies and processes to remain aligned with current best practices and the organization's risk profile.

Expected Outcome: Establishing a continuous review process for impact and likelihood assessments ensures the organization's risk evaluation methods stay current and effective. This adaptability enhances the organization's resilience against new and evolving threats by ensuring that risk management strategies are based on accurate and up-to-date assessments.

ID.RA-05: Threats, Vulnerabilities, Likelihoods, and Impacts Are Used to Understand Inherent Risk and Inform Risk Response Prioritization

Inherent risk assessment and prioritization are the bedrock of an organization's cybersecurity risk management program. This initial step involves a comprehensive analysis of potential threats and vulnerabilities, absent of any mitigating controls, to understand the raw exposure of the organization to cyber threats. Such an assessment illuminates the pure risk landscape, enabling cybersecurity professionals to quantify and prioritize risks based on their potential impact on the organization. Emphasizing inherent risk helps in crafting both efficient and effective strategies, ensuring that cybersecurity efforts are aligned with the organization's overall risk management objectives.

Integrating threat and vulnerability analysis into risk assessment requires a structured framework that systematically identifies, assesses, and combines various cyber risk components. This integration is pivotal in developing a holistic view of an organization's cybersecurity threats. Through this comprehensive approach, organizations can correlate specific vulnerabilities with potential threats, assess the likelihood of those threats being realized, and evaluate the possible impacts. Such a framework ensures that risk assessments are grounded in the organization's threat landscape and technological environment.

Calculating inherent risk based on threats, vulnerabilities, and impacts involves quantifying cybersecurity incidents' potential severity and likelihood. This calculation often leverages methodologies incorporating qualitative and quantitative analysis, enabling organizations to assign risk scores to various scenarios. These scores facilitate the comparison and prioritization of risks, guiding decision-makers in allocating resources effectively. Through this process, organizations gain insights into which risks pose the greatest threat to their operations and can tailor their risk response strategies accordingly.

Risk matrices and models for prioritizing risk responses are critical tools in the risk management arsenal. These tools provide a visual representation of the risk landscape, mapping the likelihood of an event against its potential impact. By categorizing risks, organizations can quickly identify high-priority areas that require immediate attention. Risk matrices and models are particularly useful for communicating complex risk information in an accessible manner, supporting strategic decision-making across the organization.

Developing a risk register for tracking and management is essential for maintaining an organized and actionable inventory of identified risks. This register is a central repository for all risk-related information, including risk scores, potential impacts, mitigation strategies, and responsible parties. By systematically documenting and tracking risks, organizations can ensure that risk management activities are coordinated and that accountability is maintained. Furthermore, a well-maintained risk register is invaluable for reporting risk status and progress to stakeholders.

Criteria for decision-making in risk response selection are vital for ensuring that responses are both appropriate and proportional to the risks faced. These criteria should consider factors such as the cost-effectiveness of mitigation measures, the impact on business operations, and regulatory requirements. Establishing clear criteria supports a consistent and rational approach to selecting risk responses, ensuring that cybersecurity measures align with the organization's risk appetite and business objectives.

Alignment of risk prioritization with organizational risk appetite is crucial for ensuring that cybersecurity efforts support broader business goals. Risk appetite defines the level of risk an organization is willing to accept in pursuit of its objectives. By aligning risk prioritization efforts with this appetite, cybersecurity professionals can ensure that their strategies are in harmony with the organization's overall tolerance for risk. This alignment is key to balancing security measures with operational efficiency and strategic ambitions.

Communication strategies for risk prioritization findings are necessary to ensure that all relevant stakeholders are informed and engaged in the risk management process. Effective communication involves clear, concise, and targeted messaging that conveys the significance of risk findings and the rationale behind prioritization decisions. By fostering an open dialogue around cybersecurity risks, organizations can build a culture of risk awareness and support collaborative efforts to enhance resilience against cyber threats.

Recommendations

- **Leverage Comprehensive Risk Assessment Frameworks**: Adopt structured frameworks for integrating threat and vulnerability analysis into risk assessment, ensuring a holistic view of the cybersecurity threat landscape.
- **Implement Quantitative and Qualitative Risk Analysis**: Utilize quantitative and qualitative methodologies for calculating inherent risk, enabling accurate risk scoring and prioritization.
- **Utilize Risk Matrices and Models**: Employ risk matrices and models to visually map and prioritize risk responses, facilitating strategic decision-making and communication.
- **Maintain an Organized Risk Register**: Develop and maintain a risk register for tracking and managing cybersecurity risks, ensuring coordinated risk management activities and accountability.

- **Align Risk Management with Organizational Risk Appetite**: Ensure that risk prioritization and response strategies align with the organization's defined risk appetite, supporting business objectives while managing cybersecurity risks effectively.

NIST 800-53 Controls

(Refer to our appendix for a comprehensive description of the NIST 800-53 Controls, essential for organizations adhering to strict regulatory standards.)

- PM-16,RA-02,RA-03,RA-07

Simplified Security Controls (SSC)

This section outlines our SSC, tailored for environments beyond the scope of NIST compliance. These controls offer streamlined, effective security solutions suitable for any organization seeking to enhance its cybersecurity defenses with clear, actionable measures.

Control 1: Integrated Risk Management Platform Deployment

Control Objective: To deploy an integrated risk management platform that consolidates data on threats, vulnerabilities, likelihoods, and impacts to facilitate comprehensive inherent risk assessment and inform risk response prioritization accurately.

Implementation Steps:

1.1 **Platform Selection**: Choose a risk management platform that integrates with existing threat intelligence, vulnerability management systems, and organizational databases to effectively aggregate and analyze risk data.

1.2 **Custom Configuration and Implementation**: Configure the platform to align with the organization's specific risk assessment criteria, including the ability to calculate inherent risk scores and prioritize risks based on customized matrices and models.

1.3 **Staff Training and Process Integration**: Train relevant staff on using the platform and integrate platform use into the organization's standard risk assessment and prioritization processes to ensure consistent application across all departments.

Expected Outcome: Deploying an integrated risk management platform streamlines the inherent risk assessment process, providing a centralized view of all risk data. This facilitates informed decision-making and ensures that risk response prioritization is aligned with the organization's risk appetite, thereby enhancing the overall efficiency of the risk management program.

Control 2: Development of a Dynamic Risk Register

Control Objective: To create and maintain a dynamic risk register as a living document for tracking and managing identified risks, their assessment scores, mitigation actions, and monitoring status.

Implementation Steps:

2.1 **Risk Register Creation**: Develop a risk register template that includes fields for risk description, inherent risk score (based on threats, vulnerabilities, likelihoods, and impacts), chosen risk responses, and status of mitigation actions.

2.2 **Regular Update Procedures**: Establish procedures for regularly updating the risk register with new risks, changes in risk scores following reassessment, and progress on risk mitigation actions.

2.3 **Accessibility and Accountability**: Ensure the risk register is accessible to all relevant stakeholders and assign clear accountability for updating the register and monitoring risk mitigation progress.

Expected Outcome: Creating and regularly maintaining a dynamic risk register enhances the organization's ability to track and manage risks systematically. This ensures continuous visibility into the risk landscape and the progress of mitigation efforts, supporting effective risk management and decision-making processes.

Control 3: Risk Appetite Alignment Workshops

Control Objective: To ensure that risk prioritization and response strategies are aligned with the organization's defined risk appetite through regular workshops involving key stakeholders across the organization.

Implementation Steps:

3.1 **Workshop Planning**: Organize regular workshops with key stakeholders from various departments, including executive management, IT security, risk management, and business units, to discuss and align on the organization's risk appetite.

3.2 **Alignment Activities**: During the workshops, engage in activities that compare current risk response strategies with the organization's risk appetite, identifying areas where adjustments are needed to better align risk-taking with strategic objectives.

3.3 **Documentation and Implementation of Adjustments**: Document outcomes of the workshops, including any decisions on adjusting risk response strategies or the risk appetite itself, and implement these adjustments across the organization.

Expected Outcome: Regular risk appetite alignment workshops ensure that the organization's risk response strategies remain aligned with its strategic objectives and risk-taking capacity. This fosters a risk-aware culture and enhances the risk management program's effectiveness by ensuring that resources are allocated to mitigating risks in a manner consistent with the organization's overall risk appetite.

ID.RA-06: Risk Responses Are Chosen, Prioritized, Planned, Tracked, and Communicated

Developing a comprehensive risk response strategy is integral to an organization's overall cybersecurity framework. This process involves identifying, evaluating, and implementing measures based on their assessment to mitigate, transfer, accept, or avoid risks. The essence of a robust risk response strategy lies in its alignment with organizational objectives and its capacity to protect the organization's critical assets while ensuring operational continuity. A well-formulated strategy provides a structured approach to managing cybersecurity risks, enabling organizations to respond swiftly and effectively to threats.

The criteria for selecting appropriate risk responses are multifaceted, encompassing factors such as the severity of the risk, the cost of implementation, the potential impact on business operations, and compliance requirements. These criteria ensure that the chosen responses are feasible and effective in reducing risk to an acceptable level. Decisions regarding risk responses should be made in consultation with stakeholders across the organization, including IT, security, business units, and executive management, to ensure a comprehensive understanding of the risk landscape and to foster organization-wide support for mitigation strategies.

Risk mitigation measures' planning and implementation processes require careful coordination and project management. This involves outlining the steps to implement each response, assigning responsibilities, setting deadlines, and allocating resources. Effective planning ensures that risk responses are executed efficiently and minimizes potential disruptions to business operations. Implementation may involve technical measures, such as deploying new security controls, or administrative actions, such as revising policies and procedures.

Prioritizing risk responses based on organizational objectives ensures that resources are focused on addressing the most critical risks first. This prioritization is guided by assessing each risk's potential impact on the organization's strategic goals, financial health, and operational efficiency. By aligning risk response efforts with organizational priorities, security leaders can ensure that their strategies contribute directly to the organization's resilience and long-term success.

Employing project management principles in tracking risk response progress is essential for ensuring that initiatives are completed on time and within budget. These principles provide a framework for monitoring the implementation of risk responses, identifying any deviations from the plan, and making necessary adjustments. Regular status updates and reviews help keep stakeholders informed of progress and facilitate the early detection of issues that could impede successful implementation.

Communication plans for risk response actions and outcomes are crucial for maintaining transparency and building stakeholder trust. Effective communication involves regularly updating all relevant parties on the status of risk mitigation efforts, changes to the risk landscape, and any incidents that occur. By keeping stakeholders informed, organizations can foster a culture of security awareness and ensure that risk management is perceived as a shared responsibility.

Integrating risk responses into organizational policies and procedures ensures that risk management practices are embedded into the organization's fabric. This integration helps standardize responses across the organization, making managing risks consistently and effectively easier. It also ensures that risk management considerations are incorporated into decision-making, from strategic planning to day-to-day operations.

Mechanisms for monitoring and reviewing the effectiveness of risk responses are vital for ensuring that risk management efforts achieve their intended outcomes. This involves assessing whether risk responses have effectively reduced risks to acceptable levels and identifying new or evolving risks that may require attention. Continuous monitoring and periodic reviews allow organizations to adapt their risk management strategies to the changing threat landscape, ensuring ongoing resilience against cyber threats.

Recommendations

- **Develop Clear Criteria for Response Selection**: Establish and document specific criteria for selecting risk responses, ensuring that these criteria align with organizational objectives and the severity of risks.
- **Implement Project Management for Risk Initiatives**: Utilize project management principles to oversee the planning, implementation, and tracking of risk mitigation measures, ensuring accountability and efficiency.
- **Prioritize Responses to Align with Objectives**: Ensure risk response prioritization is closely aligned with organizational goals, focusing on protecting critical assets and functions.
- **Foster Transparent Communication**: Create comprehensive communication plans to keep stakeholders informed about risk management activities, fostering a culture of awareness and collaboration.
- **Integrate and Review Risk Management Practices**: Regularly integrate risk responses into organizational policies and conduct reviews to assess their effectiveness, making adjustments as necessary to address new challenges and improve resilience.

NIST 800-53 Controls

(Refer to our appendix for a comprehensive description of the NIST 800-53 Controls, essential for organizations adhering to strict regulatory standards.)

- PM-09,PM-18,PM-30,RA-07

Simplified Security Controls (SSC)

This section outlines our SSC, tailored for environments beyond the scope of NIST compliance. These controls offer streamlined, effective security solutions suitable for any organization seeking to enhance its cybersecurity defenses with clear, actionable measures.

Control 1: Risk Response Integration System

Control Objective: To implement a comprehensive system that ensures risk responses are consistently integrated into organizational policies, procedures, and operations, aligning risk mitigation efforts with overall business strategy.

Implementation Steps:

1.1 **System Development and Configuration**: Develop and configure a centralized risk response integration system that can catalog, monitor, and manage the implementation of risk responses across various organizational departments.

1.2 **Policy and Procedure Updates**: Utilize the system to identify areas where organizational policies and procedures must be updated to incorporate chosen risk responses, ensuring that changes are systematically applied.

1.3 **Continuous Monitoring and Reporting**: Establish monitoring capabilities within the system to track the implementation status of risk responses and generate reports for senior management review, highlighting progress and any areas requiring attention.

Expected Outcome: The Risk Response Integration System ensures that all chosen risk responses are effectively integrated into the organization's policies and operations. This systematizes updating organizational practices to reflect risk mitigation measures, enhancing the overall coherence and effectiveness of the organization's risk management strategy.

Control 2: Dynamic Risk Response Training Program

Control Objective: To establish an ongoing training program that equips employees with the necessary knowledge and skills to understand their roles and responsibilities in implementing risk responses, enhancing the organization's overall capability.

Implementation Steps:

2.1 **Training Program Development**: Create a comprehensive training program that covers the principles of risk management, specific risk responses adopted by the organization, and the roles and responsibilities of employees in enacting these responses.

2.2 **Role-Specific Training Modules:** Develop role-specific training modules that address the unique responsibilities of different departments and teams in responding to risks, ensuring that training is relevant and practical.

2.3 **Regular Training Sessions and Updates**: Schedule regular training sessions and update the program content as risk management strategies evolve to maintain staff awareness and competence in risk response processes.

Expected Outcome: The Dynamic Risk Response Training Program ensures that all employees know their roles in risk response and are equipped with the skills needed to implement those responses effectively. This broadens the organization's risk management capacity and fosters a proactive risk-aware culture.

Control 3: Risk Response Prioritization Framework

Control Objective: To prioritize risk responses effectively based on their alignment with organizational objectives and the potential impact on the organization's operations and strategic goals.

Implementation Steps:

3.1 **Framework Development**: Develop a risk response prioritization framework that categorizes risks and their responses based on criteria such as impact, likelihood, resource requirements, and alignment with organizational objectives.

3.2 **Application of the Framework**: Apply the framework to all identified risks and their proposed responses, ensuring that prioritization decisions are structured and consistent.

3.3 **Review and Adjustment Process**: Establish a regular review process for the prioritization framework to ensure its ongoing relevance and effectiveness, allowing for adjustments in response to organizational or external environment changes.

Expected Outcome: The Risk Response Prioritization Framework ensures that risk responses are selected and implemented to best support the organization's strategic objectives and operational needs. This approach facilitates the efficient allocation of resources to the areas of highest priority, enhancing the organization's ability to manage risk effectively.

ID.RA-07: Changes and Exceptions Are Managed, Assessed for Risk Impact, Recorded, and Tracked

Managing changes and exceptions within an organization's IT environment is critical to a robust cybersecurity strategy. While necessary for business growth and innovation, introducing new technologies, updates, or modifications to existing systems introduces potential vulnerabilities that adversaries could exploit. A systematic approach to change and exception management is essential, encompassing thorough risk assessments, meticulous documentation, and effective communication strategies.

Managing changes and exceptions begins with clearly understanding the potential risk impact of proposed changes. This involves conducting comprehensive risk assessments that evaluate the security implications of each change or exception. Such assessments are crucial for identifying vulnerabilities that could compromise the organization's information security posture. By systematically evaluating the risks associated with changes, organizations can make informed decisions about implementing, modifying, or rejecting proposed changes.

Documentation and tracking mechanisms play a pivotal role in change and exception management. These processes ensure that all changes and associated risk assessments and decisions are accurately recorded and easily accessible. This not only facilitates accountability and traceability but also aids in continuously monitoring and managing changes throughout their lifecycle. Effective documentation practices are foundational to maintaining an audit trail that supports compliance with regulatory requirements and internal governance standards.

The criteria for approving or rejecting change requests based on risk impact must be clearly defined and communicated within the organization. These criteria should align with the organization's risk appetite and cybersecurity policies, ensuring that only changes that meet acceptable risk thresholds are implemented. This decision-making process should be transparent, involving stakeholders from relevant departments to ensure that all potential impacts are considered.

Integrating change management with IT governance frameworks is essential for ensuring that changes are aligned with the organization's overall IT strategy and risk management objectives.

This integration promotes consistency in how changes are evaluated, implemented, and reviewed, ensuring that cybersecurity considerations are embedded in all aspects of IT governance.

Communication strategies are critical for effectively conveying change decisions and rationales to all affected stakeholders, including IT staff, business units, and external partners. Clear and timely communication helps manage expectations, facilitate smooth implementation, and minimize disruptions to business operations. It also supports a culture of transparency and collaboration, which is essential to effective change management.

The review and postimplementation analysis of change impacts are vital for assessing the effectiveness of change management processes and identifying opportunities for improvement. This analysis should examine whether the implemented changes achieved their objectives without introducing unanticipated vulnerabilities or risks. Lessons learned from this review process should inform future change management practices, contributing to improving the organization's cybersecurity posture.

Training on change management policies and risk assessment procedures ensures that all personnel initiating, evaluating, or implementing changes have the necessary knowledge and skills. Regular training sessions should cover the organization's change management framework, risk assessment techniques, and the roles and responsibilities of individuals in the change management process.

- **Implement a Standardized Change Management Toolkit**: Adopt a standardized set of tools and templates for managing changes and exceptions. This includes risk assessment templates, change request forms, and tracking logs to ensure consistency and efficiency in the change management process.
- **Foster a Culture of Continuous Improvement**: Encourage a culture that values continuous improvement in change and exception management practices. This can be achieved through regular training, postimplementation reviews, and incorporating feedback mechanisms for refining processes.
- **Enhance Stakeholder Communication**: Develop comprehensive communication plans that outline how changes and their potential impacts are communicated to stakeholders. This should include mechanisms for gathering feedback and addressing concerns throughout the change lifecycle.
- **Leverage Technology for Automation**: Utilize technology solutions to automate aspects of the change management process, such as risk assessments, documentation, and tracking. Automation can improve accuracy, reduce manual errors, and save resources for more critical tasks.

NIST 800-53 Controls

(Refer to our appendix for a comprehensive description of the NIST 800-53 Controls, essential for organizations adhering to strict regulatory standards.)

- CA-07,CM-03,CM-04

Simplified Security Controls (SSC)

This section outlines our SSC, tailored for environments beyond the scope of NIST compliance. These controls offer streamlined, effective security solutions suitable for any organization seeking to enhance its cybersecurity defenses with clear, actionable measures.

Control 1: Automated Change Management and Risk Assessment System

Control Objective: To automate managing, assessing, documenting, and tracking changes and exceptions within the organization's IT environment, ensuring a systematic approach to risk management.

Implementation Steps:

1.1 **System Implementation**: Deploy an automated change management system that integrates with the organization's risk assessment tools, allowing for seamless evaluation of the risk impact of proposed changes and exceptions.

1.2 **Configuration for Risk Analysis**: Configure the system to automatically perform an initial risk assessment for each proposed change or exception based on predefined criteria, such as the sensitivity of affected assets and the nature of the change.

1.3 **Documentation and Tracking**: Ensure the system can document all change requests, risk assessments, decisions, and rationales and provide robust tracking functionalities for monitoring change implementation and outcomes.

Expected Outcome: Deploying an automated change management and risk assessment system streamlines the change management process, ensuring that all changes and exceptions are assessed for risk impact consistently and efficiently. This control minimizes the risk of unintended consequences from changes, enhancing the organization's overall security posture.

Control 2: Change Advisory Board (CAB) Formation

Control Objective: To establish a CAB responsible for reviewing, assessing, and making informed decisions on change requests based on their potential risk impact.

Implementation Steps:

2.1 **Board Formation**: Form a CAB comprising members from various departments, including IT, security, compliance, and business operations, to ensure a multidisciplinary approach to change management and risk assessment.

2.2 **Define Review Process**: Develop a structured process for reviewing change requests submitted to the CAB, including criteria for risk assessment, decision-making guidelines, and timelines for response.

2.3 **Postimplementation Review**: Incorporate a mandatory postimplementation review process to assess the actual impact of changes, validate the effectiveness of risk mitigation measures, and identify lessons learned.

Expected Outcome: The formation of a CAB ensures that changes and exceptions are evaluated by a multidisciplinary team, facilitating informed decision-making that balances operational needs with risk management objectives. This control also enhances organizational learning by systematically reviewing the outcomes of change implementations.

Control 3: Comprehensive Change Management Training Program

Control Objective: To ensure that all employees involved in initiating, reviewing, or implementing changes are adequately trained on change management policies, risk assessment procedures, and their specific roles and responsibilities.

Implementation Steps:

3.1 **Training Program Development**: Develop a comprehensive training program covering the organization's change management framework, risk assessment methodologies, documentation requirements, and adherence to established processes.

3.2 **Role-Specific Training Modules**: Create role-specific training modules to address the unique responsibilities of different stakeholders in the change management process, ensuring that each participant understands their role in managing risks associated with changes.

3.3 **Regular Training Sessions and Updates**: Conduct training sessions for new employees and refresher courses for existing staff, updating the program to reflect changes in policies, technologies, or organizational objectives.

Expected Outcome: Implementing a comprehensive change management training program ensures that all stakeholders are knowledgeable about the change management process and their roles within it. This control promotes a culture of risk awareness and compliance with change management policies, reducing the likelihood of security incidents resulting from poorly managed changes or exceptions.

ID.RA-08: Processes for Receiving, Analyzing, and Responding to Vulnerability Disclosures Are Established

Vulnerability disclosure management is critical to a robust cybersecurity strategy, ensuring that potential weaknesses in systems and applications are identified, assessed, and mitigated efficiently. An introduction to this discipline highlights the need for organizations to establish structured processes for engaging with external parties reporting vulnerabilities. This proactive approach not only enhances an organization's security posture but also fosters a collaborative relationship with the cybersecurity community. Establishing a formal VDP is the foundation for this endeavor, setting clear guidelines for reporting, evaluating, and addressing vulnerabilities.

Establishing a Formal VDP marks a significant step toward institutionalizing the vulnerability management process. A well-defined VDP outlines the mechanisms for external researchers to submit reports securely, the organization's commitment to addressing these reports, and the expected timelines for response. This clarity encourages responsible disclosure practices, providing a safe and structured channel for security researchers and the public to contribute to the organization's cybersecurity efforts.

Procedures for securely receiving and handling vulnerability reports are essential to an effective VDP. These procedures should ensure the confidentiality, integrity, and availability of the information submitted, protecting both the reporter's details and the details of the vulnerability itself. Employing secure communication channels, such as encrypted email or a dedicated reporting portal, is crucial for maintaining the trust of the reporting community and safeguarding sensitive information.

Analysis and validation processes for reported vulnerabilities are critical for determining their legitimacy and potential impact. This step involves replicating the reported vulnerability under controlled conditions to confirm its existence and assess the severity of the threat it poses. Such rigorous evaluation is necessary to prioritize response efforts and allocate resources effectively, ensuring that the most critical vulnerabilities are addressed promptly.

Coordination with external researchers and reporting entities underscores the collaborative nature of vulnerability management. Establishing a constructive dialogue with those who report vulnerabilities fosters a positive relationship with the cybersecurity community, encouraging ongoing contributions. This collaboration can also provide insights into emerging threat trends and attacker techniques, enhancing the organization's defensive capabilities.

Developing response plans for confirmed vulnerabilities is a systematic approach to mitigating risks associated with identified security weaknesses. These plans should outline the steps for patch development, testing, and deployment and any temporary measures to protect against exploitation in the interim. Effective response planning ensures that vulnerabilities are addressed promptly and efficiently, minimizing potential impacts on the organization and its stakeholders.

Communication strategies for disclosure and patch releases are vital in vulnerability management. Timely and transparent communication about vulnerabilities and the steps taken to address them is essential for maintaining stakeholder trust. This includes the immediate response to a reported vulnerability, public advisories, and patch release announcements that inform a broader audience about resolving security issues.

Integration of VDP outcomes with continuous improvement processes ensures that lessons learned from handling vulnerability disclosures are used to enhance future security practices. This integration can lead to refinements in the VDP itself, software development practices, and enhancements in security monitoring and response capabilities. Organizations can continuously evolve their cybersecurity defenses by treating vulnerability disclosure as an opportunity for learning and growth.

Recommendations

- **Establish a Comprehensive VDP**: Create a formal VDP that clearly outlines report submission, evaluation, and response procedures. This program should provide secure reporting channels and establish expectations for reporters and the organization.
- **Foster Open Communication**: Develop transparent communication strategies for engaging with external researchers and the public regarding vulnerability disclosures and patch releases. This openness builds trust and encourages more security researchers to participate in the VDP.
- **Prioritize Validation and Response Processes**: Implement rigorous processes for analyzing, validating, and prioritizing reported vulnerabilities to ensure that the most critical issues are addressed promptly and effectively.
- **Engage in Continuous Collaboration**: Maintain ongoing collaboration with the cybersecurity community to benefit from external expertise and insights, enhancing the organization's ability to identify and mitigate emerging threats.
- **Leverage Vulnerability Disclosures for Improvement**: Integrate lessons learned from vulnerability disclosures into continuous improvement processes for cybersecurity practices, including refinements to the VDP, enhancements in development practices, and improvements in security monitoring and response strategies.

NIST 800-53 Controls

(Refer to our appendix for a comprehensive description of the NIST 800-53 Controls, essential for organizations adhering to strict regulatory standards.)

- RA-05

Simplified Security Controls (SSC)

This section outlines our SSC, tailored for environments beyond the scope of NIST compliance. These controls offer streamlined, effective security solutions suitable for any organization seeking to enhance its cybersecurity defenses with clear, actionable measures.

Control 1: Secure Vulnerability Disclosure Portal

Control Objective: To provide a secure, accessible channel for external researchers and the public to report vulnerabilities, ensuring confidentiality and integrity of the information shared.

Implementation Steps:

1.1 **Portal Development and Deployment**: Develop and deploy a secure web-based vulnerability disclosure portal with encryption to protect the confidentiality of submissions. The portal should include clear report submission instructions and outline the organization's commitment to addressing vulnerabilities.

1.2 **Authentication and Access Control**: Implement robust authentication mechanisms and access controls to ensure that only authorized personnel can access the vulnerability reports submitted through the portal.

1.3 **Acknowledgment and Feedback Mechanism**: Set up an automated acknowledgment system for received reports and establish a process for providing periodic feedback to reporters on the status of their submissions.

Expected Outcome: Establishing a secure vulnerability disclosure portal enhances the organization's ability to receive detailed and confidential reports on potential vulnerabilities. This facilitates a streamlined communication process with external researchers, encouraging responsible disclosure and strengthening the organization's cybersecurity posture.

Control 2: Vulnerability Response and Patch Management Team

Control Objective: To ensure timely and effective response to reported vulnerabilities by establishing a dedicated team responsible for analyzing, prioritizing, and coordinating vulnerability remediation efforts.

Implementation Steps:

2.1 **Team Formation**: Form a cross-functional team comprising members from IT security, software development, and operations. This team assesses reported vulnerabilities, develops response plans, and oversees the patch management process.

2.2 **Standard Operating Procedures (SOPs) Development**: Develop SOPs for the vulnerability response process, including assessment, prioritization based on severity and potential impact, and developing remediation plans or patches.

2.3 **Coordination with Affected Parties**: Establish protocols for coordinating with internal teams and external entities affected by the vulnerability to ensure consistent communication and comprehensive resolution efforts.

Expected Outcome: Creating a Vulnerability Response and Patch Management Team equipped with clear SOPs ensures that vulnerabilities are addressed promptly and effectively. This control minimizes the exposure window to potential threats, reducing the risk of exploitation and enhancing system security.

Control 3: Continuous Improvement Program for Vulnerability Management

Control Objective: To integrate lessons learned from handling vulnerability disclosures into the organization's cybersecurity practices, promoting continuous improvement in vulnerability management processes.

Implementation Steps:

3.1 **Review and Analysis of Vulnerability Cases**: Regularly review resolved vulnerability cases to identify trends, assess the effectiveness of response strategies, and pinpoint areas for process improvement.

3.2 **Integration with Risk Management and Security Policies**: Use insights from case reviews to update risk management strategies and security policies, ensuring that they reflect current best practices and the evolving threat landscape.

3.3 **Stakeholder Engagement and Training**: Engage with organizational stakeholders to communicate changes and conduct training sessions to ensure that all relevant personnel are informed about updated vulnerability management practices.

Expected Outcome: Implementing a Continuous Improvement Program for Vulnerability Management ensures that the organization's handling of vulnerability disclosures is regularly refined based on experience. This proactive approach fosters a culture of learning and adaptation, leading to more robust defense mechanisms against emerging cybersecurity threats.

ID.RA-09: The Authenticity and Integrity of Hardware and Software Are Assessed Before Acquisition and Use

Assessing the authenticity and integrity of hardware and software before their acquisition and use is a pivotal step in safeguarding an organization's information technology landscape. This introduction to preacquisition assessment emphasizes the importance of verifying that IT assets originate from reputable sources and are free from tampering or malicious alterations. Such assessments are crucial for preventing the introduction of vulnerabilities and ensuring that the organization's IT infrastructure remains secure and trustworthy. A systematic approach to evaluating the security attributes of IT assets before they are integrated into the operational environment can significantly mitigate potential risks.

Criteria for evaluating hardware and software authenticity include verifying the source of the products, ensuring they have not been modified from their original state, and confirming that they meet the organization's security standards. Authenticity checks can prevent counterfeit or compromised devices and software from entering the organization's IT ecosystem, which could introduce security vulnerabilities or facilitate unauthorized access. Establishing clear criteria for these evaluations helps maintain the integrity of the IT environment and supports compliance with regulatory requirements and industry best practices.

Processes for ensuring the integrity of IT assets before deployment involve rigorous inspection and testing to detect any tampering or unauthorized modifications. This might include verifying digital signatures, conducting checksum validations, and ensuring that the hardware has not been physically altered. Such processes are essential for confirming that the assets have not been compromised during manufacturing, transit, or storage and are safe to integrate into the organization's IT infrastructure.

Vendor assurance and security certification requirements play a significant role in the preacquisition assessment of IT assets. Organizations should seek vendors that adhere to recognized security standards and can provide assurance regarding the security measures implemented in developing and distributing their products. Security certifications, such as ISO/IEC 27001 for information security management or specific product certifications, offer valuable evidence of a vendor's commitment to security.

Risk assessment for third-party hardware and software involves analyzing the potential risks of integrating these assets into the organization's IT environment. This includes evaluating the vendor's security practices, the potential impact of vulnerabilities, and the feasibility of implementing adequate controls. A thorough risk assessment helps organizations make informed decisions about acquiring third-party IT assets, balancing the benefits against the potential security implications.

Documentation and record-keeping for asset assessments are essential for maintaining a detailed history of the security evaluations conducted on IT assets. This documentation should include the assessment criteria, methodologies, findings, and any remediation actions taken. Proper record-keeping supports ongoing risk management and facilitates compliance with regulatory and industry standards.

Integrating assessment findings with procurement policies ensures that security considerations are embedded in the organization's procurement processes. This integration helps institutionalize security as a core component of the procurement lifecycle, from vendor selection to asset disposal. By aligning procurement policies with the outcomes of security assessments, organizations can ensure a consistent and comprehensive approach to securing their IT assets.

Training on secure procurement and asset onboarding practices is vital for ensuring that staff involved in acquiring and deploying IT assets are aware of the security risks and the measures necessary to mitigate them. This training should cover the assessment criteria, the importance of vendor assurance, and best practices for securely integrating new assets into the organization's IT environment.

Recommendations

- **Establish Comprehensive Assessment Criteria**: Develop detailed criteria for evaluating the authenticity and integrity of hardware and software, ensuring that these criteria are aligned with industry best practices and regulatory requirements.
- **Strengthen Vendor Assurance Processes**: Implement robust processes to assess vendor security practices and ensure that all procured IT assets meet the organization's security certification requirements.
- **Document and Maintain Assessment Records**: Keep meticulous records of all security assessments conducted on IT assets, including the methodologies used, findings, and actions taken in response.
- **Integrate Security into Procurement Policies**: Ensure that security assessments are integral to the procurement process, from vendor selection to asset disposal to embedding security considerations throughout the procurement lifecycle.
- **Provide Training on Secure Procurement Practices**: Offer comprehensive training for all staff involved in procuring and deploying IT assets, emphasizing the importance of security assessments and secure integration practices.

NIST 800-53 Controls

(Refer to our appendix for a comprehensive description of the NIST 800-53 Controls, essential for organizations adhering to strict regulatory standards.)

- SA-04,SA-05,SA-10,SA-11,SA-15,SA-17,SI-07,SR-05,SR-06,SR-10,SR-11

Simplified Security Controls (SSC)

This section outlines our SSC, tailored for environments beyond the scope of NIST compliance. These controls offer streamlined, effective security solutions suitable for any organization seeking to enhance its cybersecurity defenses with clear, actionable measures.

Control 1: Vendor Assurance and Security Certification Verification Process

Control Objective: To verify the authenticity and assess the security of hardware and software through a rigorous vendor assurance and security certification verification process before acquisition.

Implementation Steps:

1.1 **Establish Verification Criteria**: Develop criteria for vendor assurance that include verification of business legitimacy, security certifications (e.g., ISO/IEC 27001 and SOC 2), and compliance with industry standards.

1.2 **Conduct Vendor Assessments**: Perform thorough assessments of potential vendors against the established criteria, including reviews of security documentation, third-party audits, and certification statuses.

1.3 **Continuous Vendor Performance Monitoring**: Implement a process for the ongoing monitoring of vendor security practices and performance, including regular reviews of vendor-supplied security updates and changes in certification status.

Expected Outcome: Implementing a vendor assurance and security certification verification process ensures that only hardware and software from reputable and secure sources are acquired. This control significantly reduces the risk of introducing vulnerabilities into the organization's IT environment through compromised or counterfeit assets.

Control 2: Predeployment Integrity Verification System

Control Objective: To ensure the integrity of IT assets before deployment into the operational environment by establishing a system for their verification.

Implementation Steps:

2.1 **Integrity Check Protocols**: Develop and document protocols for conducting integrity checks on all new hardware and software, including checksum verification, digital signature validation, and physical inspection procedures.

2.2 **Automated Verification Tools**: Deploy automated tools capable of performing integrity checks against predefined baselines and known good configurations for software and hardware components.

2.3 **Incident Response for Integrity Failures**: Establish an incident response protocol for scenarios where IT assets fail integrity checks, including isolating the affected assets and investigation procedures.

Expected Outcome: Establishing a predeployment integrity verification system ensures that all IT assets are verified as unaltered and secure before integration into the organization's network. This control minimizes the risk of deploying compromised assets, enhancing the overall security posture.

Control 3: Integration of IT Asset Security Assessment with Procurement Processes

Control Objective: To integrate security assessments of IT assets into the procurement process, ensuring that security considerations are central to acquisition decisions.

Implementation Steps:

3.1 **Update Procurement Policies**: Revise procurement policies to require security assessments as part of the evaluation process for all IT hardware and software acquisitions.

3.2 **Cross-Departmental Collaboration**: Foster collaboration between procurement, IT security, and other relevant departments to ensure that security assessments are conducted efficiently and are considered in procurement decisions.

3.3 **Supplier Security Requirements**: Define and communicate security requirements to suppliers, including transparency in security practices and providing necessary documentation for assessment purposes.

Expected Outcome: Integrating IT asset security assessments with procurement processes ensures that security considerations are prioritized in acquisition decisions. This control leads to procuring more secure IT assets, reducing the risk of vulnerabilities introduced through new acquisitions, and enhancing the organization's overall security infrastructure.

ID.RA-10: Critical Suppliers Are Assessed Before Acquisition

In the complex ecosystem of modern business operations, the role of suppliers extends far beyond the provision of goods and services; they become integral to the organization's security posture. An introduction to supplier security assessment processes highlights the necessity of evaluating potential suppliers' cybersecurity practices as part of the procurement process. This scrutiny ensures that the organization's supply chain does not introduce unacceptable risks, safeguarding against vulnerabilities that could be exploited through third-party connections.

Developing criteria for identifying and assessing critical suppliers is a foundational step in this evaluation process. These criteria should consider the supplier's access to the organization's systems and data, the criticality of the supplied products or services to the organization's operations, and the potential impact on the organization's security posture. Organizations can prioritize assessments for suppliers that pose the most significant risk by establishing clear and comprehensive criteria.

Conducting due diligence and security assessments for potential suppliers involves thoroughly examining their cybersecurity practices, policies, and histories of security incidents. This evaluation should include assessing the supplier's ability to protect sensitive information, incident response capabilities, and compliance with relevant industry standards and regulations. Such due diligence is crucial for making informed decisions about supplier relationships and mitigating potential security risks.

Integration of supplier assessments with vendor management programs ensures a holistic approach to managing supplier risks. This integration facilitates the continuous monitoring of suppliers' security postures and the enforcement of security requirements throughout the duration of the supplier relationship. By embedding security assessments within broader vendor management processes, organizations can maintain a comprehensive view of supplier risks and implement necessary controls to mitigate those risks.

Risk management considerations in supplier selection and contracting involve evaluating the potential risks associated with a supplier relationship and implementing contractual safeguards to mitigate those risks. This may include requirements for regular security audits, incident reporting protocols, and compliance with specific security standards. By addressing these considerations during the selection and contracting phases, organizations can establish a strong foundation for secure supplier relationships.

Documentation practices for supplier assessments and decisions are critical for maintaining a record of the evaluation process and its outcomes. This documentation should detail the assessment criteria, the findings of the evaluation, and the rationale behind the decision to proceed or not proceed with a supplier. Proper documentation supports accountability and facilitates the review and continuous improvement of the supplier assessment process.

Communication of assessment requirements and findings to suppliers is essential to a transparent and collaborative relationship. Organizations can work with suppliers to address identified vulnerabilities and enhance security measures by clearly articulating security expectations and sharing assessment outcomes. This open dialogue fosters mutual trust and encourages suppliers to prioritize cybersecurity.

Continuous monitoring and review of supplier security postures recognize that cybersecurity is a dynamic field, and the security status of suppliers can change over time. Organizations must implement mechanisms for regularly reassessing their suppliers' security, including reviewing security audits, incident reports, and compliance certifications. Continuous monitoring ensures that organizations can respond promptly to changes in supplier security postures, maintaining the integrity of the supply chain.

Recommendations

- **Establish Comprehensive Assessment Criteria**: Define clear criteria for identifying and evaluating critical suppliers, focusing on the potential impact on the organization's security posture.
- **Conduct Thorough Due Diligence**: Implement rigorous due diligence processes to assess potential suppliers' cybersecurity practices and history of security incidents.
- **Integrate Assessments with Vendor Management**: Embed security assessments within vendor management programs to ensure continuous oversight of supplier risks.
- **Implement Risk-Based Supplier Selection**: Incorporate risk management considerations into supplier selection and contracting, including specific security requirements and audit protocols.
- **Maintain Detailed Documentation**: Develop robust documentation practices for recording the assessment process, findings, and decision-making rationale, supporting transparency and accountability.
- **Foster Open Communication with Suppliers**: Communicate security expectations and assessment findings to suppliers, encouraging collaborative efforts to enhance cybersecurity measures.
- **Adopt Continuous Monitoring Practices**: Establish mechanisms for the ongoing monitoring and review of supplier security postures, adapting to changes and emerging threats in the cybersecurity landscape.
- **Provide Training on Supplier Risk Management**: Offer comprehensive training on supplier risk management and assessment techniques, ensuring that procurement and vendor management staff are equipped to identify and mitigate supplier-related risks effectively.

NIST 800-53 Controls

(Refer to our appendix for a comprehensive description of the NIST 800-53 Controls, essential for organizations adhering to strict regulatory standards.)

- SR-06

Simplified Security Controls (SSC)

This section outlines our SSC, tailored for environments beyond the scope of NIST compliance. These controls offer streamlined, effective security solutions suitable for any organization seeking to enhance its cybersecurity defenses with clear, actionable measures.

Control 1: Critical Supplier Identification and Risk-Based Assessment Framework

Control Objective: To systematically identify critical suppliers and assess their security posture through a risk-based assessment framework, ensuring the security and resilience of the supply chain.

Implementation Steps:

1.1 **Establish Identification Criteria**: Define and document criteria for identifying critical suppliers based on their access to sensitive data, the criticality of supplied products or services, and potential impact on the organization's operations.

1.2 **Develop Risk-Based Assessment Framework**: Create a framework for conducting security assessments of potential and existing suppliers, incorporating factors such as compliance with security standards, past incident history, and security practices.
1.3 **Regular Review and Update**: Schedule periodic reviews of the identification criteria and assessment framework to adapt to changes in the organizational risk profile and emerging threats in the supply chain.

Expected Outcome: Implementation of a critical supplier identification and risk-based assessment framework ensures that suppliers essential to the organization's operations are thoroughly vetted for security risks. This control helps minimize potential vulnerabilities introduced through the supply chain, enhancing overall organizational resilience.

Control 2: Supplier Security Clause in Contracting and Procurement Processes

Control Objective: To integrate security considerations into supplier selection and contracting processes, ensuring suppliers meet the organization's security requirements before formal engagement.

Implementation Steps:

2.1 **Develop Security Clauses**: Draft security clauses to be included in contracts and procurement agreements, specifying requirements for data protection, incident reporting, compliance with security standards, and the right to audit.
2.2 **Integration into Procurement Processes**: Ensure that procurement teams incorporate the security clauses in all requests for proposals and contracts with suppliers, making compliance a condition for engagement.
2.3 **Supplier Acknowledgment and Compliance Verification**: Require potential suppliers to acknowledge and demonstrate compliance with these security requirements in the bidding and contract negotiations.

Expected Outcome: Incorporating security clauses in supplier contracts and procurement processes ensures that all suppliers are contractually obligated to adhere to the organization's security standards. This control mechanism significantly reduces the risk of security breaches from the supply chain and facilitates legal recourse in the event of non-compliance.

Control 3: Continuous Supplier Security Performance Monitoring Program

Control Objective: To establish an ongoing program for monitoring the security performance of suppliers, ensuring continuous compliance with the organization's security requirements.

Implementation Steps:

3.1 **Program Establishment**: Establish a continuous monitoring program that includes regular security assessments, audits, and reviews of supplier security practices and compliance with contractual obligations.
3.2 **Monitoring Tools and Techniques**: Utilize a combination of automated monitoring tools, third-party audits, and self-assessment questionnaires to evaluate supplier security postures on an ongoing basis.
3.3 **Feedback and Remediation Process**: Implement a process for providing feedback to suppliers on security performance and require corrective action plans for any identified deficiencies or non-compliance issues.

Expected Outcome: A continuous supplier security performance monitoring program ensures that suppliers maintain adherence to security requirements throughout their engagement with the organization. This ongoing oversight allows for the early identification and remediation of security risks, enhancing the security integrity of the supply chain.

13

Improvement (ID.IM)

> Adhering to the NIST framework isn't just about compliance; it's a commitment to excellence, requiring a relentless pursuit of improvement to turn the tide against sophisticated threats.

In the evolving cybersecurity landscape, continuous improvement is beneficial and essential for survival and resilience against increasingly sophisticated threats. This principle is embodied in the NIST framework's identification of improvements through rigorous evaluations, security testing, operational process analysis, and the meticulous development and maintenance of incident response plans. Each area serves as a cornerstone for strengthening an organization's cybersecurity posture. From conducting comprehensive evaluations and security exercises to integrating technological advancements and ensuring the effectiveness of incident response plans, the framework emphasizes the importance of a proactive approach. It advocates for the engagement of stakeholders, prioritization based on impact and feasibility, and the establishment of mechanisms for continuous monitoring and updating. Through these practices, organizations can safeguard against current threats and anticipate and prepare for future challenges, ensuring a robust and adaptive cybersecurity strategy.

ID.IM-01: Improvements Are Identified from Evaluations

The critical nature of evaluations in the unending enhancement of cybersecurity measures is profoundly significant. These evaluations form a structured methodology for assessing the adequacy of prevailing cybersecurity efforts, highlighting strengths while identifying areas requiring improvement. By systematically evaluating cybersecurity strategies and incorporating current threats and tactics, techniques, and procedures (TTPs), organizations can maintain pace with the changing digital environment, thereby preempting potential risks. These evaluations serve as an essential feedback loop, enabling the fine-tuning of cybersecurity approaches in light of new findings and evolving threats.

A comprehensive framework for conducting detailed evaluations of cybersecurity practices is crucial for deriving meaningful insights. This framework should include a variety of evaluative methods, such as self-assessments considering current threats and TTPs, third-party assessments, independent audits, and penetration testing. This approach provides a complete picture of an organization's cybersecurity health. Adaptability within this framework is key, allowing for integrating novel evaluative methods as they emerge. This all-encompassing strategy thoroughly scrutinizes all cybersecurity facets, laying a solid foundation for ongoing enhancement.

A Comprehensive Guide to the NIST Cybersecurity Framework 2.0: Strategies, Implementation, and Best Practice, First Edition. Jason Edwards.
© 2025 John Wiley & Sons Ltd. Published 2025 by John Wiley & Sons Ltd.

The critical analysis of evaluation outcomes is fundamental for discerning actionable insights. Analysis techniques should encompass quantitative and qualitative methods, facilitating not just the identification of vulnerabilities but also the understanding of their root causes. Through methodical evaluation result analyses, organizations can craft specific strategies to mitigate identified shortcomings, thus bolstering their cybersecurity framework.

Identifying improvement areas from evaluation findings is pivotal in the continual enhancement cycle. This entails a rigorous review of results to detect exact weaknesses, insufficient controls, or deviations from best practices and compliance standards. This acknowledgment of improvement areas is the preliminary step toward evolving a fortified cybersecurity strategy capable of withstanding contemporary and prospective cyber threats.

Improvement initiatives should be prioritized based on their impact and implementability to ensure judicious resource allocation. Each improvement's urgency and potential influence on an organization's cybersecurity stance vary. By evaluating the benefits of each initiative against its cost and intricacy, organizations can develop a prioritized enhancement roadmap that is in harmony with their strategic goals and resource limitations.

Incorporating improvement plans into the overarching cybersecurity strategy is vital for the plans' practicality and effectiveness. This amalgamation necessitates a coordinated strategy, considering the interplay between new initiatives and existing policies, procedures, and controls. Embedding improvement endeavors within the broader cybersecurity framework ensures smooth integration and minimizes operational disruptions.

Implementing mechanisms to monitor improvement progress is essential for tracking advancements and fostering accountability. Tools such as project management applications, performance dashboards, and routine progress updates are instrumental. Setting definitive metrics and benchmarks allows organizations to gauge the success of their improvement efforts, adjusting as required to meet objectives. This continuous monitoring confirms the implementation of improvements and promotes a perpetual enhancement and accountability culture.

Involving stakeholders in the evaluation and improvement framework is crucial for cultivating a cybersecurity-conscious culture and collaborative environment. This involvement should transcend the IT department, encompassing individuals from various organizational sectors alongside external partners and suppliers. By integrating diverse viewpoints into the evaluation process, organizations can ensure that their comprehensive improvement actions resonate with all stakeholder needs and priorities. Additionally, this broad participation underlines a collective commitment to cybersecurity, amplifying the strategy's overall efficacy.

Periodic reviews, evaluation criteria, and methodology updates are imperative to retain relevance and effectiveness. As cybersecurity landscapes evolve, the standards and evaluation techniques should also adapt, reflecting current threats and best practices. This perpetual refinement guarantees that evaluations accurately represent contemporary challenges and standards, arming organizations with the knowledge to stay ahead of potential security vulnerabilities.

Recommendations

- **Establish a Regular Evaluation Schedule:** Formulate a consistent timetable for conducting thorough cybersecurity evaluations, ensuring continual identification and implementation of improvements.
- **Adopt a Holistic Evaluation Framework:** Employ diverse evaluation techniques, including self-assessments, third-party assessments, and independent audits, for a complete understanding of your cybersecurity stance.

- **Prioritize Based on Risk and Resources:** Prioritize identified improvements based on their potential cybersecurity impact and implementation viability.
- **Incorporate Improvements into Strategy:** Integrate identified enhancements seamlessly into your broader cybersecurity strategy to heighten effectiveness and maintain cohesion.
- **Engage and Educate Stakeholders:** Include various organizational stakeholders in the evaluation and enhancement process, promoting a collective cybersecurity responsibility and awareness.

NIST 800-53 Controls

(Refer to our appendix for a comprehensive description of the NIST 800-53 Controls, essential for organizations adhering to strict regulatory standards.)

- AC-01,AT-01,AU-01,CA-01,CM-01,CP-01,IA-01,IR-01,MA-01,MP-01,PE-01,PL-01,PM-01,PS-01, PT-01,RA-01,SA-01,SC-01,SI-01,SR-01,CA-02,CA-05,CA-07,CA-08,CP-02,IR-04,IR-08,PL-02, RA-03,RA-05,RA-07,SA-08,SA-11,SA-17,SI-02,SI-04,SR-05

Simplified Security Controls (SSC)

This section outlines our Simplified Security Controls (SSC), tailored for environments beyond the scope of NIST compliance. These controls offer streamlined, effective security solutions suitable for any organization seeking to enhance its cybersecurity defenses with clear, actionable measures.

Control 1: Development and Implementation of a Continuous Cybersecurity Evaluation Program

Control Objective: To systematically evaluate and monitor the effectiveness of current cybersecurity practices, identify areas for improvement, and implement changes to enhance the organization's cybersecurity posture.

Implementation Steps:

1.1 **Establish Evaluation Framework:** Develop a comprehensive framework for evaluating cybersecurity practices, including assessment tools, techniques, and metrics for measuring effectiveness against predefined benchmarks.

1.2 **Conduct Regular Evaluations:** Execute regular evaluations of cybersecurity practices, utilizing a mix of internal audits, penetration testing, and employee feedback to assess the performance of existing cybersecurity measures.

1.3 **Analyze Evaluation Results and Implement Changes:** Systematically analyze the results from cybersecurity evaluations to identify weaknesses or areas for enhancement. Prioritize improvements based on impact and feasibility, and implement changes in a coordinated manner across the organization.

Expected Outcome: Establishing a continuous cybersecurity evaluation program enables the organization to identify and address security vulnerabilities proactively. This ongoing process ensures that cybersecurity measures always align with current threats and organizational needs, leading to a stronger and more resilient cybersecurity posture.

Control 2: Cybersecurity Improvement Tracking Dashboard

Control Objective: To create a comprehensive, real-time dashboard that tracks the progress of identified cybersecurity improvements, ensuring transparency and accountability in the implementation process.

Implementation Steps:

2.1 **Dashboard Development:** Design and develop a cybersecurity improvement tracking dashboard that visualizes the status, progress, and impact of ongoing improvement initiatives across the organization.

2.2 **Integration with Existing Systems:** Integrate the dashboard with existing cybersecurity management tools and databases to automate data collection related to improvement initiatives, ensuring up-to-date information.

2.3 **Regular Review and Reporting:** Schedule regular review meetings with key stakeholders to discuss the progress of improvement initiatives as reflected in the dashboard and adjust strategies as necessary to address any delays or challenges.

Expected Outcome: Implementing a cybersecurity improvement tracking dashboard provides stakeholders with a clear and current view of the status and effectiveness of improvement initiatives. This transparency facilitates better decision-making, ensures accountability, and accelerates the implementation of critical cybersecurity enhancements.

Control 3: Stakeholder Engagement and Feedback Mechanism

Control Objective: To establish a formal mechanism for engaging stakeholders in the cybersecurity evaluation and improvement process, ensuring that feedback from various perspectives is collected, considered, and integrated into cybersecurity strategies.

Implementation Steps:

3.1 **Develop Engagement Strategy:** Create a structured approach for engaging stakeholders, including regular communication channels, feedback forums, and involvement in decision-making processes related to cybersecurity improvements.

3.2 **Implement Feedback Collection Tools:** Utilize surveys, feedback forms, and interactive platforms to gather insights and suggestions from stakeholders on the effectiveness of cybersecurity measures and areas for improvement.

3.3 **Review and Integrate Feedback:** Systematically review collected feedback to identify common themes and actionable insights. Integrate this feedback into the continuous improvement process, ensuring that stakeholder perspectives contribute to enhancing cybersecurity measures.

Expected Outcome: Establishing a stakeholder engagement and feedback mechanism ensures that cybersecurity strategies are informed by diverse perspectives and insights from across the organization. This collaborative approach enhances the relevance and effectiveness of cybersecurity improvements, fostering a culture of shared responsibility and continuous enhancement in cybersecurity practices.

ID.IM-02: Improvements Are Identified from Security Tests and Exercises, Including Those Done in Coordination with Suppliers and Relevant Third Parties

Understanding the significance of security testing and exercises is fundamental to strengthening an organization's cybersecurity posture. These activities are not just about uncovering vulnerabilities; they are critical components in a broader strategy to enhance security resilience. Practical security tests and exercises simulate real-world attack scenarios, offering valuable insights into how well an organization's defenses can withstand cyber threats. Furthermore, they provide a framework for evaluating the efficiency of incident response protocols and identifying areas for improvement. Engaging in these activities regularly ensures that cybersecurity measures remain robust and responsive to evolving threats.

The planning and execution of comprehensive security tests and exercises demand meticulous attention to detail and a thorough understanding of the organization's cyber threat landscape. This involves selecting appropriate testing methodologies, such as penetration testing, red team exercises, and tabletop simulations, that align with the organization's specific vulnerabilities and threat models. Effective planning also requires setting clear objectives, defining success criteria, and ensuring the participation of all relevant stakeholders. By approaching these exercises strategically, organizations can maximize the value derived from each testing initiative, leading to more effective cybersecurity practices.

Coordination with suppliers and third parties for joint testing initiatives introduces a collaborative approach to cybersecurity. In today's interconnected digital ecosystem, an organization's security is often as strong as the weakest link in its supply chain. Joint testing initiatives help to identify and mitigate risks that could be exploited in a supply chain attack, thereby enhancing the overall security posture of all involved entities. These collaborations facilitate sharing threat intelligence, best practices, and resources, making addressing complex security challenges more efficiently possible.

Analyzing the outcomes of security tests and exercises is critical to identifying improvement opportunities. This analysis should go beyond merely cataloging vulnerabilities; it must include an assessment of the organization's response capabilities, detection times, and recovery processes. Understanding the root causes of failures or weaknesses uncovered during these exercises allows for developing targeted action plans to strengthen security measures. Additionally, this analysis helps benchmark the organization's cybersecurity maturity against industry standards and best practices.

Documenting lessons learned and best practices from security activities is crucial in continuously improving cybersecurity programs. This documentation is valuable for refining security policies, procedures, and controls. It also aids in the training and development of the cybersecurity workforce, equipping them with practical insights and knowledge derived from real-world scenarios. By systematically capturing and sharing these lessons, organizations can foster a continuous learning and adaptation culture, which is essential for keeping pace with the rapidly evolving cyber threat landscape.

Developing action plans to address identified security gaps is a proactive step toward enhancing an organization's defense mechanisms. To ensure effective implementation, these action plans should outline specific, measurable, achievable, relevant, and time-bound (SMART) goals. Prioritizing these improvements based on risk and organizational impact ensures that resources are allocated efficiently, focusing on areas that would yield the most significant benefits. Engaging stakeholders in this process helps build a collective commitment to enhancing cybersecurity.

Prioritizing improvements based on risk and organizational impact requires a strategic resource allocation and decision-making approach. This involves conducting a thorough risk assessment to evaluate the potential consequences of identified vulnerabilities and determine their exploitation likelihood. By mitigating risks that pose the greatest threat to critical assets and business operations, organizations can ensure that their cybersecurity efforts are aligned with their overall strategic objectives. This prioritization also facilitates the efficient use of limited cybersecurity resources, maximizing the return on investment in security initiatives.

Communicating findings and improvement plans to all stakeholders is essential for fostering a shared understanding of cybersecurity risks and the measures to address them. Effective communication strategies should include regular updates on the progress of improvement implementation, highlighting successes, and learning from any challenges encountered. This transparency builds trust and encourages a collaborative approach to cybersecurity, with stakeholders across the organization playing an active role in safeguarding against cyber threats. Additionally, open communication channels facilitate the exchange of ideas and feedback, further enhancing the effectiveness of the cybersecurity program.

Continuous monitoring of improvement implementation progress ensures that action plans are executed as intended and deliver the expected outcomes. This monitoring should involve regular reviews of key performance indicators and metrics that reflect the effectiveness of implemented changes. Based on this ongoing evaluation, adjustments to the action plans may be necessary, allowing for a flexible and responsive approach to cybersecurity management. Organizations can adapt to new threats and vulnerabilities by focusing on continuous improvement, ensuring their cybersecurity measures remain effective.

Recommendations

- **Engage in Regular, Comprehensive Security Testing:** Schedule and conduct security tests and exercises routinely to identify vulnerabilities and assess the effectiveness of current cybersecurity measures.
- **Foster Collaboration with Suppliers and Third Parties:** Establish partnerships for joint security testing initiatives to address supply chain risks and enhance the overall security posture.
- **Implement a Structured Analysis and Feedback Loop:** After each testing exercise, thoroughly analyze outcomes to identify improvement opportunities and continuously document lessons learned to refine security practices.
- **Prioritize Action Plans Based on Risk Assessment:** Develop and execute action plans to address identified security gaps, with prioritization based on the potential impact and risk to the organization.
- **Promote Continuous Monitoring and Adjustment:** Ensure ongoing monitoring of the implementation of improvements, making necessary adjustments based on evolving threats and organizational changes.

NIST 800-53 Controls

(Refer to our appendix for a comprehensive description of the NIST 800-53 Controls, essential for organizations adhering to strict regulatory standards.)

- AC-01,AT-01,AU-01,CA-01,CM-01,CP-01,IA-01,IR-01,MA-01,MP-01,PE-01,PL-01,PM-01,PS-01, PT-01,RA-01,SA-01,SC-01,SI-01,SR-01,CA-02,CA-05,CA-07,CA-08,CP-02,CP-04,IR-03,IR-04, IR-08,PL-02,PM-04,PM-31,RA-03,RA-05,RA-07,SA-08,SA-11,SI-02,SI-04,SR-05

Simplified Security Controls (SSC)

This section outlines our SSC, tailored for environments beyond the scope of NIST compliance. These controls offer streamlined, effective security solutions suitable for any organization seeking to enhance its cybersecurity defenses with clear, actionable measures.

Control 1: Establishment of a Joint Cybersecurity Testing Framework with Suppliers and Third Parties

Control Objective: To develop a collaborative framework for conducting joint cybersecurity tests and exercises with suppliers and third-party vendors to identify and mitigate security vulnerabilities across the supply chain.

Implementation Steps:

1.1 **Framework Development:** Create a comprehensive joint testing framework that outlines the objectives, scope, methodologies, and roles and responsibilities of all involved parties.

1.2 **Engagement and Agreement:** Engage with suppliers and third-party vendors to agree on the framework, ensuring mutual understanding and commitment to the joint testing initiatives.

1.3 **Execution and Coordination:** Coordinate the planning and execution of joint cybersecurity tests and exercises, leveraging shared resources and expertise to maximize the coverage and effectiveness of security assessments.

Expected Outcome: Establishing a joint cybersecurity testing framework with suppliers and third parties enhances the security posture of the organization and its extended network. This collaborative approach leads to a more resilient supply chain capable of identifying and addressing security vulnerabilities more effectively.

Control 2: Comprehensive Security Testing and Exercise Documentation Process

Control Objective: To implement a systematic process for documenting all security tests and exercises, including planning, execution, outcomes, lessons learned, and best practices identified.

Implementation Steps:

2.1 **Documentation Standards and Templates:** Develop standardized templates and guidelines for documenting security tests and exercises to ensure consistency and comprehensiveness.

2.2 **Documentation Training:** Train relevant personnel on how to document security tests and exercises effectively, emphasizing the importance of capturing detailed and actionable insights.

2.3 **Review and Update Mechanism:** Establish a mechanism for regularly reviewing and updating documentation to incorporate new findings, lessons learned, and best practices from recent security activities.

Expected Outcome: Implementing a comprehensive security testing and exercise documentation process ensures that valuable insights are captured systematically and are readily available for reference. This facilitates continuous improvement in cybersecurity practices and enhances the organization's ability to respond to emerging threats.

Control 3: Dynamic Security Improvement Tracking System

Control Objective: To create a dynamic system for tracking the implementation progress of security improvements identified from tests and exercises, ensuring timely and effective resolution of identified vulnerabilities.

Implementation Steps:

3.1 **System Development:** Develop an integrated tracking system to monitor the status, responsible parties, deadlines, and outcomes of improvement initiatives.

3.2 **Integration with Existing Systems:** Ensure the tracking system is fully integrated with existing security information and event management (SIEM) systems and project management tools to enable real-time monitoring and updates.

3.3 **Regular Review Meetings:** Schedule regular meetings with key stakeholders to review the progress of improvement initiatives, address any barriers to implementation, and adjust priorities as necessary.

Expected Outcome: Establishing a dynamic security improvement tracking system provides real-time visibility into the status of security improvements, fostering accountability and ensuring that critical vulnerabilities are addressed promptly. This system supports the organization's ability to adapt and respond to the dynamic cybersecurity landscape, strengthening its overall security posture.

ID.IM-03: Improvements Are Identified from Execution of Operational Processes, Procedures, and Activities

Operational excellence in cybersecurity is not merely about deploying advanced technologies or having a set of processes in place. It encapsulates a culture of continuous improvement, where every operational activity, process, and procedure is scrutinized for efficiency and effectiveness. This commitment to excellence ensures that cybersecurity operations are robust and adaptable to the evolving cyber threat landscape. Achieving operational excellence requires a meticulous approach to reviewing existing operations, identifying potential improvements, and implementing changes that enhance the overall cybersecurity posture of an organization.

Continuous monitoring and review of operational cybersecurity processes are foundational to identifying areas for improvement. This ongoing vigilance allows organizations to detect real-time inefficiencies, vulnerabilities, and emerging threats. Utilizing tools and methodologies for continuous monitoring, such as SIEM systems, can provide valuable insights into the operational landscape. When systematically reviewed, these insights form the basis for making informed decisions about where and how to refine cybersecurity operations to maintain an optimal level of security.

Identifying inefficiencies and gaps in current processes and procedures is critical to operational improvement. This requires a thorough analysis of the existing cybersecurity framework, including examining incident response times, detection capabilities, and the effectiveness of existing security controls. Identifying these inefficiencies and gaps is not a one-time task but a continuous process that demands regular audits and assessments. By pinpointing these areas, organizations can focus on making targeted improvements that enhance their cybersecurity defenses.

Gathering and analyzing feedback from process execution is essential for a well-rounded understanding of operational effectiveness. Feedback can come from various sources, including system logs, user reports, and incident response outcomes. Analyzing this feedback provides direct insights into how processes perform under real-world conditions, highlighting areas where adjustments are needed. This iterative feedback loop is crucial for refining operational procedures, ensuring they are theoretically sound and practical.

Implementing process improvements to enhance efficiency and effectiveness is the natural progression from identifying gaps and gathering feedback. This might involve streamlining workflows, adopting new security technologies, or revising response strategies to improve agility. Implementing these improvements requires careful planning to integrate seamlessly with existing operations and contribute positively to the organization's overall cybersecurity posture. It is also essential to measure the effectiveness of these improvements to ensure they deliver the intended outcomes.

Integrating technological advancements into operational practices represents a proactive approach to maintaining a strong cybersecurity posture. As cyber threats evolve, so must the technologies and methodologies used to combat them. This could involve adopting next-generation firewalls, advanced endpoint detection and response systems, or leveraging artificial intelligence and machine learning for threat detection. However, integrating new technologies requires a strategic approach to complement existing processes and not introduce new vulnerabilities.

Training and awareness programs are vital components of supporting operational improvements. These programs ensure that all organization members know the latest cybersecurity practices, policies, and procedures. Regular training sessions can help to instill a culture of security awareness, ensuring that cybersecurity is not just the responsibility of the IT department but of every employee. Moreover, specialized training for IT staff can equip them with the skills and knowledge needed to implement and manage operational improvements effectively.

Evaluation of the impact of implemented improvements on operations is necessary to understand their effectiveness. This evaluation should be comprehensive, considering the direct effects on cybersecurity posture and the broader implications for business operations. Assessing the impact of improvements helps to validate the changes made and guides future efforts in operational enhancement. It also ensures that cybersecurity excellence aligns with the organization's strategic objectives and operational realities.

Recommendations

- **Establish a Continuous Improvement Framework:** Implement a structured framework for continuously monitoring, reviewing, and improving cybersecurity operational processes. This framework should include regular assessments, feedback mechanisms, and a clear process for implementing improvements.
- **Leverage Technology Strategically:** Integrate the latest cybersecurity technologies into operational practices, ensuring they complement existing processes and enhance the organization's ability to respond to cyber threats.
- **Foster a Culture of Continuous Learning:** Develop comprehensive training and awareness programs that promote a culture of security awareness across the organization and equip IT staff with the skills necessary for operational excellence.
- **Measure and Evaluate Improvement Impact:** Regularly evaluate the impact of implemented operational improvements, assessing their effectiveness in enhancing cybersecurity and aligning with business objectives.
- **Prioritize Scalability and Sustainability:** Focus on implementing improvements that are effective in the short term and scalable and sustainable, ensuring they can adapt to future cyber threats.

NIST 800-53 Controls

(Refer to our appendix for a comprehensive description of the NIST 800-53 Controls, essential for organizations adhering to strict regulatory standards.)

- AC-01,AT-01,AU-01,CA-01,CM-01,CP-01,IA-01,IR-01,MA-01,MP-01,PE-01,PL-01,PM-01,PS-01, PT-01,RA-01,SA-01,SC-01,SI-01,SR-01,CA-02,CA-05,CA-07,CA-08,CP-02,IR-04,IR-08,PL-02, PM-04,PM-31,RA-03,RA-05,RA-07,SA-04,SA-08,SA-11,SI-02,SI-04,SR-05

Simplified Security Controls (SSC)

This section outlines our SSC, tailored for environments beyond the scope of NIST compliance. These controls offer streamlined, effective security solutions suitable for any organization seeking to enhance its cybersecurity defenses with clear, actionable measures.

Control 1: Operational Cybersecurity Process Optimization Program

Control Objective: To systematically enhance the efficiency and effectiveness of operational cybersecurity processes through continuous identification, analysis, and improvement of inefficiencies and gaps.

Implementation Steps:

1.1 **Program Establishment:** Launch an operational cybersecurity process optimization program, dedicating resources and personnel to continuously identify and address inefficiencies within cybersecurity operations.

1.2 **Process Mapping and Analysis:** Conduct comprehensive mapping and analysis of existing cybersecurity processes to identify inefficiencies, redundancies, and gaps in process execution.

1.3 **Implementation of Optimized Processes:** Develop and implement optimized processes based on the analysis, incorporating streamlined workflows, automation, and best practices to enhance operational efficiency and effectiveness.

Expected Outcome: Implementing an operational cybersecurity process optimization program results in more efficient and effective cybersecurity operations. By systematically identifying and addressing inefficiencies, the organization can improve its cybersecurity posture while achieving greater operational agility and resilience.

Control 2: Cybersecurity Innovation Integration Initiative

Control Objective: To ensure the continuous integration of technological advancements and innovations into operational cybersecurity practices, enhancing the organization's ability to respond to emerging threats.

Implementation Steps:

2.1 **Innovation Scouting Team Formation:** Form a cross-functional team responsible for continuously scouting and evaluating new cybersecurity technologies and innovations.

2.2 **Evaluation and Pilot Testing:** Conduct thorough evaluations of identified technologies and innovations, followed by pilot testing within the organization to assess their applicability and impact on operational cybersecurity.

2.3 **Full-Scale Integration:** For technologies and innovations that demonstrate significant value during pilot testing, proceed with full-scale integration into operational cybersecurity practices, including necessary training and awareness activities for staff.

Expected Outcome: The Cybersecurity Innovation Integration Initiative ensures that the organization remains at the forefront of cybersecurity by continuously integrating cutting-edge technologies and innovations. This proactive approach enhances the organization's defensive capabilities and operational efficiency, strengthening its security posture.

Control 3: Continuous Cybersecurity Training and Awareness Program

Control Objective: To support operational improvements and integrate new technologies within cybersecurity practices through ongoing training and awareness programs for all relevant personnel.

Implementation Steps:

3.1 **Program Development:** Develop a comprehensive cybersecurity training and awareness program that addresses current cybersecurity challenges, operational processes, and new technologies and innovations.

3.2 **Regular Training Sessions:** Conduct regular training sessions for all relevant personnel, including IT staff, operational teams, and executives, to ensure they are aware of and proficient in updated cybersecurity practices and technologies.

3.3 **Feedback and Program Adjustment:** Collect participant feedback and continuously adjust the training and awareness program to address new threats, technological advancements, and operational changes.

Expected Outcome: The Continuous Cybersecurity Training and Awareness Program ensures that all personnel are informed and proficient in the latest cybersecurity practices and technologies. This ongoing education supports the effective implementation of operational improvements and the integration of innovations, contributing to a robust cybersecurity posture and a culture of continuous improvement.

ID.IM-04: Incident Response Plans and Other Cybersecurity Plans That Affect Operations Are Established, Communicated, Maintained, and Improved

Developing comprehensive incident response plans is paramount for organizations to manage and mitigate cybersecurity incidents effectively. These plans provide a structured approach for responding to security breaches, minimizing their impact on operations. A well-crafted incident response plan outlines clear roles and responsibilities, escalation procedures, and communication strategies to ensure a coordinated and efficient response. It also incorporates criteria for classifying the severity of incidents to prioritize responses accordingly. This foundational element of cybersecurity strategy empowers organizations to navigate the complexities of incident management with confidence and resilience.

The framework for creating and maintaining cybersecurity plans emphasizes a systematic approach to developing, implementing, reviewing, and updating cybersecurity measures. This framework ensures that cybersecurity plans remain relevant, comprehensive, and aligned with current and emerging threats. Key components include risk assessment, control implementation, performance monitoring, and continuous improvement. Regularly updating these plans in response to new insights, technologies, and threats keeps an organization's cybersecurity posture robust and proactive.

Effective communication strategies for cybersecurity plan dissemination are essential to ensure that all stakeholders are informed and prepared. These strategies should cater to diverse audiences within the organization, from executive leadership to operational staff, tailoring messages to be clear and actionable. Utilizing multiple communication channels, such as emails, intranet postings, and training sessions, enhances the reach and effectiveness of these messages. Ensuring everyone understands their role in cybersecurity defense mechanisms strengthens the collective response to threats.

Regular testing and reviewing incident response and cybersecurity plans are crucial for verifying their effectiveness and identifying areas for enhancement. Simulated attack exercises, such as tabletop exercises and red teaming, provide practical insights into the operational readiness of an organization to manage real-world incidents. These exercises reveal gaps in plans and practice, offering valuable opportunities for refinement. Incorporating feedback from these tests into plan updates ensures that strategies evolve in line with dynamic cybersecurity landscapes.

Incorporation of lessons learned into plan updates and improvements is a vital aspect of the continuous improvement cycle. Analyzing both successful and unsuccessful elements of incident response efforts provides critical insights that can enhance future responses. This process involves documenting incidents comprehensively, conducting postincident reviews, and updating plans to reflect a new understanding. Organizations can elevate their cybersecurity resilience and response capabilities by systematically integrating lessons learned.

Coordination with internal and external stakeholders in plan development fosters a collaborative approach to cybersecurity. Engaging various departments within the organization ensures that cybersecurity plans are comprehensive and account for the nuances of different operational areas. Collaborating with external partners, such as law enforcement and cybersecurity firms, enriches plans with broader insights and best practices. This multifaceted approach ensures that cybersecurity strategies benefit from a wide range of expertise and perspectives.

Integration of cybersecurity plans with overall business continuity strategies ensures that organizations can maintain critical operations despite cybersecurity incidents. This integration involves aligning incident response plans with broader business continuity and disaster recovery plans, providing a cohesive approach to managing disruptions. By considering cybersecurity as an integral component of overall business resilience, organizations can ensure they are prepared to sustain operations under various scenarios, safeguarding assets and reputation.

Training and exercise programs to enhance plan effectiveness are vital for preparing teams to respond effectively to incidents. These programs should cover various scenarios, from common threats to sophisticated cyberattacks, ensuring that response teams have the skills and knowledge to manage diverse challenges. Regular training sessions and practical exercises reinforce the principles outlined in cybersecurity plans and improve the organization's overall preparedness.

Recommendations

- **Develop Dynamic Incident Response Plans:** Create comprehensive incident response plans that outline clear procedures and are regularly updated to reflect the evolving cybersecurity landscape.
- **Implement Regular Testing and Review Cycles:** Conduct periodic testing and review of cybersecurity plans to ensure they remain effective and relevant, incorporating lessons learned from exercises and actual incidents.
- **Foster Comprehensive Communication Strategies:** Establish robust communication strategies to ensure widespread understanding and adherence to cybersecurity plans across all levels of the organization.
- **Promote Cross-Departmental Collaboration:** Engage with stakeholders and external partners to develop and refine cybersecurity plans, leveraging a broad spectrum of insights and expertise.
- **Integrate Cybersecurity with Business Continuity:** Align cybersecurity planning closely with business continuity strategies to ensure seamless operations during and after cybersecurity incidents.

NIST 800-53 Controls

(Refer to our appendix for a comprehensive description of the NIST 800-53 Controls, essential for organizations adhering to strict regulatory standards.)

- CP-02,IR-08,PL-02,SR-02

Simplified Security Controls (SSC)

This section outlines our SSC, tailored for environments beyond the scope of NIST compliance. These controls offer streamlined, effective security solutions suitable for any organization seeking to enhance its cybersecurity defenses with clear, actionable measures.

Control 1: Comprehensive Incident Response Plan Development and Maintenance Protocol

Control Objective: To ensure the creation, ongoing maintenance, and improvement of comprehensive incident response plans capable of effectively managing and mitigating cybersecurity incidents.

Implementation Steps:

1.1 **Development of Comprehensive Plans:** Draft detailed incident response plans that clearly outline steps for identification, containment, eradication, and recovery from cybersecurity incidents, including roles and responsibilities.

1.2 **Regular Maintenance Schedule:** Establish a routine schedule for reviewing and updating the incident response plans to ensure they remain relevant and effective against new and evolving cyber threats.

1.3 **Incorporation of Feedback and Lessons Learned:** Systematically integrate feedback and lessons learned from incident response drills, actual incidents, and changes in the cybersecurity landscape into the plans to ensure continuous improvement.

Expected Outcome: Establishing a protocol for developing and maintaining incident response plans ensures that the organization is prepared to respond swiftly and effectively to cybersecurity incidents, minimizing potential damage and downtime.

Control 2: Cybersecurity Communication and Awareness Program

Control Objective: To effectively communicate cybersecurity plans and policies across the organization and ensure all employees know and understand their roles in maintaining cybersecurity.

Implementation Steps:

2.1 **Development of Communication Strategies:** Create comprehensive communication strategies that utilize multiple channels (e.g., email, intranet, and meetings) to disseminate cybersecurity plans and updates to all stakeholders.

2.2 **Implementation of an Awareness Program:** Roll out a continuous cybersecurity awareness program that educates employees on the importance of cybersecurity, their specific roles within the incident response plan, and how to report incidents.

2.3 **Feedback Mechanism for Improvement:** Establish a feedback mechanism that allows employees to voice concerns, suggest improvements, and report on the effectiveness of the communication strategies.

Expected Outcome: An effective cybersecurity communication and awareness program ensures that all employees are informed, vigilant, and prepared to play their role in the organization's cybersecurity efforts, strengthening the overall cybersecurity posture.

Control 3: Integration of Cybersecurity and Business Continuity Planning

Control Objective: To seamlessly integrate cybersecurity incident response plans with overall business continuity strategies, ensuring that cybersecurity incidents are managed to minimize impact on business operations.

Implementation Steps:

3.1 **Cross-functional Planning Teams:** Form cross-functional teams comprising members from cybersecurity, business continuity planning, and key operational departments to align cybersecurity incident response plans with business continuity strategies.

3.2 **Development of Integrated Response Strategies:** Develop response strategies that address the cybersecurity aspects of incidents and their potential impact on business operations, ensuring that plans are comprehensive and aligned.

3.3 **Regular Joint Exercises:** Conduct regular joint exercises that simulate cybersecurity incidents affecting business operations to test the integrated plans' effectiveness and identify improvement areas.

Expected Outcome: The integration of cybersecurity and business continuity planning ensures that the organization can maintain critical operations during and after a cybersecurity incident, minimizing operational and financial impact while providing a swift return to normal operations.

14

Identity Management, Authentication, and Access Control (PR.AA)

> The strategic orchestration of identity and credential management is not merely a procedural necessity; it embodies the essence of trust and control in the digital realm, ensuring access is as much about safeguarding as enabling.

In cybersecurity, managing identities and credentials is a cornerstone of organizational security, underpinning the safeguarding of users, services, and hardware. Central to this effort is establishing a comprehensive identity management system meticulously designed to assign and manage credentials while enforcing robust security measures to protect against unauthorized access. Regular audits, integrated training programs, and the adept handling of compromised credentials ensure that the organization stays ahead of potential security breaches. This framework is not static; it demands ongoing review and adaptation, incorporating the latest multifactor authentication (MFA) and encryption technologies to maintain a resilient defense against the ever-evolving cybersecurity landscape. Through a concerted effort that marries technology with best practices, organizations can create a secure environment where access is both a privilege and a right, carefully granted and vigilantly protected.

PR.AA-01: Identities and Credentials for Authorized Users, Services, and Hardware Are Managed by the Organization

Identity and credential management is a cornerstone in safeguarding an organization's digital assets. The process begins with a robust understanding of what constitutes identity within the digital realm and the critical role credentials play in securing access to resources. Organizations must prioritize the management of identities and credentials to protect against unauthorized access, ensuring that only authorized users, services, and hardware can interact with sensitive data and systems.

Establishing a centralized identity management system is the first critical step toward securing an organization's digital environment. Such a system is the foundation for efficiently assigning and managing digital identities and credentials. By centralizing identity management, organizations gain a consolidated view of user access across the board, facilitating easier monitoring and management of access rights. This approach streamlines administrative processes and enhances security by providing a single source of truth for identity-related information.

A Comprehensive Guide to the NIST Cybersecurity Framework 2.0: Strategies, Implementation, and Best Practice, First Edition. Jason Edwards.
© 2025 John Wiley & Sons Ltd. Published 2025 by John Wiley & Sons Ltd.

Key to this framework is the development of comprehensive processes for assigning and managing credentials for users, services, and hardware. These processes must be meticulously designed to ensure access rights are granted according to the principles of least privilege and need-to-know. Each credential should be uniquely tied to an individual user, service, or hardware component, eliminating generic accounts and reducing the risk of unauthorized access. This granularity in access control further strengthens the security posture of an organization.

Security measures for protecting identities and credentials are paramount. These include implementing strong password policies, MFA, and encryption of credentials at rest and in transit. Organizations can significantly reduce the risk of compromised credentials by adopting such measures. Furthermore, regular security audits and penetration testing can help identify vulnerabilities in the identity management process, enabling timely remediation before adversaries can exploit them.

The lifecycle of credentials necessitates regular review and updating to adapt to changes within the organization, such as role changes, departures, or changes in access requirements. This continuous monitoring ensures that access rights remain aligned with current job responsibilities and that outdated or unnecessary credentials are promptly revoked. Such diligence prevents the accumulation of excess privileges, a common vector for security breaches.

Integrating identity management with access control systems further streamlines granting and managing access. This integration allows for automated enforcement of access policies based on the credentials and roles of users, services, and hardware. Real-time adjustments to access rights can be made as circumstances change, enhancing the responsiveness of the security framework to evolving threats and organizational dynamics.

Handling lost or compromised credentials promptly and effectively minimizes potential security breaches. Organizations must have clear procedures for reporting and responding to such incidents. This includes the immediate revocation of compromised credentials, conducting a thorough investigation to determine the scope of the breach, and taking corrective action to prevent future occurrences.

Training programs on secure credential use and management are essential for reinforcing the importance of cybersecurity hygiene among employees. Through regular training and awareness programs, employees can be educated on the best practices for managing and safeguarding their credentials and recognizing and responding to potential security threats. Such programs are vital for fostering a culture of security within the organization.

Auditing and logging credential management activities provide a detailed record of access and changes, which is critical for detecting unauthorized activity and investigating security incidents. These logs should be securely stored, regularly reviewed, and analyzed for signs of suspicious behavior. Through diligent monitoring and analysis, organizations can swiftly identify and address security issues related to identity and credential management.

Recommendations

- **Develop a Comprehensive Credential Management Policy**: Establish clear guidelines and procedures for the creation, distribution, storage, and revocation of credentials. This policy should cover all aspects of credential management, from assigning access rights to handling compromised credentials.
- **Implement Multifactor Authentication (MFA) Across All Systems**: Enhance the security of your identity and credential management system by requiring MFA for access to sensitive systems and data. MFA adds a layer of protection, significantly reducing the risk of unauthorized access.

- **Conduct Regular Access Reviews**: Review and audit user access rights across the organization. This ensures that employees only have access to the resources necessary for their current roles and responsibilities, minimizing the potential for internal and external security breaches.
- **Establish a Rapid Response Mechanism for Compromised Credentials**: Create a dedicated process for quickly responding to lost or compromised credentials reports. This includes immediate revocation of access, investigation of the incident, and corrective measures to prevent recurrence.
- **Invest in Employee Training and Awareness Programs**: Educate employees on the importance of secure credential management and their role in protecting the organization's assets. Regular training sessions can help instill good cybersecurity practices and prepare employees to respond effectively to potential security threats.

NIST 800-53 Controls

(Refer to our appendix for a comprehensive description of the NIST 800-53 Controls, essential for organizations adhering to strict regulatory standards.)

- AC-01,AC-02,AC-14,IA-01,IA-02,IA-03,IA-04,IA-05,IA-06,IA-07,IA-08,IA-09,IA-10,IA-11

Simplified Security Controls (SSC)

This section outlines our Simplified Security Controls (SSC), tailored for environments beyond the scope of NIST compliance. These controls offer streamlined, effective security solutions suitable for any organization seeking to enhance its cybersecurity defenses with clear, actionable measures.

Control 1: Centralized Identity Management System Deployment

Control Objective: To create and implement a centralized system for managing the identities and credentials of authorized users, services, and hardware, ensuring consistency, efficiency, and security across the organization.

Implementation Steps:

1.1 **System Selection**: Conduct a thorough assessment of available identity management systems, selecting one that best fits the organization's size, complexity, and specific security requirements.

1.2 **Integration Planning**: Develop a comprehensive plan for integrating the chosen identity management system with existing IT infrastructure, including user databases, access control systems, and other relevant applications.

1.3 **Rollout and Migration**: Execute a phased rollout of the identity management system, migrating users, services, and hardware credentials in a controlled manner to minimize disruption and ensure a smooth transition.

Expected Outcome: The deployment of a centralized identity management system streamlines the process of managing identities and credentials, enhancing security by ensuring consistent application of policies and procedures across the organization. This system also simplifies the tasks of onboarding, modifying, and revoking access, contributing to a more efficient and secure operational environment.

Control 2: Comprehensive Credential Security Measures Implementation

Control Objective: To safeguard identities and credentials from unauthorized access and compromise by adopting robust security measures, including encryption, MFA, and regular security audits.

Implementation Steps:

2.1 **Encryption of Stored Credentials**: Implement strong encryption for credentials stored in databases and directories, using industry-standard algorithms to protect against data breaches.

2.2 **Deployment of Multifactor Authentication**: Introduce MFA for accessing sensitive systems and information, requiring users to provide two or more verification factors.

2.3 **Conduct Regular Security Audits**: Conduct regular security audits of credential management practices, promptly identifying and addressing potential vulnerabilities.

Expected Outcome: Implementing comprehensive security measures significantly reduces the risk of credential compromise, ensuring that identities and access rights are protected. This approach enhances the organization's overall security posture by making unauthorized access more difficult and providing mechanisms for quick detection and response to potential breaches.

Control 3: Regular Training and Awareness Programs on Secure Credential Management

Control Objective: To enhance the organization's security culture by providing employees with regular training and awareness programs focused on the importance of secure credential use and management practices.

Implementation Steps:

3.1 **Develop Comprehensive Training Material**: Create detailed training materials that cover best practices for secure credential management, including password complexity, the importance of regular updates, and the handling of phishing attempts.

3.2 **Schedule Regular Training Sessions**: Organize periodic training sessions for all employees, making attendance mandatory to ensure everyone is informed about the latest security practices and threats.

3.3 **Assess and Update Training Content**: Continually assess the effectiveness of training programs and update content as necessary to address new threats, technologies, and security practices.

Expected Outcome: Regular training and awareness programs significantly increase employees' understanding and adherence to secure credential management practices. This proactive approach reduces the risk of security breaches due to human error and strengthens the organization's overall security culture.

PR.AA-02: Identities Are Proofed and Bound to Credentials Based on the Context of Interactions

Understanding the intricacies of identity proofing and the secure binding of identities to credentials is fundamental to the architecture of robust cybersecurity strategies. This process is essential for verifying that a person is who they claim to be before granting them access to sensitive information or systems. This requires a thorough analysis of the context in which interactions occur. Contextual analysis aids in determining the appropriate level of identity proofing required, considering factors such as the sensitivity of the information accessed and the potential risks involved.

Procedures for securely binding identities to credentials are paramount. These procedures involve establishing a reliable association between the individual and the credentials they use, often employing methods such as biometric verification or the issuance of secure tokens. This binding process is critical for preventing identity theft and ensuring that credentials cannot be misused if they fall into the wrong hands. It requires a careful balance of security measures to protect against threats while ensuring legitimate users can access their accounts without undue friction.

The use of MFA significantly enhances the security of this process. MFA requires users to provide two or more verification factors to gain access to a resource, such as something they know (a password), something they have (a security token), or something they are (a biometric verification). This layered approach to security makes it much more difficult for unauthorized parties to breach accounts, even if they have compromised one element of the authentication process.

Incorporating mechanisms for context-aware authentication processes further strengthens security measures. These mechanisms adjust the authentication requirements based on the access request context, such as the user's location, the device used, or the time of access. For example, access attempts from unfamiliar locations or devices might trigger additional authentication challenges. This dynamic approach to authentication ensures a higher level of security for sensitive operations while maintaining user convenience for routine access.

Organizations must periodically reassess their identity-proofing and credential-binding practices. This periodic review ensures that these practices remain effective against evolving threats and adapt to changes in technology and user behavior. Continuous improvement in these areas is crucial for maintaining the integrity of the identity management system and protecting against unauthorized access.

Integrating identity proofing with onboarding processes is crucial for establishing secure access. When new users are added to the system—whether employees, customers, or partners—the initial identity verification and credential issuance set the stage for all future interactions; streamlining this integration enhances security and improves the user experience by making the onboarding process smoother and more efficient.

Strategies for managing identity across different contexts require a flexible and adaptive approach. As users interact with systems in various capacities—such as switching between customer and employee roles or accessing resources from different devices—the identity management system must dynamically adjust to these changes. This adaptability ensures that security measures remain effective and relevant, regardless of the scenario.

Implementing dynamic authentication methods based on risk assessment further refines the security landscape. Authentication requirements can be adjusted in real time to mitigate potential threats by evaluating the risk associated with each access request. This risk-based approach allows for stricter authentication measures in high-risk situations while streamlining access in lower-risk scenarios, thus optimizing both security and user experience.

Recommendations

- **Establish a Contextual Analysis Framework**: Implement a framework for performing contextual analysis to determine the appropriate level of identity proofing required for different scenarios. This will ensure that security measures are aligned with the potential risks and the sensitivity of the information being accessed.
- **Enhance Security with Advanced MFA Solutions**: Invest in advanced MFA solutions that offer a range of verification options, including biometric verification, to strengthen the security of your identity and credential-binding processes.
- **Regularly Update and Review Identity Proofing Practices**: Schedule periodic reviews of your identity proofing and credential binding practices to ensure they remain effective against new and evolving threats. This should include updating technologies and methods in line with best practices.

- **Seamlessly Integrate Identity Proofing in User Onboarding**: Design your onboarding processes to incorporate secure identity proofing measures, ensuring a secure and efficient initiation for new users into your systems.
- **Adopt Dynamic Authentication Based on Risk Assessment**: Implement authentication methods that adjust dynamically based on the risk associated with each access request. This approach balances the need for security with the user experience by applying stricter authentication for higher-risk scenarios and simplifying it for lower-risk situations.

NIST 800-53 Controls

(Refer to our appendix for a comprehensive description of the NIST 800-53 Controls, essential for organizations adhering to strict regulatory standards.)

- IA-12

Simplified Security Controls (SSC)

This section outlines our SSC, tailored for environments beyond the scope of NIST compliance. These controls offer streamlined, effective security solutions suitable for any organization seeking to enhance its cybersecurity defenses with clear, actionable measures.

Control 1: Contextual Identity Proofing System Integration

Control Objective: To implement a system that integrates contextual analysis into the identity proofing process, ensuring that the proofing and credential binding level are appropriate for the risks associated with different types of interactions.

Implementation Steps:

1.1 **Develop Risk-Based Profiles**: Create profiles that categorize types of interactions by risk level, determining the appropriate identity proofing requirements for each.

1.2 **Integrate Contextual Analysis Tools**: Deploy tools capable of analyzing the context of access requests in real time, matching them to the predefined risk profiles.

1.3 **Automate Identity Proofing Adjustments**: Implement automation rules within the identity management system to adjust the proofing requirements based on the interaction context.

Expected Outcome: Integrating a contextual identity proofing system ensures that verifying identities is dynamically adjusted based on the level of risk associated with each interaction. This approach enhances security by applying stricter identity proofing for higher-risk scenarios while streamlining the user experience for lower-risk interactions.

Control 2: Comprehensive Multifactor Authentication (MFA) Deployment

Control Objective: To enhance the security of identity proofing and credential binding across the organization by deploying comprehensive MFA for all users, considering the context of their interactions.

Implementation Steps:

2.1 **Evaluate and Select MFA Solutions**: Assess available MFA solutions, selecting those that offer flexibility, scalability, and compatibility with the organization's IT infrastructure.

2.2 **Define MFA Policies**: Establish policies that dictate when and how MFA should be applied, including criteria for different levels of access and interaction contexts.

2.3 **Implement and Monitor MFA Deployment**: Roll out the selected MFA solution according to the defined policies and monitor its effective operation and user compliance.

Expected Outcome: The comprehensive deployment of MFA across the organization significantly strengthens the security of identity proofing and credential binding. Requiring multiple forms of verification greatly reduces the risk of unauthorized access, especially in high-risk interactions.

Control 3: Dynamic Authentication Method Implementation Based on Risk Assessment

Control Objective: To implement dynamic authentication methods that adjust the authentication process in real time based on a risk assessment of each access request or interaction.

Implementation Steps:

3.1 **Develop a Risk Assessment Model**: Create a model for assessing the risk of access requests, incorporating factors such as user behavior, access location, and the sensitivity of the requested resource.

3.2 **Integrate Risk Assessment with Authentication Processes**: Ensure the identity management system can apply the risk assessment model in real time to each access request, adjusting the authentication requirements accordingly.

3.3 **Continuous Improvement and Adaptation**: Regularly review and update the risk assessment model and authentication processes to adapt to new threats, technological advances, and changes in organizational policies.

Expected Outcome: Implementing dynamic authentication methods based on risk assessment provides a flexible and adaptive approach to security. This system enhances the organization's ability to protect sensitive information and systems by applying more stringent authentication measures in higher-risk scenarios while maintaining user convenience in lower-risk situations.

PR.AA-03: Users, Services, and Hardware Are Authenticated

Authentication forms the bedrock of security protocols, ensuring access to an organization's digital environment is securely controlled and monitored. The authentication process verifies the identity of users, services, and hardware, granting access only to those with legitimate credentials. The formulation of strong authentication policies and procedures is paramount.

These policies dictate the methods and protocols to be used in verifying the identities of users and devices, setting the stage for a secure authentication environment. Effective policies are comprehensive, addressing various scenarios and types of access, and include guidelines for password management, MFA, and handling authentication failures. Establishing these guidelines ensures a standardized approach across the organization, significantly reducing the risk of unauthorized access.

The deployment of MFA for sensitive systems elevates security measures, requiring users to provide multiple pieces of evidence of their identity. MFA combines something the user knows (like a password), something the user has (such as a mobile device or security token), and something the user is (verified through biometrics). This method considerably diminishes the likelihood of unauthorized access, as the compromise of one factor alone is insufficient to breach the system.

Authentication extends beyond users to encompass services and hardware, recognizing that machines also access and exchange information within networks. Securing these non-human elements of the IT infrastructure is critical, necessitating unique credentials and authentication protocols. This holistic approach ensures that all components within the network are authenticated, thereby safeguarding against vulnerabilities that could be exploited through automated processes or compromised devices.

The advent of advanced authentication technologies, including biometrics and security tokens, has introduced new layers of security. Biometric authentication utilizes unique physical characteristics, such as fingerprints or facial recognition, offering a high level of protection due to the difficulty of replication. Security tokens, whether hardware-based or software-generated, provide a dynamically changing code that must be entered alongside traditional passwords, adding a barrier to unauthorized access.

Regular review and enhancement of authentication systems are crucial for maintaining their effectiveness. As technology evolves and new threats emerge, authentication systems must adapt to address these changes. Regular audits and updates ensure that authentication mechanisms remain robust against sophisticated attacks, incorporating the latest security advancements and addressing identified vulnerabilities.

Integrating authentication protocols with system and network security controls creates a unified defense strategy. This integration ensures that authentication mechanisms are supported by and complement other security measures, such as firewalls, intrusion detection systems, and encryption protocols. Such a cohesive security infrastructure provides more comprehensive protection against cyber threats, making it more difficult for attackers to exploit weaknesses.

Training and awareness programs are essential in educating users about authentication best practices and their critical role in maintaining security. Users become active participants in the organization's security posture by understanding the importance of strong passwords, the benefits of MFA, and the procedures for reporting suspicious activity. These programs should be ongoing, reflecting the latest cyber threats and authentication technology trends.

Monitoring and logging authentication attempts and failures provide valuable insights into security threats. Analyzing these logs can reveal patterns indicative of attempted breaches, enabling timely intervention. Furthermore, detailed logging is instrumental in forensic analysis following a security incident, offering clues to the methods and origins of the attack. Effective monitoring and analysis of authentication activities are vital to a proactive security strategy.

Recommendations

- **Enhance Security with Adaptive Multifactor Authentication (MFA)**: Implement adaptive MFA that adjusts the level of authentication required based on contextual factors, such as the user's location or device security posture, to balance security with usability.
- **Regularly Update Authentication Technologies**: Stay abreast of advancements in authentication technologies and regularly update your systems to incorporate the latest security features. This includes upgrading to more secure forms of biometric verification and exploring next-generation authentication tokens.
- **Conduct Regular Authentication System Audits**: Schedule periodic audits of your authentication systems to identify and remediate any vulnerabilities or weaknesses. These audits should also assess compliance with established authentication policies and procedures.
- **Integrate Authentication Training into Cybersecurity Awareness Programs**: Develop comprehensive training programs focusing on authentication best practices, the importance of secure password management, and the effective use of MFA. Encourage a culture where security is everyone's responsibility.
- **Implement Real-Time Monitoring and Alerting for Authentication Systems**: Set up real-time monitoring and alerting mechanisms for authentication attempts and failures to quickly identify and respond to potential security incidents. Use advanced analytics to detect patterns indicating a breach or attempted attack.

NIST 800-53 Controls

(Refer to our appendix for a comprehensive description of the NIST 800-53 Controls, essential for organizations adhering to strict regulatory standards.)

- AC-07,AC-12,IA-02,IA-03,IA-05,IA-07,IA-08,IA-09,IA-10,IA-11

Simplified Security Controls (SSC)

This section outlines our SSC, tailored for environments beyond the scope of NIST compliance. These controls offer streamlined, effective security solutions suitable for any organization seeking to enhance its cybersecurity defenses with clear, actionable measures.

Control 1: Comprehensive Multifactor Authentication Implementation

Control Objective: To significantly enhance the security of user, service, and hardware authentication by implementing MFA across all sensitive systems and applications within the organization.

Implementation Steps:

1.1 **MFA Requirement Analysis**: Conduct an analysis to identify which systems, services, and hardware are considered sensitive and require MFA based on the data they handle or their criticality to the organization.

1.2 **MFA Solution Selection and Deployment**: Choose an MFA solution compatible with the organization's technology stack and user needs, ensuring it supports various authentication methods (e.g., SMS, authenticator apps, and hardware tokens). Deploy the solution across identified sensitive systems.

1.3 **User and Administrator Training**: Provide comprehensive training for users and administrators on the importance of MFA, how to use it, and troubleshooting common issues to ensure smooth adoption and operation.

Expected Outcome: The deployment of MFA across all sensitive systems significantly reduces the risk of unauthorized access, ensuring that even if passwords are compromised, the additional authentication factors provide a critical security layer.

Control 2: Advanced Authentication Technologies Adoption

Control Objective: To adopt and integrate advanced authentication technologies, such as biometrics and security tokens, to provide stronger and more user-friendly authentication mechanisms for accessing corporate resources.

Implementation Steps:

2.1 **Technology Assessment and Selection**: Evaluate various advanced authentication technologies, considering security, usability, cost, and compatibility with existing systems. Select technologies that best meet the organization's security and operational requirements.

2.2 **Pilot Testing**: Before a full-scale rollout, conduct pilot testing of the selected technologies with a small, controlled group of users to identify potential issues and gather user feedback on usability.

2.3 **Organization-Wide Implementation**: Following successful pilot testing, implement the chosen authentication technologies across the organization, providing necessary training and support to ensure user adoption and system compatibility.

Expected Outcome: The adoption of advanced authentication technologies enhances the security and usability of the organization's authentication system, making it more difficult for attackers to gain unauthorized access while improving the user experience.

Control 3: Authentication System Review and Enhancement Program

Control Objective: To establish an ongoing program to regularly review and enhance authentication systems, ensuring they remain effective against evolving threats and meet the organization's security and usability needs.

Implementation Steps:

3.1 **Establish a Review Schedule**: Set a regular schedule for reviewing the effectiveness, security, and user satisfaction with the organization's authentication systems, including MFA and advanced technologies.

3.2 **Conduct Security and Usability Assessments**: Periodically assess the authentication systems for security vulnerabilities and usability issues, incorporating feedback from users and IT staff.

3.3 **Implement Enhancements**: Based on the review findings, implement enhancements to the authentication systems, including upgrading technologies, adjusting policies, or providing additional user training.

Expected Outcome: Establishing a regular review and enhancement program ensures that the organization's authentication systems remain robust against new threats and efficient for users, supporting ongoing security and operational needs.

PR.AA-04: Identity Assertions Are Protected, Conveyed, and Verified

Securely managing identity assertions is a critical component of modern cybersecurity practices. Identity assertions—statements made by one digital entity about the identity of another—are fundamental in enabling secure communications and access control across distributed systems.

Techniques for securely protecting identity assertions are varied and must be carefully selected based on the specific needs and contexts of the systems in use. These techniques include the application of digital signatures to validate the origin and integrity of the assertion, as well as encryption to maintain confidentiality. By securely protecting identity assertions, organizations can prevent unauthorized access and ensure that the digital identities represented by these assertions are accurately and securely managed.

The methods for safely conveying identity assertions across systems play a vital role in the interconnected landscape of digital services. Secure transmission protocols, such as transport layer security (TLS), ensure that assertions are not intercepted or altered during transit. Additionally, secure token services can facilitate the safe exchange of assertions by issuing tokens that represent the identity information securely, minimizing the exposure of sensitive data.

Processes for the accurate verification of identity assertions are equally crucial. These processes involve checking the validity of the assertions against trusted sources or directories and ensuring that the assertion has not been tampered with. Verification mechanisms often rely on public key infrastructure or similar cryptographic methods to ensure the authenticity of the digital signatures attached to the assertions.

Implementing secure communication channels for assertion transfer is essential to protect the data between systems. Establishing these channels involves deploying encryption at every step of the communication process and ensuring that all parties involved in the exchange are authenticated and authorized to participate. This protects the data in transit and helps maintain the privacy and security of the individuals or entities the data represents.

Encryption and other security measures for assertion protection are foundational elements of a robust identity management strategy. Beyond encrypting the data in transit, it's important to secure it at rest, employing encryption standards such as advanced encryption standard to safeguard stored identity assertions. Other security measures, including access controls and the principle of least privilege, ensure that only authorized personnel can access or manage these assertions.

Regular audits of assertion protection, conveyance, and verification practices help organizations identify potential weaknesses in their identity management processes and make necessary adjustments. These audits should assess the effectiveness of the implemented security measures, compliance with relevant regulations and standards, and the ability of the processes to resist emerging threats. Continuous improvement based on audit findings is key to maintaining the security and integrity of identity assertions.

Integration with federated identity management systems offers a scalable and efficient way to manage identity assertions across multiple domains and services. Federated identity management enables the sharing of identity information among different organizations in a secure and controlled manner, reducing the need for multiple accounts and passwords. This approach simplifies the user experience while enhancing security by centralizing the management of identity assertions.

Training on the importance and management of identity assertions is indispensable for ensuring that all stakeholders understand their roles in safeguarding these critical pieces of information. Regular training sessions should cover the best practices for handling identity assertions, the potential risks associated with their misuse, and the security protocols that must be followed. Empowering users and administrators with this knowledge strengthens the organization's overall security posture.

Recommendations

- **Implement Comprehensive Encryption Standards**: Adopt advanced encryption standards for both data in transit and at rest to ensure the secure protection of identity assertions. Regularly review and update these encryption practices to align with security advancements and threats.
- **Leverage Federated Identity Management Solutions**: Integrate with federated identity management systems to streamline the secure management and sharing of identity assertions across different domains and services, enhancing security and user convenience.
- **Conduct Regular Security Audits**: Schedule and conduct regular audits of your identity assertion management practices, including protection, conveyance, and verification processes. Use these audits to identify and address vulnerabilities, ensuring continuous improvement in security measures.
- **Educate Stakeholders on Security Best Practices**: Develop and deliver comprehensive training programs on the importance of secure identity assertion management. These programs should target all relevant stakeholders, including users, IT staff, and management, to foster a culture of security awareness and compliance.
- **Deploy Secure Transmission Protocols**: Use secure transmission protocols, such as TLS, to convey identity assertions. Regularly update these protocols to incorporate the latest security features and protection mechanisms.

NIST 800-53 Controls

(Refer to our appendix for a comprehensive description of the NIST 800-53 Controls, essential for organizations adhering to strict regulatory standards.)

- IA-13

Simplified Security Controls (SSC)

This section outlines our SSC, tailored for environments beyond the scope of NIST compliance. These controls offer streamlined, effective security solutions suitable for any organization seeking to enhance its cybersecurity defenses with clear, actionable measures.

Control 1: Secure Assertion Transfer Protocol Implementation

Control Objective: To safeguard the integrity and confidentiality of identity assertions as they are conveyed across systems, ensuring that assertions remain protected from interception or alteration during transit.

Implementation Steps:

1.1 **Protocol Selection**: Evaluate and select secure transfer protocols that offer robust encryption and integrity checks, such as TLS or Internet protocol security, tailored to the organization's technical infrastructure and specific needs.

1.2 **Configuration and Deployment**: Configure the chosen protocols to enforce strong encryption standards and deploy them across all systems, conveying identity assertions and ensuring compatibility and optimal security settings.

1.3 **Continuous Monitoring and Updating**: Establish a process for continuously monitoring the secure transfer protocols to detect and respond to potential security threats and regularly update the protocols to leverage the latest security enhancements and patches.

Expected Outcome: Implementing secure assertion transfer protocols ensures that identity assertions are protected during conveyance between systems, mitigating the risk of interception or tampering and enhancing the overall security of the identity management framework.

Control 2: Federated Identity Management System Integration

Control Objective: To streamline the management of identity assertions and enhance security by integrating federated identity management systems, enabling seamless and secure access across different systems and organizational boundaries.

Implementation Steps:

2.1 **System Evaluation and Selection**: Assess available federated identity management solutions, selecting one that aligns with the organization's security requirements, technical environment, and interoperability needs with external partners.

2.2 **Integration Planning and Execution**: Develop a detailed plan for integrating the federated identity management system with existing identity and access management (IAM) infrastructure, ensuring secure configuration and seamless operation across all systems.

2.3 **Partnership and Collaboration**: Work with external partners and service providers to establish trust relationships and secure communication channels, facilitating the secure exchange and verification of identity assertions.

Expected Outcome: Integrating a federated identity management system simplifies the secure management and exchange of identity assertions across diverse systems and organizational boundaries, reducing complexity and enhancing security through centralized oversight and consistent application of security policies.

Control 3: Comprehensive Training Program on Identity Assertion Management

Control Objective: To raise awareness and improve the understanding of the importance and secure management of identity assertions among all stakeholders, including IT staff, users, and management.

Implementation Steps:

3.1 **Program Development**: Develop a comprehensive training program covering the principles of identity assertion management, the importance of protecting identity assertions, and best practices for secure handling and verification.

3.2 **Delivery and Engagement**: Deliver the training program through multiple formats (e.g., workshops, e-learning modules, and seminars) to ensure broad accessibility and engagement, tailoring content to different roles and responsibilities within the organization.

3.3 **Feedback and Continuous Improvement**: Collect feedback from training participants to assess the program's effectiveness, adjusting and updating the content and delivery methods based on input and evolving security practices.

Expected Outcome: The delivery of a comprehensive training program on identity assertion management fosters a culture of security awareness within the organization, ensuring that all stakeholders understand their role in protecting identity assertions and are equipped with the knowledge to implement best practices effectively.

PR.AA-05: Access Permissions, Entitlements, and Authorizations Are Defined in a Policy, Managed, Enforced, and Reviewed, and Incorporate the Principles of Least Privilege and Separation of Duties

The development and documentation of access control policies form the backbone of a secure access management system. These policies define who can access specific resources within the organization, under what circumstances, and with what kind of permissions. A well-documented access control policy provides a clear framework for implementing security measures, guiding the deployment of technologies and practices that enforce these policies effectively.

Implementing the least privilege and separation of duties principles is essential in minimizing the potential for unauthorized access or insider threats. The principle of least privilege ensures that individuals have only the access necessary to perform their job functions, nothing more. Separation of duties divides critical functions among different individuals or teams to prevent any single point of vulnerability from being exploited. Together, these principles form a robust defense against external and internal threats.

Processes for managing and enforcing access permissions and entitlements must be rigorous and dynamic. This involves the initial assignment of access rights and the ongoing management of those rights as individuals' roles or circumstances change. Effective access management systems allow for the quick modification, revocation, or escalation of access rights, ensuring that the access control policies remain relevant and are enforced consistently across the organization.

Regularly reviewing and updating access policies and authorizations is critical to adapting to the changing threat landscape and organizational needs. This entails periodic audits of access rights, ensuring that they align with users' current roles and responsibilities. Such reviews help identify redundant or obsolete access permissions that may pose a security risk, allowing for timely remediation.

The use of access control technologies and solutions facilitates the enforcement of access policies. These technologies, including IAM systems, MFA, and role-based access control systems, provide the technical means to implement and enforce access control policies efficiently and effectively. The choice of technologies should reflect the organization's unique needs and security requirements.

Training on access control policies and procedures is indispensable for ensuring all employees understand their roles and responsibilities in maintaining security. This training should cover the basic principles of access control, the specific policies and procedures of the organization, and the correct use of access control technologies. An informed and vigilant workforce is a critical line of defense in the security posture of any organization.

Auditing access control implementations for compliance and effectiveness is essential to verify that access controls function as intended and comply with relevant regulations and standards. Regular audits provide an opportunity to assess the effectiveness of access control measures, identify areas for improvement, and ensure that access control practices align with the overall security objectives of the organization.

Access control integration with other security policies and practices ensures a cohesive and comprehensive security strategy. Access control measures should work with other security protocols, such as data encryption, network security, and incident response plans, to provide layered defense against threats. This holistic approach maximizes the effectiveness of security measures and provides a more secure environment for the organization's operations.

Recommendations

- **Establish Dynamic Access Control Policies**: Develop access control policies adaptable to changing organizational roles and the evolving threat landscape. Ensure these policies incorporate the principles of least privilege and separation of duties.
- **Invest in Advanced Access Control Technologies**: Leverage the latest access control technologies, such as IAM and MFA, to enforce policy decisions dynamically and securely. These technologies should support seamless real-time access rights and permissions adjustments.
- **Conduct Regular Access Rights Reviews**: Implement a regular schedule for reviewing and updating access rights and permissions to ensure they align with current job roles and minimize unnecessary access that could pose a security risk.
- **Enhance Training Programs on Access Control**: Develop comprehensive training programs for all employees on access control policies and procedures, emphasizing the importance of security and each employee's role in safeguarding the organization's assets.
- **Perform Comprehensive Access Control Audits**: Schedule and conduct thorough audits of access control measures to assess their effectiveness, identify areas for improvement, and ensure compliance with relevant regulations and security standards.

NIST 800-53 Controls

(Refer to our appendix for a comprehensive description of the NIST 800-53 Controls, essential for organizations adhering to strict regulatory standards.)

- AC-01,AC-02,AC-03,AC-05,AC-06,AC-10,AC-16,AC-17,AC-18,AC-19,AC-24,IA-13

Simplified Security Controls (SSC)

This section outlines our SSC, tailored for environments beyond the scope of NIST compliance. These controls offer streamlined, effective security solutions suitable for any organization seeking to enhance its cybersecurity defenses with clear, actionable measures.

Control 1: Development and Implementation of a Comprehensive Access Control Policy

Control Objective: To establish a well-defined access control policy that clearly outlines access permissions, entitlements, and authorization processes, incorporating the principles of least privilege and separation of duties to ensure secure and efficient access management within the organization.

Implementation Steps:

1.1 **Policy Formulation**: Assemble a task force comprising representatives from IT, security, legal, and human resources departments to develop a comprehensive access control policy. This policy should define roles, responsibilities, and procedures for granting, reviewing, and revoking access.

1.2 **Policy Communication and Training**: Communicate the organization's newly developed access control policy through various channels. Conduct training sessions for all employees to ensure they understand the policy, their roles within it, and the importance of compliance.

1.3 **Policy Enforcement and Compliance Monitoring**: Implement mechanisms and tools to continuously enforce the access control policy and monitor compliance. Use automated systems where possible to manage access rights efficiently and ensure adherence to the policy.

Expected Outcome: The organization will have a clear and comprehensive access control policy that all employees understand and follow. This policy will facilitate secure and efficient access permissions and entitlements management, significantly reducing the risk of unauthorized access and data breaches.

Control 2: Regular Access Rights Review and Recertification Process

Control Objective: To ensure that access rights and entitlements remain aligned with users' current roles and responsibilities, reflecting the principle of least privilege through regular reviews and recertification of access permissions.

Implementation Steps:

2.1 **Schedule Regular Reviews**: Establish a regular schedule for the review and recertification of access rights, such as semi-annually or annually, depending on the sensitivity of the information and systems.

2.2 **Automated Review and Recertification Tools**: Deploy automated tools to streamline the review and recertification process, enabling managers and security teams to assess and adjust access rights as needed efficiently.

2.3 **Documentation and Audit Trail**: Ensure that all changes to access rights are documented and auditable, providing a clear trail of who approved the access, the rationale for access, and when it was reviewed or modified.

Expected Outcome: Regularly reviewing and recertifying access rights will ensure that users have only the access necessary for their current job functions, minimizing the risk of excessive privileges that could lead to security vulnerabilities. The process will also ensure compliance with internal policies and external regulations.

Control 3: Integration of Access Control Systems with IT and Security Operations

Control Objective: To enhance security and operational efficiency by integrating access control systems with other IT and security operations, ensuring a cohesive and automated approach to access management and security monitoring.

Implementation Steps:

3.1 **Identify Integration Opportunities**: Map out current IT and security systems and identify key areas where integration with access control systems could improve security and operational efficiency, such as incident response, user behavior analytics, and identity management.

3.2 **Implement Integration Solutions**: Develop or procure integration solutions that enable the seamless flow of information between access control systems and other IT and security operations, ensuring compatibility and secure data exchange.

3.3 **Continuous Improvement and Evolution**: Regularly review the integration setup to identify improvements and adapt to evolving security challenges and technological advancements, ensuring the integrated system remains effective and up-to-date.

Expected Outcome: Integrating access control systems with other IT and security operations will create a more unified and efficient approach to security management, enabling faster response to security incidents, more effective monitoring and reporting, and streamlined access management processes.

PR.AA-06: Physical Access to Assets Is Managed, Monitored, and Enforced Commensurate with Risk

Physical access control systems (PACS) are crucial for securing an organization's physical assets and protecting sensitive information from unauthorized access. These systems serve as the first line of defense, preventing unauthorized personnel from accessing critical infrastructure and data centers. Implementing robust PACS involves a comprehensive approach, starting with a risk assessment to identify potential vulnerabilities and determine the appropriate level of security required to mitigate these risks.

Risk assessment for physical security involves evaluating the potential threats to an organization's physical assets, including hardware, facilities, and personnel. This assessment helps prioritize security measures based on the risk associated with different assets and locations. By understanding the specific security requirements, organizations can tailor their PACS to effectively protect against identified risks, ensuring that resources are allocated efficiently to where they are needed most.

Implementing PACS to manage and monitor physical access is a complex process that requires careful planning and execution. This includes deploying physical barriers, such as gates, turnstiles, and secured doors, as well as electronic systems like card readers, biometric scanners, and surveillance cameras. These measures work together to ensure that only authorized individuals can access secured areas while also providing a means to monitor and record access events for future review.

Procedures for granting, modifying, and revoking physical access rights are essential to an effective PACS. These procedures must be clearly defined and strictly followed to prevent unauthorized access. The process for granting access should include verification of the individual's identity and authorization level. In contrast, modifications and revocations of access rights should be promptly executed to reflect changes in employment status or roles within the organization.

Integrating physical and logical access control systems enhances overall security by providing a cohesive approach to access management. This integration ensures that physical access controls align with cyber security policies, creating a comprehensive security posture that addresses physical and digital threats. For example, an individual's physical access to a data center can be linked to network access, ensuring access rights are consistent across both domains.

Monitoring and logging physical access events and incidents are critical for detecting and responding to security breaches. These records provide valuable insights into access patterns and can help identify suspicious activities. Effective monitoring involves real-time surveillance and automated alerts for unauthorized access attempts, allowing security personnel to respond quickly to potential security incidents.

Regular testing and maintenance of physical security measures are necessary to ensure their effectiveness and reliability. This includes routine checks of access control devices, physical barriers, and surveillance equipment to identify and rectify any malfunctions or vulnerabilities. Regular testing ensures the operational integrity of physical security measures and helps adapt to new threats and technologies.

Training programs on physical security awareness and procedures are vital in fostering a security culture within the organization. Employees should be educated on the importance of physical security, the proper use of access control systems, and the protocols for reporting suspicious activities or security breaches. This training empowers individuals to act as an additional layer of security, complementing the technical and procedural measures in place.

Reviewing and updating physical access controls based on evolving risks ensures that security measures remain effective. As threats evolve and new vulnerabilities emerge, organizations must continuously assess and adjust their physical security strategies to address these changes. This ongoing review and improvement process is essential for maintaining a secure and resilient physical environment.

Recommendations

- **Conduct Comprehensive Risk Assessments Regularly**: Perform regular risk assessments to identify and evaluate potential threats to physical assets. Use these assessments to inform and prioritize the implementation of security measures.
- **Integrate Physical and Logical Security Systems**: Develop an integrated security strategy encompassing physical and digital access controls. This approach ensures a unified defense against a wide range of threats.
- **Implement Advanced Monitoring and Surveillance Technologies**: Leverage the latest surveillance and monitoring technologies to enhance your physical security system's detection and response capabilities.
- **Establish Rigorous Access Control Procedures**: Define and enforce strict procedures for granting, modifying, and revoking physical access. Ensure these procedures are followed consistently to maintain the integrity of your access control system.
- **Invest in Physical Security Training and Awareness Programs**: Provide regular training and awareness programs for all employees on the importance of physical security and their role in maintaining it. Empower your workforce to contribute to a secure working environment.

NIST 800-53 Controls

(Refer to our appendix for a comprehensive description of the NIST 800-53 Controls, essential for organizations adhering to strict regulatory standards.)

- PE-02,PE-03,PE-04,PE-05,PE-06,PE-08,PE-18,PE-19,PE-20

Simplified Security Controls (SSC)

This section outlines our SSC, tailored for environments beyond the scope of NIST compliance. These controls offer streamlined, effective security solutions suitable for any organization seeking to enhance its cybersecurity defenses with clear, actionable measures.

Control 1: Comprehensive Risk Assessment for Physical Security

Control Objective: To ensure that physical access controls are appropriately aligned with the assessed risks to the organization's assets, preventing unauthorized access to critical infrastructure and sensitive areas.

Implementation Steps:

1.1 **Conduct a Comprehensive Risk Assessment**: Perform a detailed risk assessment focusing on physical security vulnerabilities, considering factors like location security, potential threats, asset criticality, and previous security incidents.

1.2 **Prioritize Assets and Areas for Enhanced Security**: Based on the risk assessment outcomes, prioritize assets and areas that require enhanced physical security measures, directing resources where they are most needed.

1.3 **Develop Risk Mitigation Strategies**: For identified risks, develop specific mitigation strategies, including implementing additional physical security controls or enhancing existing ones.

Expected Outcome: The organization will clearly understand its physical security risks and a prioritized plan for addressing them. This ensures that physical access controls are commensurate with the identified risks, effectively protecting critical assets and sensitive areas from unauthorized access.

Control 2: Integration of Physical and Logical Access Control Systems

Control Objective: To create a cohesive security posture by integrating PACS with logical access controls, ensuring synchronized access rights management across physical and digital domains.

Implementation Steps:

2.1 **Assess Integration Requirements**: Evaluate the existing physical and logical access control systems to identify integration requirements and potential challenges, ensuring compatibility and seamless data exchange.

2.2 **Implement Integrated Access Control Solution**: Deploy an integrated access control solution that synchronizes physical and logical access rights, allowing for a unified access management and monitoring approach.

2.3 **Continuous Monitoring and Adjustment**: Establish processes for continuously monitoring the integrated access control system, ensuring that any discrepancies between physical and logical access rights are quickly identified and rectified.

Expected Outcome: Integrating physical and logical access control systems will ensure a holistic security approach, where changes in one domain are automatically reflected in the other. This enhances the organization's ability to manage and monitor access rights, improving security and operational efficiency.

Control 3: Regular Training and Awareness Programs for Physical Security

Control Objective: To enhance the overall security culture of the organization by providing regular training and awareness programs on physical security awareness and procedures, ensuring that all employees understand their role in maintaining a secure environment.

Implementation Steps:

3.1 **Develop Comprehensive Training Material**: Create engaging and informative training materials that cover key physical security concepts, policies, procedures, and the importance of vigilance in maintaining security.

3.2 **Conduct Regular Training Sessions**: Schedule regular training sessions for all employees, including specific training for security personnel on the operation and monitoring of PACS.

3.3 **Implement an Awareness Campaign**: Launch an ongoing physical security awareness campaign using posters, emails, and intranet resources to reinforce key messages and maintain high levels of awareness across the organization.

Expected Outcome: Implementing regular training and awareness programs will result in a well-informed workforce that understands the importance of physical security and is vigilant in detecting and reporting potential security threats or breaches. This contributes to a stronger security posture and a safer working environment.

15

Awareness and Training (PR.AT)

> Cultivating a culture of cybersecurity awareness is akin to arming every individual with the knowledge and tools to act as both shield and sentinel, transforming the collective workforce into a formidable line of defense against the ceaseless tide of cyber threats.

In cybersecurity, the imperative of equipping all personnel with the necessary awareness and skills cannot be understated. This is achieved by establishing comprehensive cybersecurity awareness and training programs as the foundation for cultivating a vigilant and informed workforce. These programs are meticulously designed to integrate fundamental cybersecurity practices into the fabric of general training sessions, utilizing interactive methods to foster engagement and enhance the learning experience. With the landscape of cyber threats constantly evolving, the curriculum undergoes regular updates to reflect the latest emerging threats and trends. Moreover, the initiative extends beyond mere training, embedding a culture of security throughout the organization through mandatory sessions for all employees, continuous awareness campaigns, and a structured feedback system to assess these programs' effectiveness, laying the groundwork for ongoing improvement in cybersecurity practices.

PR.AT-01: Personnel Are Provided with Awareness and Training So That They Possess the Knowledge and Skills to Perform General Tasks with Cybersecurity Risks in Mind

Cybersecurity awareness and training within any organization cannot be overstated. Personnel at all levels need the knowledge and skills to recognize and mitigate cybersecurity risks as part of their daily responsibilities. This requirement is underscored by the NIST Cybersecurity Framework, specifically through the Protect function's aspect PR.AT-01 emphasizes that personnel are provided with awareness and training. This section delves into the multifaceted approach to developing and implementing a robust cybersecurity awareness and training program that aligns with PR.AT-01 requirements.

Implementing cybersecurity awareness and training programs is fundamental for all personnel across the organization. An introductory program serves as the cornerstone for building a secure organizational culture, ensuring that every member understands the basics of cybersecurity. This includes recognizing potential cyber threats, understanding the importance of maintaining digital hygiene, and being aware of the organization's policies and procedures regarding information

A Comprehensive Guide to the NIST Cybersecurity Framework 2.0: Strategies, Implementation, and Best Practice, First Edition. Jason Edwards.
© 2025 John Wiley & Sons Ltd. Published 2025 by John Wiley & Sons Ltd.

security. A well-structured introduction sets the stage for more detailed and role-specific training programs.

Developing a comprehensive cybersecurity awareness curriculum is critical in reinforcing the organization's security posture. This curriculum should cover various topics, from password security and phishing awareness to advanced threat detection techniques. Tailoring the content to the audience's level of expertise ensures that the training is accessible and engaging. Moreover, incorporating real-world scenarios and recent case studies can significantly enhance the learning experience, making the theoretical knowledge applicable to everyday situations.

Incorporating basic cybersecurity practices into general training sessions ensures that all employees, regardless of their role, understand how their actions can impact the organization's cybersecurity. These sessions can cover topics such as secure password creation, the importance of regular software updates, and the correct handling of sensitive information. Embedding these practices into the training framework cultivates a baseline security awareness among all personnel.

The use of interactive and engaging training methods is crucial to engage participants and enhance the retention of cybersecurity concepts. Gamification, simulations, and scenario-based training exercises can transform traditional learning experiences into dynamic and interactive sessions. These methods increase engagement and allow participants to practice their response to cybersecurity incidents in a controlled environment, fostering a deeper understanding of the material.

Keeping the training content up-to-date with the latest cybersecurity threats and trends is vital for maintaining an effective awareness and training program. The cyber threat landscape continuously evolves, with new vulnerabilities and attack vectors emerging regularly. Regular updates to the training content ensure that the organization's personnel are aware of the latest threats and understand how to protect against them.

Implementing mandatory training sessions for new and existing employees guarantees that everyone within the organization understands cybersecurity risks and best practices. These sessions can be part of the onboarding process for new hires and an ongoing requirement for all personnel, ensuring that the organization's cybersecurity knowledge base remains strong and up-to-date.

Regular cybersecurity awareness campaigns and activities can supplement formal training sessions, keeping cybersecurity in employees' minds. These campaigns can include newsletters, posters, quizzes, and even mock phishing exercises. Such activities reinforce the training and help create a culture of security awareness that permeates the entire organization.

Finally, implementing assessment and feedback mechanisms is crucial for evaluating the effectiveness of cybersecurity awareness and training programs. Surveys, quizzes, and practical assessments can provide valuable insights into how well the personnel understand and apply their cybersecurity knowledge. Feedback from participants can also guide the continuous improvement of the training content and delivery methods, ensuring that the program remains effective and relevant.

Recommendations

- **Develop Role-Specific Training Modules**: Tailor cybersecurity training programs to address different organizational roles' unique requirements and risks. This ensures that each employee receives relevant and practical knowledge to safeguard against cybersecurity threats in their work area.
- **Leverage Behavioral Analytics to Identify Training Needs**: Utilize behavioral analytics to monitor employee actions and identify common security mistakes. This data can inform targeted

training interventions, focusing efforts where needed to improve the organization's cybersecurity posture.

- **Establish a Cybersecurity Mentoring Program**: Pair less experienced employees with cybersecurity mentors who can provide guidance, share experiences, and offer advice on best practices. This peer-learning approach can enhance the overall effectiveness of the cybersecurity training program.
- **Integrate Cybersecurity Awareness into Daily Operations**: Embed cybersecurity awareness into the fabric of daily work routines. For example, incorporate reminders about secure practices in system login messages or email signatures to keep cybersecurity in mind for all personnel.
- **Regularly Review and Update the Cybersecurity Training Curriculum**: Establish a regular review cycle for the training curriculum to ensure it remains current with the latest cybersecurity trends and threats. Involving cybersecurity professionals in this process can provide valuable insights into emerging threats and the latest defensive tactics.

NIST 800-53 Controls

(Refer to our appendix for a comprehensive description of the NIST 800-53 Controls, essential for organizations adhering to strict regulatory standards.)

- AT-02,AT-03

Simplified Security Controls (SSC)

This section outlines our simplified security controls (SSC), tailored for environments beyond the scope of NIST compliance. These controls offer streamlined, effective security solutions suitable for any organization seeking to enhance its cybersecurity defenses with clear, actionable measures.

Control 1: Integration of Cybersecurity Awareness into Employee Onboarding and Continuous Education Programs

Control Objective: To ensure that all employees, from the moment they join the organization, are equipped with the necessary cybersecurity awareness to function effectively and safely within the digital workplace. This control aims to establish a foundational understanding of cybersecurity best practices and maintain this awareness through continuous education.

Implementation Steps:

1.1 **Develop Onboarding Cybersecurity Modules**: Create comprehensive cybersecurity awareness modules for the employee onboarding process. These modules should cover essential cybersecurity practices, organizational policies, and identifying and reporting potential cyber threats.

1.2 **Incorporate Regular Cybersecurity Training Updates**: Establish a schedule for regular updates to the cybersecurity training programs. This schedule should account for emerging threats and evolving best practices, ensuring all employees receive up-to-date information and training.

1.3 **Implement Continuous Cybersecurity Education**: Introduce mechanisms for continuous cybersecurity education, such as annual refresher courses, cybersecurity newsletters, and participation in cybersecurity awareness months or events.

Expected Outcome: Employees will begin their tenure with a strong foundation in cybersecurity awareness, which is maintained and enhanced through ongoing education. This continuous learning environment fosters a culture of security awareness across the organization, significantly reducing the risk of security breaches caused by human error.

Control 2: Creation of a Cybersecurity Feedback and Incident Reporting System

Control Objective: To develop an efficient, user-friendly system for reporting cybersecurity incidents and providing feedback on the organization's cybersecurity policies and training programs. This system aims to facilitate a rapid response to incidents and continuously improve cybersecurity awareness initiatives.

Implementation Steps:

2.1 **Implement an Incident Reporting Platform**: Deploy a secure, easily accessible platform where employees can report cybersecurity incidents and suspicious activities without fear of reprisal. This platform should ensure the confidentiality and anonymity of the reporter, if necessary.

2.2 **Establish a Feedback Mechanism for Training Programs**: Create a feedback system for employees to provide insights and suggestions on cybersecurity training and awareness programs. This could be through surveys, suggestion boxes, or digital feedback forms.

2.3 **Analyze Feedback and Incident Reports for Continuous Improvement**: Regularly review the feedback and incident reports to identify trends, areas for improvement, and the effectiveness of current cybersecurity measures. Use this analysis to inform future training content and cybersecurity policies.

Expected Outcome: Establishing a feedback and incident reporting system will promote a proactive security culture where employees feel empowered to contribute to the organization's cybersecurity health. This system enables the organization to respond swiftly to threats and refine its cybersecurity awareness efforts based on direct employee input, enhancing the overall effectiveness of its security posture.

Control 3: Regular Cybersecurity Simulations and Drills

Control Objective: To assess and enhance the organization's preparedness for real-world cybersecurity incidents through the regular execution of simulations and drills. These exercises aim to test the effectiveness of current security protocols and the employees' ability to respond to cyber threats.

Implementation Steps:

3.1 **Plan and Schedule Regular Cybersecurity Drills**: Organize periodic cybersecurity simulations tailored to mimic recent cyber threat trends, such as phishing exercises or breach response drills. These drills should be scheduled at regular intervals and involve all levels of the organization.

3.2 **Analyze Performance and Identify Areas for Improvement**: Conduct a thorough analysis of the organization's response after each simulation. Identify strengths, weaknesses, and areas needing improvement, focusing on technical defenses and employee actions.

3.3 **Provide Targeted Training Based on Drill Outcomes**: Develop targeted training sessions or modules to address the specific weaknesses identified during the simulations. This training should bolster the organization's cybersecurity defenses by improving employee knowledge and response capabilities.

Expected Outcome: Regularly executing cybersecurity simulations and drills prepares the organization for real cyber incidents by testing and improving its response protocols and employee readiness. This proactive approach reduces the potential impact of actual cyberattacks and strengthens the organization's overall cybersecurity posture.

PR.AT-02: Individuals in Specialized Roles Are Provided with Awareness and Training So That They Possess the Knowledge and Skills to Perform Relevant Tasks with Cybersecurity Risks in Mind

The nuanced demands of cybersecurity necessitate that individuals in specialized roles receive tailored awareness and training. This specificity ensures that these key personnel comprehend the broader cybersecurity landscape and possess deep insights into the risks and protocols pertinent to their specific functions. The NIST Cybersecurity Framework underscores this need through PR.AT-02, focusing on the development and delivery of targeted training initiatives. This section outlines strategies for identifying roles that require specialized training, developing role-specific modules, integrating advanced topics, and ensuring the continuous evolution of these educational programs.

Identifying specialized roles that require advanced cybersecurity knowledge is the initial step toward crafting a tailored training program. This process thoroughly analyzes the organization's structure, the critical functions supporting its core mission, and the specific cybersecurity risks associated with these roles. Positions that handle sensitive information, manage critical infrastructure, or are involved in the organization's cybersecurity defenses are typically prioritized for advanced training. Recognizing these roles allows for the development of focused training objectives aligned with the organization's cybersecurity strategy.

The development of role-specific cybersecurity training modules follows the identification of specialized roles. These modules address each role's unique responsibilities and risks, going beyond general cybersecurity awareness to cover in-depth topics such as advanced threat analysis, incident response tactics, and secure coding practices. Effective training modules are structured to balance theoretical knowledge with practical applications, making the learning experience informative and engaging for the participants.

Integrating advanced cybersecurity topics relevant to specialized roles is crucial for the efficacy of the training program. This integration ensures that the content covers basic cybersecurity principles and delves into complex issues such as cryptographic protocols, network security architectures, and forensic analysis techniques. The choice of topics should reflect the latest developments in the field of cybersecurity, ensuring that the training remains current and relevant.

Coordination with external experts for specialized training sessions can enhance the depth and breadth of the training program. Bringing in professionals with cutting-edge knowledge and real-world experience adds a valuable dimension to the training, exposing participants to current trends and emerging threats. Collaborations with academic institutions, cybersecurity firms, and industry consortia can provide access to a wider range of expertise and resources.

Customization of training programs to address the specific cybersecurity challenges faced by the organization is essential for maximizing the impact of the training. This customization involves adapting the training content to reflect the organization's risk profile, regulatory requirements, and technological infrastructure. Customized training ensures that participants can directly apply what they learn to mitigate risks and enhance the security of their specific operational environments.

Providing hands-on training and simulation exercises is key to effective learning for specialized roles. Practical exercises such as penetration testing, incident response drills, and secure coding workshops allow participants to apply theoretical knowledge in a controlled, realistic environment. This experiential learning approach reinforces concepts and skills, making it easier for participants to translate training into effective cybersecurity practices in their daily work.

Regular review and update of specialized training content are critical to maintaining the relevance and effectiveness of the program. The dynamic nature of cybersecurity threats and technologies necessitates continuously adapting training materials to incorporate new knowledge, best practices, and case studies. This iterative process ensures that the training remains aligned with current challenges and advances in the field.

Certification and continuous education opportunities for specialized personnel underscore the importance of ongoing professional development in cybersecurity. Offering pathways for certification in specific cybersecurity domains not only validates the expertise of the individuals but also encourages a culture of continuous learning and improvement. Continuous education initiatives, such as workshops, seminars, and conferences, provide platforms for sharing knowledge, networking with peers, and staying abreast of innovations in cybersecurity.

Recommendations

- **Foster Collaboration with Industry and Academic Partners**: Establish partnerships with leading cybersecurity organizations and academic institutions to access cutting-edge research, training resources, and expert instructors. This collaboration can significantly enhance the quality and relevance of specialized training programs.
- **Implement Adaptive Learning Platforms**: Utilize adaptive learning technologies to personalize the training experience for specialized roles. These platforms can assess individual learning styles and knowledge gaps, adjusting the training content accordingly to maximize engagement and retention.
- **Encourage Participation in Cybersecurity Competitions**: Support and encourage specialized personnel to participate in cybersecurity competitions and capture-the-flag events. These competitive environments offer unique challenges that test skills in a practical, high-pressure context, fostering problem-solving abilities and innovation.
- **Leverage Virtual Reality (VR) for Immersive Training**: Integrate VR technology into training programs to create immersive simulation environments. This approach can simulate complex cybersecurity scenarios in a safe and controlled setting, providing hands-on experience in tackling sophisticated threats.
- **Establish a Continuous Feedback Loop**: Create mechanisms for ongoing feedback from training participants and stakeholders. This feedback loop allows for the timely adjustment of training content and methods, ensuring that the program effectively addresses the evolving needs of specialized roles and the broader cybersecurity landscape.

NIST 800-53 Controls

(Refer to our appendix for a comprehensive description of the NIST 800-53 Controls, essential for organizations adhering to strict regulatory standards.)

- AT-03

Simplified Security Controls (SSC)

This section outlines our SSC, tailored for environments beyond the scope of NIST compliance. These controls offer streamlined, effective security solutions suitable for any organization seeking to enhance its cybersecurity defenses with clear, actionable measures.

Control 1: Development of Specialized Cybersecurity Training Labs

Control Objective: To create dedicated training environments that simulate real-world cybersecurity scenarios, enabling individuals in specialized roles to gain practical, hands-on experience. These labs complement theoretical knowledge by applying advanced cybersecurity practices in a controlled setting.

Implementation Steps:

1.1 **Identify Key Scenarios and Technologies**: Work with cybersecurity experts to identify critical scenarios and technologies relevant to the organization's specialized roles. This could include scenarios like advanced persistent threats, incident response, and secure coding practices.

1.2 **Lab Setup and Development**: Develop training labs with the necessary software, hardware, and network configurations to simulate the identified scenarios. Ensure these labs are isolated from the organization's live environment to prevent potential risks.

1.3 **Integrate Labs into Training Modules**: Seamlessly integrate these labs into the specialized training modules, ensuring participants can apply what they learn in a realistic environment. Schedule regular lab sessions as part of the ongoing training curriculum.

Expected Outcome: Developing specialized cybersecurity training labs provides a dynamic learning environment where individuals can apply theoretical knowledge to practical situations. This hands-on experience is crucial for understanding the complexities of cybersecurity threats and responses, enhancing the skill set of those in specialized roles, and improving the organization's overall security posture.

Control 2: Cybersecurity Role-Based Competency Framework

Control Objective: To establish a structured framework that defines the competencies and skills required for each specialized cybersecurity role within the organization. This framework aims to standardize specialized personnel's training and development paths, ensuring a consistent level of expertise across all cybersecurity functions.

Implementation Steps:

2.1 **Define Role-Specific Competencies**: Collaborate with department leads and cybersecurity experts to define each specialized role's essential competencies and skills. This includes both technical skills and soft skills related to cybersecurity.

2.2 **Develop Assessment Mechanisms**: Create mechanisms for assessing the current competency levels of personnel in specialized roles. This could involve tests, simulations, or performance evaluations.

2.3 **Align Training Programs with Competency Needs**: Tailor the cybersecurity training programs to address the identified competency gaps. This includes updating existing modules and developing new content focused on the specific needs of each role.

Expected Outcome: Implementing a cybersecurity role-based competency framework ensures that all specialized personnel possess the necessary skills and knowledge to perform their roles effectively. This standardization leads to a more robust cybersecurity posture, with a clear path for continuous professional development and skill enhancement within the organization.

Control 3: Partnership for Cybersecurity Innovation and Learning

Control Objective: To foster a culture of continuous learning and innovation in cybersecurity by establishing partnerships with academic institutions, research organizations, and industry leaders. This control aims to provide specialized personnel access to cybersecurity research, trends, and technologies.

Implementation Steps:

3.1 **Identify and Establish Partnerships**: Identify potential academic and industry partners that lead in cybersecurity research and innovation. Formalize partnerships through agreements that provide access to research materials, experts, and advanced training technologies.

3.2 **Incorporate Cutting-Edge Knowledge into Training**: Leverage the insights and materials from these partnerships to enrich the cybersecurity training programs. This includes guest lectures, access to advanced cybersecurity tools, and participation in joint research projects.

3.3 **Facilitate Knowledge Exchange and Collaboration**: Organize regular workshops, seminars, and forums that allow specialized personnel to interact with external experts, share experiences, and discuss the latest developments in cybersecurity.

Expected Outcome: Establishing partnerships for cybersecurity innovation and learning ensures that the organization remains at the forefront of cybersecurity developments. This access to external expertise and resources enriches the training programs for specialized roles, fostering a culture of innovation and continuous improvement in the organization's cybersecurity practices.

16

Data Security (PR.DS)

> Securing data in all its states isn't just a technical necessity; it's a foundational pillar of trust and resilience, ensuring that every byte of information remains under our guardianship, irrespective of its phase in the digital lifecycle.

In cybersecurity, protecting data across different states—whether at rest, in transit, or use—is fundamental to ensuring the overall security of an organization's informational assets. The confidentiality, integrity, and availability of data at rest are safeguarded through encryption techniques, stringent access control measures, and regular security audits, ensuring that sensitive information stored on devices remains secure from unauthorized access or alterations. Similarly, data-in-transit requires robust protection facilitated by encryption protocols like transport layer security (TLS) and secure sockets layer (SSL), secure authentication mechanisms, and vigilant monitoring systems to prevent interception or breaches during transmission. For data-in-use, applying access controls, employing end-to-end encryption for processing sensitive information, and implementing secure coding practices are critical to safeguarding information in applications and processes. Finally, a comprehensive approach to data backups—characterized by regular, encrypted, and redundant backup processes—ensures that data can be recovered following unexpected incidents, maintaining the continuity and resilience of business operations.

Author's Note: PR.DS-03 through PR.DS-09 were deprecated from the CSF 2 and are not included in this chapter.

PR.DS-01: The Confidentiality, Integrity, and Availability of Data at Rest Are Protected

The safeguarding of data at rest encompasses a comprehensive strategy aimed at protecting data stored, as opposed to data in transit. This involves ensuring the stored information's confidentiality, integrity, and availability. Data at rest typically includes information stored on physical devices such as hard drives, SSDs, USBs, or backup tapes. Protecting this data is crucial as it often contains sensitive information, including personal details, financial records, or intellectual property. A breach or unauthorized access to such data can lead to significant financial losses, legal consequences, and damage to an organization's reputation.

Encryption is one of the most effective methods for securing data at rest. By converting data into an unreadable coded format without the proper decryption key, encryption helps ensure

A Comprehensive Guide to the NIST Cybersecurity Framework 2.0: Strategies, Implementation, and Best Practice, First Edition. Jason Edwards.
© 2025 John Wiley & Sons Ltd. Published 2025 by John Wiley & Sons Ltd.

that, even if data storage devices are lost or stolen, the information remains protected from unauthorized access. Techniques such as full disk encryption (FDE) or self-encrypting drives provide comprehensive protection, automatically securing all data on a device. Encryption solutions should be selected based on their robustness, key management practices, and compliance with relevant industry standards and regulations.

Access control is another pivotal aspect of protecting data at rest. Stringent access control measures ensure that only authorized users can access or modify stored data. This typically involves multifactor authentication, user role definition, and access logs. By clearly defining who has access to what data and tracking access, organizations can significantly reduce the risk of internal threats and unauthorized access incidents. Access control systems should be integrated with the security infrastructure to ensure cohesive protection measures.

Maintaining data integrity at rest is crucial to prevent unauthorized modifications from compromising data quality or leading to fraudulent activities. This can be achieved by implementing checksums, cryptographic hashes, and digital signatures, which verify that data has not been altered. Regular integrity checks should be part of the organization's routine security procedures to promptly detect and mitigate any unauthorized changes to data.

Redundancy solutions, such as RAID configurations or off-site backups, are essential to ensure data availability during hardware failure, natural disasters, or cyberattacks. By creating copies of data and storing them in multiple locations, organizations can ensure that they can recover critical data quickly and continue operations with minimal downtime. Regular testing of backup and recovery procedures is crucial to ensure they are effective when needed.

Regular security audits and assessments are vital for identifying vulnerabilities and gaps in protecting data at rest. These audits should thoroughly examine encryption practices, access control measures, and redundancy solutions. External and internal audits objectively review security practices and help ensure that data-at-rest protections are up-to-date and aligned with best practices and compliance requirements.

Documentation of data storage policies and procedures is essential for ensuring that all employees understand their roles and responsibilities in protecting data at rest. These documents should outline the security measures in place, procedures for accessing and handling data, and steps to take in the event of a security breach. Clear, accessible documentation helps foster a culture of security awareness and compliance within the organization.

Employee training on best practices for handling data at rest is crucial. Regular training sessions should cover topics such as recognizing phishing attempts, proper data handling procedures, and following security protocols. Empowering employees with the knowledge and tools to protect sensitive information is critical to a comprehensive data-at-rest security strategy.

Recommendations

- **Develop and Implement a Comprehensive Encryption Strategy**: Establish robust encryption practices for all stored data, focusing on key management and adopting industry-standard encryption technologies.
- **Enhance Access Control Measures**: Implement strict access control policies and procedures, including multifactor authentication and role-based access, to ensure that only authorized individuals can access sensitive data.
- **Regularly Conduct Data Integrity Checks**: Schedule and execute regular integrity checks on stored data to detect and promptly address any unauthorized alterations.

- **Establish a Redundant Data Storage System**: Implement and maintain a robust data redundancy solution, including regular testing of backup and recovery processes, to ensure data availability during system failures or disasters.
- **Foster a Culture of Security Awareness**: Create comprehensive training programs for all employees, emphasizing the importance of protecting data at rest and the practices necessary to maintain security.

NIST 800-53 Controls

(Refer to our appendix for a comprehensive description of the NIST 800-53 Controls, essential for organizations adhering to strict regulatory standards.)

- CA-03,CP-09,MP-08,SC-04,SC-07,SC-12,SC-13,SC-28,SC-32,SC-39,SC-43,SI-03,SI-04,SI-07

Simplified Security Controls (SSC)

This section outlines our simplified security controls (SSC), tailored for environments beyond the scope of NIST compliance. These controls offer streamlined, effective security solutions suitable for any organization seeking to enhance its cybersecurity defenses with clear, actionable measures.

Control 1: Implementation of FDE for All Data Storage Devices

Control Objective: To ensure the confidentiality and integrity of data at rest by implementing FDE across all data storage devices within the organization, rendering data unreadable to unauthorized users without the correct decryption key.

Implementation Steps:

1.1 **Inventory and Assessment**: Conduct a comprehensive inventory of all data storage devices used within the organization to identify those requiring encryption.

1.2 **Encryption Software Selection and Deployment**: Select an industry-standard FDE solution that meets organizational security requirements and deploys it across all identified devices.

1.3 **Key Management and Access Control**: Implement a secure key management process to ensure that decryption keys are accessible only to authorized personnel and integrate the encryption solution with existing access control mechanisms.

Expected Outcome: All data stored on the organization's devices is encrypted, significantly reducing the risk of unauthorized data access and ensuring that data remains confidential and integral even in physical theft or loss.

Control 2: Development and Enforcement of Comprehensive Access Control Policies

Control Objective: To limit access to sensitive data at rest to authorized users only, reducing the risk of unauthorized data exposure and modification.

Implementation Steps:

2.1 **Access Needs Analysis**: Identify and document the access requirements for all data at rest based on organizational roles.

2.2 **Policy Creation and Implementation**: Develop comprehensive access control policies that include the principles of least privilege and role-based access control (RBAC).

2.3 **Regular Access Reviews**: Establish a routine schedule for reviewing and updating access controls and permissions to ensure they remain appropriate and that unauthorized access is quickly identified and rectified.

Expected Outcome: Access to data at rest is strictly controlled, limiting exposure to unauthorized individuals and reducing the risk of data breaches while ensuring that legitimate users can access the data they need for their roles.

Control 3: Establishment of Routine Data-at-Rest Security Audits

Control Objective: To identify and address vulnerabilities in protecting data at rest through regular security audits and assessments, thereby maintaining the integrity, confidentiality, and availability of stored data.

Implementation Steps:

3.1 **Audit Planning and Scheduling**: Develop a comprehensive audit plan for data-at-rest security, including schedules for regular audits and criteria for unexpected audits triggered by security incidents.

3.2 **Execution of Audits**: Perform thorough security audits of all systems and storage devices holding data at rest, utilizing internal and external audit teams for unbiased assessments.

3.3 **Remediation and Improvement**: Address identified vulnerabilities based on audit findings, prioritize fixes based on risk, and re-audit as necessary to confirm resolution.

Expected Outcome: Regular audits provide ongoing insight into the security posture of data at rest, leading to continuous improvement in data protection measures and reduced risk of data breaches or loss.

PR.DS-02: The Confidentiality, Integrity, and Availability of Data in Transit Are Protected

Ensuring data confidentiality, integrity, and availability in transit is critical in today's interconnected digital world. Data-in-transit refers to data moving from one location to another, such as across the internet or a private network. As data travels, it becomes vulnerable to interceptions, unauthorized access, and tampering. Protecting this data ensures that sensitive information, such as personal details, financial information, or corporate secrets, remains private and unaltered during transmission. This section delves into the strategies and technologies essential for securing data-in-transit, underlining their importance in safeguarding information flows in various network environments.

The use of encryption protocols is fundamental in securing data in transit. Protocols such as TLS, SSL, and Internet Protocol Security (IPSec) provide a secure channel over which data can travel, ensuring that it remains incomprehensible and secure from eavesdroppers and hackers. The protocol choice depends on the specific requirements and context of data transmission. Effective implementation of these encryption protocols is crucial; they should be kept up to date to protect against emerging threats and vulnerabilities.

Secure authentication mechanisms play a pivotal role in safeguarding data in transit. These mechanisms ensure that only authorized users can access and transmit data. Implementations may include multifactor authentication, digital certificates, and biometric verification. Secure authentication helps prevent unauthorized access and ensures that data in transit is accessible only to intended recipients. It is the first line of defense against unauthorized data breaches.

Implementing robust network security controls is essential for protecting the channels through which data transits. These controls can include firewalls, intrusion detection systems (IDS), and intrusion prevention systems (IPS), which monitor and regulate the data passing through network points. Network security also involves securing endpoints, ensuring secure configurations, and isolating sensitive data transmissions within protected network segments.

Monitoring and detection systems are critical for identifying and responding to data breaches in transit. These systems analyze data flows for unusual patterns or signs of malicious activity, enabling rapid detection and mitigation of potential violations. Effective monitoring requires continuously examining network traffic and implementing advanced detection algorithms to identify possible threats accurately.

Virtual private network (VPN) solutions provide a secure means for remote access to data, ensuring that data in transit is encrypted and protected from interception. VPNs create a secure, encrypted tunnel between the user's device and the network, safeguarding data as it travels across public or less secure networks. This is especially important for remote workers and those accessing company resources outside the corporate network.

Developing and enforcing policies for secure data sharing and transfer is fundamental to data-in-transit security. These policies should outline acceptable methods and protocols for data transmission, identify authorized users and devices, and specify the data types that can be transmitted under different circumstances. Clear, comprehensive policies help establish a shared understanding of data protection practices among all stakeholders.

Training employees on secure communication and data transfer practices is crucial. Regular training sessions should cover the importance of data security, encryption, secure file transfer methods, and the dangers of unsecured networks. Well-informed employees are less likely to make mistakes that could lead to data breaches, making training a critical component of an overall data protection strategy.

Regular testing of data-in-transit security measures is essential to ensure their effectiveness. Security measures should be audited and tested regularly to identify vulnerabilities and ensure all systems are current. Penetration testing, security audits, and vulnerability assessments can help organizations identify weaknesses and take corrective action before attackers can exploit them.

Recommendations

- **Implement End-to-End Encryption**: Adopt robust encryption protocols for all data-in-transit, ensuring that data is encrypted from its origin to its destination point, leaving no gaps for potential interception.
- **Strengthen Authentication Processes**: Enhance security by implementing strong authentication mechanisms, such as multifactor authentication and digital certificates, to ensure that only authorized users can access and transmit sensitive data.
- **Regularly Update and Patch Systems**: Maintain the security of network and encryption systems by regularly updating and patching software to protect against known vulnerabilities and threats.
- **Develop Comprehensive Security Policies**: Create detailed policies for data transmission and sharing, clearly defining acceptable practices, authorized users, and approved methods and protocols for secure data transfer.
- **Conduct Regular Security Training and Testing**: Provide ongoing security training for employees to foster awareness and understanding of secure data handling practices. Additionally, regular tests and audits of data-in-transit security measures should be conducted to identify and address potential vulnerabilities.

NIST 800-53 Controls

(Refer to our appendix for a comprehensive description of the NIST 800-53 Controls, essential for organizations adhering to strict regulatory standards.)

- AU-16,CA-03,SC-04,SC-07,SC-08,SC-11,SC-12,SC-13,SC-16,SC-40,SC-43,SI-03,SI-04,SI-07

Simplified Security Controls (SSC)

This section outlines our SSC, tailored for environments beyond the scope of NIST compliance. These controls offer streamlined, effective security solutions suitable for any organization seeking to enhance its cybersecurity defenses with clear, actionable measures.

Control 1: Deployment of Advanced Encryption for Data-in-Transit

Control Objective: To safeguard the confidentiality and integrity of data as it moves across networks by implementing advanced encryption protocols, ensuring that data remains secure and unreadable to unauthorized individuals during transmission.

Implementation Steps:

1.1 **Protocol Selection**: Choose robust encryption protocols such as TLS 1.3 for web traffic and IPSec for network communications, considering the specific needs and context of the organization's data transmission.

1.2 **Configuration and Implementation**: Properly configure encryption settings on all transmitting devices and channels, ensuring that only strong ciphers and keys are used and weak or deprecated protocols are disabled.

1.3 **Continuous Monitoring and Updating**: Regularly monitor the effectiveness of encryption protocols and perform updates to adapt to emerging threats and maintain compliance with current standards.

Expected Outcome: With the correct implementation and maintenance of advanced encryption protocols, data-in-transit is protected against eavesdropping and interception, ensuring that sensitive information remains confidential and integral during transmission.

Control 2: Implementation of Secure Access Controls for Data Transmission

Control Objective: To ensure that only authorized users can access and transmit sensitive data, thus preventing unauthorized data breaches and maintaining the integrity of data-in-transit.

Implementation Steps:

2.1 **User Identification and Authentication**: Implement secure authentication mechanisms such as multifactor authentication (MFA) for all users accessing the network to send or receive data.

2.2 **Role-based Access Control (RBAC)**: Define user roles and assign access rights based on the least privilege principle, ensuring users can only access and transmit data necessary for their role.

2.3 **Regular Access Review and Adjustment**: Conduct periodic reviews of user access rights and make necessary adjustments to account for changes in job roles, departures, or data sensitivity.

Expected Outcome: By implementing stringent access controls and ensuring regular reviews, the organization significantly reduces the risk of unauthorized data access and ensures that data-in-transit is accessed and transmitted securely by authorized personnel.

Control 3: Establishment of Continuous Monitoring and Detection Systems

Control Objective: To detect and respond to security threats and breaches affecting data-in-transit in a timely manner, thereby minimizing the impact of any unauthorized access or data leakage.

Implementation Steps:

3.1 **Monitoring System Implementation**: Deploy advanced monitoring and detection systems to analyze data flows and identify suspicious activities or deviations from normal traffic patterns.

3.2 **Alert Configuration and Response Planning**: Configure alerts for unusual activities and establish clear procedures for responding to potential data breaches or security incidents.
3.3 **Regular Review and Updates**: Periodically review the monitoring and detection systems for efficiency and update detection mechanisms to adapt to new threats and tactics attackers use.

Expected Outcome: With a robust monitoring and detection system, the organization can quickly identify and respond to security incidents affecting data-in-transit, reducing the potential damage from breaches and ensuring continuous protection of sensitive information.

PR.DS-10: The Confidentiality, Integrity, and Availability of Data-in-Use Are Protected

Understanding and implementing a comprehensive data backup strategy is a cornerstone of modern cybersecurity defenses. This involves copying data and ensuring its availability and integrity for when it is most needed—during a recovery after a data loss event. A robust backup strategy is essential for recovering from various incidents such as cyberattacks, hardware failures, or natural disasters. Ensuring that backups are created regularly, stored securely, and can be recovered reliably is fundamental to maintaining business continuity and protecting against data loss.

Creating regular, automated backup processes is central to a robust data protection strategy. Automation ensures that backups are performed consistently and reduces the risk of human error, a common cause of incomplete or missed backups. It is crucial to determine the right frequency for backups based on the criticality and dynamism of the data—more vital or frequently changed data requires more frequent backups. Additionally, implementing versioning can protect against ransomware and other malware by preserving multiple historical states of data.

Encrypting backup data is essential for protecting its confidentiality and integrity, whether stored on-site or off-site. Encryption ensures that backup data remains inaccessible to unauthorized users, even if physical media is lost or stolen. Secure storage also involves physically protecting backup media and ensuring only authorized personnel can access it. This could mean employing biometric access controls, secure off-site storage facilities, or cloud storage solutions with robust security measures.

Implementing redundant backup solutions across multiple locations helps protect against site-specific risks, such as natural disasters or localized system failures. This approach, often called the 3-2-1 rule (three total copies of your data, two of which are local but on different devices, and one copy off-site), ensures that there is always an accessible backup, even if one is compromised. Redundancy should be built into the backup strategy to ensure data can be recovered from different sources.

Access control and authentication for backup data retrieval are as important as primary data systems. Only authorized users can restore data from backups; access should be strictly controlled and monitored. Employing multifactor authentication and maintaining detailed access logs can help ensure that attackers or malicious insiders do not exploit backup systems.

Regular maintenance and updates to backup systems are necessary to ensure they remain effective and secure. This includes updating the backup software to patch known vulnerabilities, checking the physical health of backup media, and ensuring that backup systems are compatible with the data they're protecting. Regular reviews and updates help mitigate the risk of backup failures and ensure that data can be restored when needed.

Periodic testing of backup restorations is critical to ensure they can be executed correctly in a real-world recovery scenario. Regular testing helps identify any issues in the backup process and allows for the refinement of recovery procedures. It is essential to ensure that data can be restored quickly and accurately, with minimal impact on business operations.

Documentation of backup policies and recovery procedures is essential for ensuring clarity and consistency in managing backups. This documentation should be detailed, easy to understand, and readily accessible. It should cover aspects such as the scope of backups, frequency, storage locations, security measures, and step-by-step recovery procedures.

Employee training on backup and recovery processes is crucial for ensuring all staff understand their roles and responsibilities during data recovery. Training should cover how to perform backups correctly, recognize signs of data loss, and initiate a recovery process. A well-informed team is critical in effectively executing a backup and recovery plan.

Recommendations

- **Establish Regular, Automated Backup Schedules**: Create a consistent, automated backup schedule that reflects the importance and frequency of data changes, incorporating versioning to protect against malware and accidental deletion.
- **Secure and Encrypt All Backup Data**: Implement robust encryption for all backup data, both in transit and at rest, and ensure physical and digital security measures are in place to protect backup media.
- **Implement Redundant, Multilocation Backups**: Adopt a 3-2-1 backup strategy to ensure redundancy and availability, storing backups in multiple physical locations or using cloud services to mitigate site-specific risks.
- **Regularly Test and Update Backup Systems**: Conduct routine tests of backup and recovery processes to ensure data integrity and system functionality, updating systems as needed to address new threats and technological changes.
- **Develop and Maintain Comprehensive Backup Documentation**: Create detailed, up-to-date documentation of backup policies and procedures and ensure that all relevant employees are trained on proper backup and recovery processes.

NIST 800-53 Controls

(Refer to our appendix for a comprehensive description of the NIST 800-53 Controls, essential for organizations adhering to strict regulatory standards.)

- AC-02,AC-03,AC-04,AU-09,AU-13,CA-03,CP-09,SA-08,SC-04,SC-07,SC-11,SC-13,SC-24,SC-32, SC-39,SC-40,SC-43,SI-03,SI-04,SI-07,SI-10,SI-16

Simplified Security Controls (SSC)

This section outlines our SSC, tailored for environments beyond the scope of NIST compliance. These controls offer streamlined, effective security solutions suitable for any organization seeking to enhance its cybersecurity defenses with clear, actionable measures.

Control 1: Establishment of Automated and Versioned Backup Procedures

Control Objective: To ensure consistent, reliable, and secure backups of essential data by establishing automated and versioned backup procedures, reducing the risk of data loss, and facilitating efficient data recovery in the event of an incident.

Implementation Steps:

1.1 **Define Backup Requirements**: Identify critical data sets requiring regular backups and determine the frequency of these backups based on the data's sensitivity and importance.

1.2 **Implement Automation and Versioning**: Deploy backup software that supports automated scheduling and versioning, ensuring backups occur without manual intervention and multiple historical data states are preserved.

1.3 **Validate Backup Procedures**: Regularly test the backup process to ensure data is accurately and completely backed up as scheduled.

Expected Outcome: The organization achieves a systematic and fail-safe backup routine, significantly reducing the likelihood of data loss. Versioned backups ensure recovery options simultaneously, enhancing resilience against ransomware and accidental deletions.

Control 2: Development and Enforcement of Comprehensive Backup Encryption Policies

Control Objective: To protect the confidentiality and integrity of backup data by implementing and enforcing robust encryption policies, ensuring that backup data remains secure during storage and transit.

Implementation Steps:

2.1 **Policy Formulation**: Develop a comprehensive backup encryption policy covering all aspects of backup data security, including encryption standards, key management, and access controls.

2.2 **Encryption Implementation**: Apply strong encryption standards to all backup data in transit to backup locations and at rest within storage environments.

2.3 **Policy Training and Compliance Monitoring**: Train all relevant personnel on encryption policies and monitor compliance to ensure ongoing adherence to established guidelines.

Expected Outcome: Backup data is consistently encrypted, safeguarding against unauthorized access and ensuring that sensitive information remains confidential, regardless of physical or virtual location.

Control 3: Regular Testing and Validation of Backup Integrity and Restore Processes

Control Objective: To ensure the reliability and effectiveness of backup and recovery systems by regularly testing and validating the integrity of backups and the efficiency of restore processes.

Implementation Steps:

3.1 **Develop Testing Schedule**: Establish a routine schedule for testing the integrity of backup data and the effectiveness of restore procedures.

3.2 **Conduct Restore Tests**: Perform regular restore tests from backups to different systems to ensure data can be successfully recovered in various scenarios.

3.3 **Analyze and Document Test Results**: Document the outcomes of restore tests, analyze any failures or issues, and implement improvements to the backup and recovery processes based on test results.

Expected Outcome: The organization ensures that backups are up-to-date, complete, and fully restorable, guaranteeing data availability and continuity in the face of disruptions.

PR.DS-11: Backups of Data Are Created, Protected, Maintained, and Tested

The essence of data backup strategies is in safeguarding the integrity, confidentiality, and availability of information within an organization. Data backups are a fundamental pillar in cybersecurity, providing a reliable recovery solution for data loss due to cyberattacks, hardware failures, or natural disasters. An effective backup strategy encompasses creating and storing copies and ensuring these backups are secure, accessible, and capable of being restored accurately and promptly.

Creating regular, automated backup processes forms the backbone of a robust data protection plan. Automation minimizes human error and ensures data is backed up consistently and according to a predetermined schedule. It is critical to assess the nature and frequency of data changes to determine the appropriate backup intervals. High-value or frequently modified data may require daily or hourly backups, while less critical information might be adequately served with weekly backups. This strategic approach ensures that the most recent and relevant data can be swiftly recovered in the event of a data loss incident.

Encrypting backup data adds a vital layer of security, safeguarding sensitive information against unauthorized access. Whether stored on-site or in the cloud, encrypted backups ensure that data remains confidential and intact, only accessible to individuals with the requisite decryption keys. Encrypting backup data should align with best practices and industry standards, employing strong algorithms and secure key management practices to prevent potential vulnerabilities.

Implementing redundant backup solutions across multiple locations fortifies an organization's data recovery capabilities. Adhering to the 3-2-1 backup rule—maintaining three total copies of data, with two on different media and one off-site—safeguards against localized disasters and system failures. This redundancy ensures that irrespective of the failure of a single backup source, data remains recoverable from alternative locations, thereby enhancing resilience and continuity.

Access control and authentication mechanisms are imperative for securing backup data retrieval processes. Implementing robust access controls ensures that only authorized personnel can retrieve and restore backups, preventing unauthorized access and potential data breaches. Multifactor authentication, role-based access controls, and rigorous access logging should be integral components of the backup system, aligning with the organization's broader data security policies.

Regular maintenance and updates to backup systems are necessary to ensure their continued effectiveness and security. This includes updating backup software, verifying storage locations' physical and digital security, and ensuring compatibility with the protected data. Routine maintenance identifies potential issues before they escalate into critical failures, ensuring the reliability of backup systems.

Periodic testing of backup restorations is essential to validate the integrity and availability of backed-up data. Regular testing procedures help identify any flaws or inefficiencies in the backup and restoration process, providing an opportunity to rectify issues before a real-world data recovery scenario arises. Effective testing ensures that, in the event of data loss, recovery processes are efficient, accurate, and capable of reinstating critical operations with minimal downtime.

Documentation of backup policies and recovery procedures is crucial for maintaining clear, consistent, and accessible guidelines for data protection. This documentation should outline the backup strategy, including what data is backed up, how often, where it is stored, and the procedures for data recovery. Having well-documented policies not only aids in compliance and audit trails but also serves as an educational tool for staff involved in data management and recovery processes.

Recommendations

- **Develop and Implement a Detailed Backup Schedule**: Create a tailored backup schedule that reflects the criticality and frequency of data updates, ensuring that all vital data is backed up regularly and automatically.

- **Ensure Comprehensive Data Encryption**: Implement encryption for all backup data, applying stringent security measures in transit and at rest to protect against unauthorized access and breaches.
- **Adopt and Maintain a Redundant Backup System**: Utilize a multilocation, redundant backup strategy to safeguard against local disasters and system failures, ensuring data availability.
- **Conduct Regular Backup System Audits**: Schedule routine audits and maintenance of backup systems to ensure their integrity, security, and compatibility with current data formats and structures.
- **Foster a Culture of Data Recovery Preparedness**: Provide ongoing training for employees on the importance of backups, the organization's backup procedures, and their role in maintaining data integrity and security.

NIST 800-53 Controls

(Refer to our appendix for a comprehensive description of the NIST 800-53 Controls, essential for organizations adhering to strict regulatory standards.)

- CP-06,CP-09

Simplified Security Controls (SSC)

This section outlines our SSC, tailored for environments beyond the scope of NIST compliance. These controls offer streamlined, effective security solutions suitable for any organization seeking to enhance its cybersecurity defenses with clear, actionable measures.

Control 1: Automated Backup Verification System

Control Objective: To ensure the reliability and effectiveness of backup processes by implementing an automated system to verify the success and integrity of each backup operation.

Implementation Steps:

1.1 **Develop and Deploy Verification Tools**: Implement automated tools that verify each backup's success immediately after completion, checking for completeness and integrity.

1.2 **Integrate Alerts and Reporting**: Configure the system to send immediate alerts to the IT team in case of a failed or incomplete backup, along with detailed reports for successful backups.

1.3 **Schedule Regular Review Meetings**: Set up monthly meetings to review backup success rates, discuss any issues identified through alerts and reports, and assess the overall health of the backup system.

Expected Outcome: Establishing an automated backup verification system ensures that all backups are consistently checked for integrity and completeness, significantly reducing the risk of data loss due to failed backup processes and enhancing the organization's data resilience.

Control 2: Enhanced Backup Data Encryption Protocol

Control Objective: To safeguard backup data against unauthorized access and ensure its confidentiality and integrity by implementing an enhanced encryption protocol for all backup activities.

Implementation Steps:

2.1 **Selection of Advanced Encryption Standards**: Choose industry-recognized encryption standards (such as AES-256) for encrypting all backup data, whether in transit or at rest.

2.2 **Key Management and Access Control**: Develop a secure key management policy, ensuring that encryption keys are stored separately from backups and are accessible only to authorized personnel.

2.3 **Regular Encryption Audits**: Conduct quarterly audits to maintain encryption standards across all backup processes and update outdated encryption protocols.

Expected Outcome: By implementing enhanced encryption protocols for backup data, the organization ensures the security and confidentiality of its backups, protecting against data breaches and unauthorized access even in the event of physical theft or cyber intrusion.

Control 3: Comprehensive Backup Recovery Drill Program

Control Objective: To confirm the organization's capability to restore operations from backup swiftly and effectively, minimizing downtime in the event of data loss.

Implementation Steps:

3.1 **Develop a Recovery Drill Schedule**: Create a structured schedule for conducting full-scale recovery drills, varying scenarios to cover different data loss and disaster scenarios.

3.2 **Document and Implement Drill Procedures**: Establish detailed recovery procedures for each scenario, including steps for data restoration, system checks, and operational validation.

3.3 **Analyze and Refine Post-Drill**: Conduct a thorough analysis following each drill to identify any weaknesses or failures in the recovery process and refine the backup and recovery strategies accordingly.

Expected Outcome: Regular execution of comprehensive backup recovery drills ensures the organization is well-prepared to recover critical data and resume operations quickly after a loss event, thereby minimizing downtime and associated costs.

17

Platform Security (PR.PS)

> In the complex tapestry of IT security, the discipline of Configuration Management stands as a sentinel, guiding the organization through the ever-shifting sands of technology with a steady hand, ensuring not just compliance but a fortress of reliability in the digital age.

Establishing and applying robust configuration management (CM) practices form the bedrock of a secure digital environment, ensuring that every aspect of an organization's IT infrastructure is systematically managed and safeguarded. This begins with developing a comprehensive CM plan that sets forth standards, procedures, and responsibilities, complemented by a meticulous inventory of assets critical to the CM process. Employing automated tools for continuous monitoring and managing configurations, alongside establishing and maintaining baseline configurations, fortifies the organization's defense mechanisms. Regular reviews and updates of configuration settings, coupled with strict change management processes, ensure that modifications to system configurations are effectively controlled. Additionally, meticulous documentation and auditing, combined with targeted training programs, underscore the organization's commitment to continuous improvement in CM, thereby bolstering its overall security posture.

PR.PS-01: Configuration Management Practices Are Established and Applied

CM practices are vital for maintaining the integrity, consistency, and security of information technology systems. They involve identifying, documenting, and managing changes in hardware, software, and other system elements. Proper CM practices help organizations prevent unauthorized changes and reduce the risk of security breaches. They are fundamental in establishing a controlled IT environment where updates, modifications, and access are meticulously managed and recorded.

Developing a CM plan is critical to establishing effective CM practices. This plan should clearly outline the standards, procedures, and responsibilities of managing and controlling IT system configurations. It serves as a roadmap for the organization, detailing how configuration changes should be managed, who is responsible for different aspects of the CM process, and how compliance with the plan will be monitored and enforced. The CM plan ensures that all organization members understand their roles and responsibilities in maintaining system security and integrity.

An inventory of assets is essential for the effective application of CM. This involves identifying and documenting all IT assets, including hardware, software, networking components, and

A Comprehensive Guide to the NIST Cybersecurity Framework 2.0: Strategies, Implementation, and Best Practice, First Edition. Jason Edwards.
© 2025 John Wiley & Sons Ltd. Published 2025 by John Wiley & Sons Ltd.

other relevant system elements. The inventory should be regularly updated to reflect new or decommissioned assets. This comprehensive understanding of what assets exist and how they are configured provides the foundation for managing system changes and maintaining security.

Implementing automated configuration monitoring and management tools significantly enhances CM practices' effectiveness. These tools can automatically track changes, enforce policies, and report discrepancies in system configurations. Automation reduces the likelihood of human error and ensures a consistent and timely approach to CM. By leveraging technology, organizations can ensure that their IT environments remain secure, compliant, and aligned with business objectives.

A fundamental CM component is establishing and maintaining a baseline configuration for systems and networks. This involves defining a standard, secure, and approved set of configurations that serve as the benchmark for all systems and networks within the organization. The baseline ensures that all configurations meet security requirements and are consistent across the organization. It also serves as a reference point for detecting unauthorized changes and managing updates effectively.

Regularly reviewing and updating configuration settings are crucial to ensure ongoing security and compliance. As the threat landscape evolves and business needs change, configuration settings may need to be adjusted to maintain the security and functionality of systems. Regular reviews allow organizations to identify necessary changes and ensure that configurations align with best practices and regulatory requirements.

Change management processes play a critical role in controlling modifications to system configurations. These processes ensure that all changes are reviewed, approved, and documented before implementation. By managing changes in a structured way, organizations can minimize the risk of introducing vulnerabilities or disrupting system operations. Effective change management also facilitates traceability and accountability, allowing organizations to understand the impact of changes on system security and performance.

Finally, documentation and auditing of configuration changes and states are essential for maintaining control over IT systems. Detailed records of changes, including who made the change, when it was made, and why, help organizations track the history of their systems and verify compliance with CM policies. Regular audits can identify unauthorized or improper changes, ensuring that systems remain secure and configurations align with organizational standards.

Recommendations

- **Implement a Comprehensive CM Policy**: Establish clear policies and procedures for CM that are aligned with industry best practices and regulatory requirements. This ensures a structured approach to managing system changes and maintaining security.
- **Invest in Configuration Management Tools**: Leverage automated tools to enhance the accuracy and efficiency of CM. These tools can help monitor configurations, enforce policies, and provide alerts on unauthorized changes.
- **Conduct Regular Configuration Audits**: Regularly audit system configurations to ensure compliance with the established baseline and to detect unauthorized changes. This helps maintain the integrity and security of IT systems.
- **Provide Training and Awareness Programs**: Offer training and awareness programs to educate staff about the importance of CM and their specific roles and responsibilities. This promotes a culture of security and compliance throughout the organization.

- **Establish a Configuration Control Board**: Create a cross-functional team responsible for overseeing configuration changes, ensuring that all changes are reviewed and approved in line with organizational policies and security standards. This fosters collaboration and ensures that changes are made in a controlled and secure manner.

NIST 800-53 Controls

(Refer to our appendix for a comprehensive description of the NIST 800-53 Controls, essential for organizations adhering to strict regulatory standards.)

- CM-01,CM-02,CM-03,CM-04,CM-05,CM-06,CM-07,CM-08,CM-09,CM-10,CM-11

Simplified Security Controls (SSC)

This section outlines our simplified security controls (SSC), tailored for environments beyond the scope of NIST compliance. These controls offer streamlined, effective security solutions suitable for any organization seeking to enhance its cybersecurity defenses with clear, actionable measures.

Control 1: Establishment of Comprehensive Configuration Management Plan

Control Objective: To develop and enforce a CM plan that outlines the standards, procedures, and responsibilities necessary to manage system configurations and ensure organizational security effectively.

Implementation Steps:

1.1 **Plan Development**: Assemble a cross-departmental team to develop a CM plan, incorporating industry best practices and organizational requirements. The plan should detail the configuration standards, procedures for updates and changes, roles and responsibilities, and compliance requirements.

1.2 **Plan Approval and Dissemination**: Submit the CM Plan for executive approval. Once approved, distribute the plan to all relevant stakeholders and departments, ensuring everyone knows their responsibilities and procedures.

1.3 **Regular Review and Updates**: Schedule and conduct regular reviews (e.g., annually) of the CM Plan to incorporate new technologies, address emerging threats, and reflect changes in organizational structure or strategy.

Expected Outcome: Developing and enforcing a CM plan ensures a standardized approach to managing system configurations across the organization. This results in enhanced security, improved compliance with regulatory requirements, and better overall system integrity and performance.

Control 2: Implementation of Automated Configuration Management Tools

Control Objective: To utilize automated tools to continuously monitor and manage system configurations, ensuring that all devices adhere to the organization's configuration policies and standards.

Implementation Steps:

2.1 **Tool Selection and Deployment**: Identify, select, and deploy automated CM tools that align with the organization's technical requirements and CM Plan. Ensure these tools have capabilities for real-time monitoring, alerting, and reporting.

2.2 **Tool Configuration and Integration**: Configure the tools to monitor critical system configurations and integrate them with existing IT infrastructure for comprehensive coverage. Define baseline configurations and set alerts for any deviations.

2.3 **Continuous Monitoring and Response**: Establish processes for responding to alerts generated by the automated tools. Ensure that any unauthorized changes are investigated, documented, and rectified promptly.

Expected Outcome: Implementing automated CM tools provides continuous oversight of system configurations, quickly identifying and rectifying unauthorized changes. This reduces the risk of security breaches and ensures that all systems comply with established standards.

Control 3: Configuration Change Management and Documentation

Control Objective: To establish a formal change management process for controlling modifications to system configurations, ensuring that all changes are reviewed, approved, documented, and audited to maintain security and compliance.

Implementation Steps:

3.1 **Change Management Process Development**: Develop and document a formal change management process, which includes steps for requesting, reviewing, approving, implementing, and documenting changes to system configurations.

3.2 **Change Review Board Establishment**: Form a Change Review Board comprising representatives from relevant departments to oversee and approve proposed configuration changes. This board will ensure that all changes comply with organizational policies and do not introduce security vulnerabilities.

3.3 **Documentation and Auditing**: Ensure that all configuration changes are thoroughly documented, including the rationale for the change, the approval process, and the impact assessment. Conduct regular audits of configuration changes to ensure compliance with the change management process.

Expected Outcome: Establishing a rigorous change management process ensures that all configuration changes are systematically reviewed and documented, reducing the risk of unauthorized changes and maintaining the integrity and security of the system configurations.

PR.PS-02: Software Is Maintained, Replaced, and Removed Commensurate with Risk

The management of software throughout its lifecycle is a critical aspect of an organization's cybersecurity strategy. This encompasses the processes involved from the initial acquisition or development of the software to its eventual retirement. Proper software lifecycle management ensures that applications remain efficient, secure, and aligned with business needs. It involves regular updates, timely replacement, and secure decommissioning of software, mitigating potential vulnerabilities that could lead to security breaches.

Understanding and assessing software risks based on vulnerabilities, threats, and operational impact is fundamental. An organization must evaluate the potential risks associated with its software inventory, considering the nature of the data it processes and the environment in which it operates. This assessment helps identify the criticality of different software applications and prioritizes them based on their potential impact on the organization's cybersecurity posture. A thorough risk assessment informs decision-making regarding software updates, replacements, and removals.

Procedures for regular software updates and patch management are essential components of software lifecycle management. Organizations should establish standardized processes for the timely application of patches and updates to software applications. This not only addresses known vulnerabilities but also helps maintain the operational efficiency of the software. Effective patch management requires a systematic approach, ensuring that all software, especially critical ones, is up-to-date against emerging threats.

Criteria for software replacement should include considerations for end-of-life (EOL). Software that has reached its EOL is no longer supported by its manufacturer, meaning no new updates or patches will be released. Using EOL software poses significant security risks, as vulnerabilities discovered post-EOL will not be remedied. Organizations should have clear guidelines for replacing outdated software with modern, supported alternatives that offer improved security features and compatibility with current systems.

Secure removal or decommissioning processes for obsolete software are just as crucial as the procedures for its installation and maintenance. When software is no longer required or has been replaced, it should be securely removed from all systems to prevent unauthorized access or use. This involves uninstalling the software and ensuring that associated data is securely deleted or migrated. Proper decommissioning helps avoid security gaps that attackers could exploit.

Documentation of software versions, patches, and change histories is vital for maintaining visibility and control over the software landscape of an organization. This documentation records the software's evolution and current status, supporting effective management and security compliance. It also facilitates audits and investigations, enabling organizations to demonstrate their diligence in maintaining software security and compliance with relevant regulations and standards.

Integrating software management with the enterprise risk management framework ensures that software-related risks are considered in the organization's overall risk landscape. This holistic approach enables better alignment of software management practices with organizational objectives and risk tolerance. By embedding software lifecycle management within the enterprise risk management processes, organizations can ensure that software decisions are informed and support their broader risk management strategies.

Continuous monitoring of software performance and security posture is imperative. By regularly reviewing the effectiveness and security of software applications, organizations can identify and address potential issues before they lead to significant problems. Continuous monitoring enables the early detection of suspicious activities, performance degradation, or compliance deviations, allowing for timely interventions.

Recommendations

- **Establish a Comprehensive Software Risk Assessment Process**: Implement a structured approach to assess the risks associated with software, considering factors such as vulnerabilities, threats, and operational impacts. This enables informed decision-making regarding software updates, replacements, and decommissioning.
- **Develop and Enforce a Patch Management Policy**: Create a policy that outlines the procedures for regular software updates and patch management. This policy should ensure that critical software is prioritized and patches are applied promptly to mitigate vulnerabilities.
- **Set Clear Criteria for Software Replacement**: Define clear guidelines for when software should be replaced, particularly with regard to EOL considerations. Ensure the transition to new software is planned and executed securely, minimizing operational disruptions.
- **Implement Secure Decommissioning Practices for Obsolete Software**: Establish procedures for the secure removal of obsolete software, including the secure deletion or migration of associated data. This helps to prevent security vulnerabilities associated with unused or outdated software.

- **Integrate Software Lifecycle Management into the Enterprise Risk Management Framework**: Ensure that software management practices are aligned with the organization's broader risk management strategies. This facilitates a comprehensive approach to managing software-related risks and enhances the organization's security posture.

NIST 800-53 Controls

(Refer to our appendix for a comprehensive description of the NIST 800-53 Controls, essential for organizations adhering to strict regulatory standards.)

- CM-11,MA-03,SA-10,SI-02,SI-07

Simplified Security Controls (SSC)

This section outlines our SSC, tailored for environments beyond the scope of NIST compliance. These controls offer streamlined, effective security solutions suitable for any organization seeking to enhance its cybersecurity defenses with clear, actionable measures.

Control 1: Comprehensive Software Risk Assessment Process

Control Objective: To systematically identify, analyze, and prioritize risks associated with software, guiding informed decision-making related to updates, replacements, and decommissioning to enhance organizational security.

Implementation Steps:

1.1 **Develop Assessment Criteria**: Create a set of criteria that considers software vulnerabilities, threats, and operational impacts to evaluate the security risks associated with each software asset.

1.2 **Conduct Risk Assessments**: Regularly perform risk assessments on all software assets, utilizing the developed criteria to identify and prioritize risks based on their potential impact on the organization's cybersecurity posture.

1.3 **Review and Act**: Analyze the results of risk assessments to make informed decisions about software updates, replacements, and decommissioning and implement necessary actions to mitigate identified risks.

Expected Outcome: Establishing a comprehensive software risk assessment process enables the organization to make informed decisions regarding the management of software assets, effectively reducing the likelihood and impact of security incidents.

Control 2: Patch Management Policy Development and Enforcement

Control Objective: To maintain the integrity, availability, and confidentiality of information assets by ensuring timely application of patches and updates to software applications, thereby addressing known vulnerabilities and enhancing system security.

Implementation Steps:

2.1 **Policy Formulation**: Draft a patch management policy that outlines the procedures for identifying, assessing, and applying updates and patches to software assets, particularly those classified as critical.

2.2 **Implement Patch Management Procedures**: Establish mechanisms for regularly reviewing, testing, and applying necessary software patches and updates as dictated by the policy.

2.3 **Monitoring and Compliance**: Set up systems for monitoring compliance with the patch management policy and auditing the effectiveness of patch management practices.

Expected Outcome: The creation and strict enforcement of a patch management policy ensure that all software applications are kept up-to-date, significantly reducing the organization's exposure to known vulnerabilities and cyber threats.

Control 3: Secure Decommissioning and Software Replacement Guidelines

Control Objective: To prevent security vulnerabilities associated with outdated or unnecessary software by establishing clear guidelines for the secure replacement and decommissioning software applications.

Implementation Steps:

3.1 **Develop Replacement Criteria**: Define clear, documented criteria for determining when software should be replaced, focusing on security, operational efficiency, and EOL considerations.

3.2 **Decommissioning Process**: Establish a standard process for the secure removal of obsolete software, which includes data backup, secure deletion, and documentation.

3.3 **Transition and Monitoring**: Manage the transition from old to new software securely, ensuring minimal disruption to operations, and monitor the decommissioning process to confirm it meets all security and compliance requirements.

Expected Outcome: Implementing secure decommissioning and software replacement guidelines minimizes security risks associated with outdated or unused software and ensures that transitions to new software are executed securely and efficiently, maintaining operational continuity and compliance.

PR.PS-03: Hardware Is Maintained, Replaced, and Removed Commensurate with Risk

Properly managing hardware throughout its lifecycle is essential for maintaining a robust cybersecurity posture. This process, known as Hardware Lifecycle Management (HLM), encompasses the acquisition, operation, maintenance, and eventual disposal of hardware assets. Effective HLM ensures that hardware does not become a weak link in the security chain, protecting against physical and cyber threats. It requires a strategic approach that aligns with the organization's overall risk management framework and cybersecurity goals.

Risk assessment for hardware security is a crucial step, addressing both physical and operational risks. Organizations must evaluate the vulnerability of their hardware assets to unauthorized access, damage, and theft, as well as the risks associated with their operational environment. This assessment should consider the lifecycle stage of each asset, from deployment to decommissioning, and prioritize risks based on the potential impact on the organization's operations and data security. Understanding these risks allows for developing tailored security measures and maintenance schedules.

Scheduled maintenance and inspection procedures for hardware are vital for ensuring ongoing security and operational efficiency. Regular checks can identify and mitigate emerging risks, such as hardware malfunctions, wear and tear, and vulnerabilities. These procedures should be systematic and documented, with clear frequency, scope, and responsibility guidelines. Maintenance schedules must be adhered to rigorously to prevent security gaps and ensure hardware assets continue to perform as required.

Guidelines for hardware replacement should consider both performance and security factors. As hardware ages, it may no longer meet the organization's operational needs or comply with current security standards. Replacement guidelines should outline criteria for evaluating hardware performance, assessing security vulnerabilities, and determining the appropriate time for upgrade or replacement. These decisions should be based on a comprehensive analysis of the hardware's condition, its role in the organization, and the evolving threat landscape.

Secure decommissioning and disposal of hardware assets are critical to prevent unauthorized access to residual data and ensure environmental compliance. Procedures for decommissioning should include data erasure, physical destruction, or secure repurposing of hardware. Organizations must ensure that all sensitive information is completely removed or securely transferred before disposal or repurposing of any hardware component. Compliance with legal and environmental standards during disposal is essential to mitigate legal and reputational risks.

Inventory and documentation of hardware assets and their security posture provide the foundation for effective HLM. This includes maintaining detailed records of all hardware components, configurations, security measures, and maintenance history. A comprehensive hardware inventory supports risk management by enabling accurate risk assessments, facilitating the tracking of hardware changes, and aiding in investigating security incidents. Documentation should be kept up-to-date and securely stored.

Incorporating hardware management into the organization's overall risk framework ensures that hardware-related risks are addressed within the broader context of organizational risk management. This alignment enables a holistic approach to security, ensuring that hardware security measures support overall business objectives and risk tolerance. Integration with the enterprise risk management framework facilitates coordinated action and resource allocation to address hardware vulnerabilities effectively.

The use of automated systems for monitoring hardware health and security can significantly enhance HLM. Automated monitoring tools can provide real-time alerts on hardware malfunctions, potential security breaches, and other relevant issues. These systems enable proactive management of hardware assets, allowing for timely responses to possible security threats and operational issues. Automation also supports efficient resource utilization and can reduce the likelihood of human error.

Recommendations

- **Develop a Comprehensive Hardware Risk Assessment Framework**: Establish a structured approach for assessing the security risks associated with hardware assets, considering physical and operational threats. This framework should guide the prioritization of assets and inform security measures throughout the hardware lifecycle.
- **Implement Systematic Maintenance and Inspection Protocols**: Define clear, consistent procedures for the regular maintenance and inspection of hardware assets. Ensure these protocols are followed diligently to identify and mitigate risks promptly.
- **Establish Criteria for Hardware Replacement and Decommissioning**: Set clear, objective criteria for determining when hardware should be replaced or decommissioned, focusing on security, performance, and compliance considerations. This ensures hardware remains effective and does not pose undue risk to the organization.
- **Securely Manage the Decommissioning and Disposal of Hardware**: Create secure and environmentally responsible procedures for the decommissioning and disposal of hardware, ensuring that all sensitive data is properly erased or migrated and that hardware is disposed of in compliance with legal standards.

- **Integrate Hardware Lifecycle Management into Organizational Risk Management**: Align hardware management practices with the organization's broader risk management strategies. This ensures a coordinated, holistic approach to mitigating hardware-related risks and enhancing overall security posture.

NIST 800-53 Controls

(Refer to our appendix for a comprehensive description of the NIST 800-53 Controls, essential for organizations adhering to strict regulatory standards.)

- CM-07,SA-10,SC-03,SC-39,SC-49,SC-51

Simplified Security Controls (SSC)

This section outlines our SSC, tailored for environments beyond the scope of NIST compliance. These controls offer streamlined, effective security solutions suitable for any organization seeking to enhance its cybersecurity defenses with clear, actionable measures.

Control 1: Establishment of a Hardware Risk Assessment Protocol

Control Objective: To systematically evaluate and prioritize the security risks associated with each hardware asset to inform protective measures and lifecycle decisions, enhancing the organization's cybersecurity posture.

Implementation Steps:

1.1 **Develop Assessment Criteria**: Create comprehensive criteria for evaluating the security risks associated with hardware, considering physical vulnerabilities, operational impacts, and exposure to cyber threats.

1.2 **Conduct Risk Assessments**: Perform regular risk assessments on all hardware assets using the established criteria, documenting each asset's vulnerabilities and potential impact on the organization's security.

1.3 **Prioritize and Mitigate Risks**: Use the findings from the risk assessments to prioritize hardware assets for updates, replacements, or enhanced security measures based on their risk levels.

Expected Outcome: A systematic hardware risk assessment protocol enables the organization to identify, prioritize, and mitigate risks associated with hardware assets effectively, reducing potential security breaches and operational disruptions.

Control 2: Comprehensive Maintenance and Inspection Schedule

Control Objective: To maintain operational efficiency and security of hardware assets through systematic and documented maintenance and inspection procedures.

Implementation Steps:

2.1 **Schedule Development**: Develop a comprehensive maintenance and inspection schedule for all hardware assets, defining the frequency, scope, and responsible personnel for each task.

2.2 **Procedure Implementation**: Implement the scheduled maintenance and inspection procedures, ensuring they are carried out diligently and documented accurately.

2.3 **Review and Update Processes**: Regularly review and update maintenance and inspection procedures to address emerging vulnerabilities and hardware lifecycle changes.

Expected Outcome: Establishing and adhering to a comprehensive maintenance and inspection schedule ensures that hardware assets remain secure and operational, mitigating the risk of security breaches and hardware failures.

Control 3: Hardware Decommissioning and Disposal Policy

Control Objective: To ensure that all hardware assets are decommissioned and disposed of securely and responsibly, preventing unauthorized access to residual data and complying with environmental standards.

Implementation Steps:

3.1 **Policy Development**: Develop a detailed policy for the secure decommissioning and disposal of hardware assets, outlining steps for data erasure, physical destruction, and disposal certification.

3.2 **Implementation and Training**: Implement the decommissioning policy across the organization and provide training for relevant staff on secure disposal practices and procedures.

3.3 **Monitoring and Compliance**: Establish monitoring mechanisms to ensure compliance with the decommissioning policy and conduct regular audits to assess the effectiveness of disposal procedures.

Expected Outcome: A robust hardware decommissioning and disposal policy ensures that all decommissioned hardware is disposed of securely and environmentally responsibly, preventing data breaches and legal violations.

PR.PS-04: Log Records Are Generated and Made Available for Continuous Monitoring

Understanding the role of log management within cybersecurity is fundamental to safeguarding an organization's digital assets. Log records, generated by various systems and applications, offer invaluable insights into network activities, user behaviors, and potential security threats. They serve as the factual foundation upon which security analyses are built, enabling the detection, investigation, and prevention of cybersecurity incidents.

Defining log generation policies is critical to ensuring that key systems and applications capture relevant and actionable data. These policies dictate what information should be recorded, how it should be formatted, and how frequently it should be updated. Ensuring that these policies encompass all critical assets within the organization is essential for creating a comprehensive security overview.

The configuration of logging mechanisms must be tailored to capture the data deemed necessary by the log generation policies. This involves adjusting system and application settings to ensure that logs are comprehensive, detailed, and structured in a way that supports effective analysis. Proper configuration helps distinguish normal activities from potential security incidents.

Implementing secure log storage solutions is crucial for maintaining the integrity and confidentiality of log data. Secure storage prevents unauthorized access and tampering, ensuring that log records remain reliable and unaltered. This is vital for forensic analyses and compliance with regulatory requirements, as compromised log data can severely hinder an organization's ability to respond to and recover from security incidents.

Utilizing automated tools for log analysis and anomaly detection transforms raw data into actionable intelligence. These tools can sift through vast quantities of log entries in real time, identifying patterns and anomalies that may indicate a security threat. The capability to rapidly analyze logs allows quicker responses to potential security incidents, reducing the overall impact on the organization.

Regular review and analysis of log files are imperative for maintaining an up-to-date understanding of the organization's security posture. Scheduled analyses can uncover subtle patterns and trends that automated tools might overlook, providing deeper insights into the organization's security dynamics. Regular reviews also help validate current security measures' effectiveness and identify areas for improvement.

Integrating log management with incident response processes enhances the organization's ability to react swiftly and effectively to detected threats. By establishing procedures that leverage log data during the response to an investigation of incidents, organizations can reduce response times and improve the accuracy of their investigations. This integration ensures that every response is informed by the most complete and relevant information available.

Compliance with legal and regulatory requirements for log retention is a matter of legal obedience and a component of sound cybersecurity strategy. Maintaining logs for the required duration supports thorough investigations, aids in detecting long-term patterns, and ensures that the organization meets its legal obligations. Understanding and adhering to these requirements is essential for any effective log management strategy.

Finally, training on log management tools and analysis techniques empowers employees to utilize log data in their security roles effectively. Providing staff with the knowledge and skills to interpret log data and respond to the insights it provides strengthens the organization's overall security framework.

Recommendations

- **Implement a Tailored Log Generation Policy**: Develop and enforce a log generation policy for your organization's systems and applications, ensuring comprehensive coverage and relevant data capture.
- **Regularly Audit and Update Log Management Practices**: Conduct periodic audits of your log management practices to identify and rectify gaps or inefficiencies, keeping your strategies aligned with evolving cybersecurity landscapes.
- **Enhance Log Analysis with Advanced Tools**: Leverage advanced log analysis tools and technologies to automate the detection of anomalies and streamline the analysis process, enhancing your security team's ability to identify and respond to threats swiftly.
- **Integrate Log Management with Wider Security Practices**: Ensure that log management seamlessly integrates with your organization's broader security and incident response strategies, creating a unified and effective defense mechanism.
- **Foster Continuous Learning**: Encourage ongoing education and training for your team in log management best practices and analysis techniques, fostering a culture of continuous improvement and cybersecurity excellence.

NIST 800-53 Controls

(Refer to our appendix for a comprehensive description of the NIST 800-53 Controls, essential for organizations adhering to strict regulatory standards.)

- AU-02,AU-03,AU-06,AU-07,AU-11,AU-12

Simplified Security Controls (SSC)

This section outlines our SSC, tailored for environments beyond the scope of NIST compliance. These controls offer streamlined, effective security solutions suitable for any organization seeking to enhance its cybersecurity defenses with clear, actionable measures.

Control 1: Establishment of Comprehensive Log Generation Guidelines

Control Objective: To ensure comprehensive and effective log data capture across all critical systems and applications, facilitating detailed security monitoring and analysis.

Implementation Steps:

1.1 **Policy Development**: Draft a detailed log generation policy that specifies the types of data to be logged, log format, and retention periods tailored to the organization's systems and security requirements.

1.2 **Policy Implementation**: Deploy log generation policies across all critical systems and applications, ensuring they are configured to capture all relevant data as defined by the policy.

1.3 **Continuous Policy Review and Update**: Regularly review and update the log generation policies to adapt to changes in the organization's IT environment, emerging threats, and compliance requirements.

Expected Outcome: A comprehensive set of log generation guidelines ensures that all necessary log data is consistently captured across the organization's IT environment, enhancing the ability to monitor, analyze, and respond to potential security threats.

Control 2: Secure Log Storage and Management

Control Objective: To maintain log data's integrity, confidentiality, and availability through secure storage solutions and effective log management practices.

Implementation Steps:

2.1 **Secure Storage Implementation**: Implement secure log storage solutions that provide encryption, access controls, and integrity checks to protect log data against unauthorized access and tampering.

2.2 **Access Control and Monitoring**: Define and enforce strict access controls for log data and establish monitoring mechanisms to detect and respond to unauthorized access attempts.

2.3 **Regular Security Assessments**: Conduct regular security assessments of log storage and management practices to identify and remediate potential vulnerabilities or policy violations.

Expected Outcome: Secure log storage and stringent management practices ensure that log data remains protected from unauthorized access and tampering, preserving its value for security monitoring, incident response, and compliance purposes.

Control 3: Integration of Log Management into Incident Response and Compliance Frameworks

Control Objective: To enhance the organization's incident response capabilities and ensure compliance with legal and regulatory log retention requirements through effective log management integration.

Implementation Steps:

3.1 **Develop Integration Processes**: Establish processes for integrating log management with the organization's incident response framework, ensuring timely access to log data during security incidents.

3.2 **Compliance Alignment**: Align log retention and management practices with applicable legal and regulatory requirements, ensuring that log data is retained for the required duration and in compliance with relevant standards.

3.3 **Continuous Improvement and Training**: Implement continuous improvement processes for log management and incident response integration and regularly train relevant personnel on log-related responsibilities and procedures.

Expected Outcome: Effective integration of log management with incident response and compliance frameworks enhances the organization's ability to swiftly respond to and recover from security incidents while meeting all legal and regulatory requirements for log retention and management.

PR.PS-05: Installation and Execution of Unauthorized Software Are Prevented

Preventing unauthorized software installation and execution is critical to an organization's cybersecurity defenses. Unauthorized software can introduce vulnerabilities, facilitate data breaches, and undermine the integrity of corporate IT environments. This section explores strategies and practices to mitigate these risks, ensuring a secure and controlled IT landscape.

Application whitelisting and blacklisting emerge as fundamental techniques in the fight against unauthorized software. Whitelisting allows only preapproved applications to run, offering a proactive defense by default and denying unauthorized software. Conversely, blacklisting prohibits specific applications known to be malicious or unnecessary, though it operates reactively, addressing known threats while potentially leaving new, unidentified ones unchallenged.

The configuration of endpoint security solutions plays a pivotal role in detecting and blocking unauthorized software. Effective solutions monitor real-time system activities, identifying and stopping unauthorized executions. These tools, integral to an organization's endpoint security strategy, must be continually updated to recognize the latest threats and maintain robust defense mechanisms.

Establishing comprehensive policies and procedures for software approval and installation is essential. These policies should outline clear criteria for software evaluation, approval processes, and installation procedures, ensuring consistency and control across the organization. Such formalized processes help to prevent unauthorized software from penetrating the corporate environment, reducing associated security risks.

Regularly auditing systems is necessary to identify and remove unauthorized software, ensuring compliance with established policies. Audits can reveal discrepancies, unauthorized installations, and outdated software, providing an opportunity to reinforce policies and enhance security. Scheduled, systematic reviews maintain the integrity of the IT environment, helping to mitigate potential vulnerabilities.

Network access control (NAC) systems complement these efforts by limiting device access based on compliance with security policies. By enforcing access controls, NAC systems prevent non-compliant devices from connecting to the network, reducing the risk of unauthorized software introductions. These controls are particularly effective in environments with BYOD policies or diverse device ecosystems.

Employee training on the risks of unauthorized software and compliance with organizational policies is crucial. Awareness programs should educate employees about the dangers of downloading and installing unapproved software, emphasizing personal and organizational consequences. Well-informed employees become a proactive line of defense, reducing the likelihood of unauthorized software breaches.

Incident response plans tailored explicitly to unauthorized software detection are vital. These plans should outline clear procedures for responding to detections of unauthorized software, detailing steps for investigation, mitigation, and reporting. Quick and effective responses to such incidents can significantly reduce their impact, restoring normal operations with minimal disruption.

Finally, continuous monitoring and updating of security controls are essential to address new threats and adapt to evolving IT landscapes. As cyber threats evolve, so too must the defenses against them. Regularly reviewing and updating security measures ensures protections remain effective against current and emerging threats.

Recommendations

- **Implement Robust Application Control**: Establish and maintain strict application whitelisting and blacklisting policies to control software execution and prevent unauthorized applications from operating within the network.
- **Strengthen Endpoint Security**: Deploy comprehensive endpoint security solutions regularly updated to detect and block unauthorized software effectively.
- **Formalize Software Management Policies**: Develop clear policies and procedures for software approval, installation, and removal to ensure consistent and secure software management across the organization.
- **Conduct Regular System Audits**: Perform scheduled audits of all systems to identify and rectify instances of unauthorized software, ensuring compliance with organizational policies.
- **Foster a Culture of Security Awareness**: Provide ongoing employee training on the risks associated with unauthorized software and the importance of adhering to company policies, enhancing the organization's overall security posture.

NIST 800-53 Controls

(Refer to our appendix for a comprehensive description of the NIST 800-53 Controls, essential for organizations adhering to strict regulatory standards.)

- CM-07, SC-34

Simplified Security Controls (SSC)

This section outlines our SSC, tailored for environments beyond the scope of NIST compliance. These controls offer streamlined, effective security solutions suitable for any organization seeking to enhance its cybersecurity defenses with clear, actionable measures.

Control 1: Enhanced Application Whitelisting and Blacklisting Mechanisms

Control Objective: To ensure only authorized applications run within the network, reducing the risk of unauthorized software leading to security breaches.

Implementation Steps:

1.1 **Define Application Lists**: Develop comprehensive lists of approved (whitelisting) and unauthorized (blacklisting) applications based on security assessments and business needs.

1.2 **Enforce Policy Compliance**: Implement technical controls to enforce these lists, preventing the execution of unauthorized applications and ensuring only approved software runs on network devices.

1.3 **Regular Review and Update**: Schedule regular reviews of the whitelists and blacklists to add new applications, remove outdated ones, and adjust to changing security landscapes and business requirements.

Expected Outcome: A robust application control mechanism is in place, significantly reducing the risk of unauthorized software installations and executions, thereby enhancing the organization's overall security posture.

Control 2: Establishment of a Secure Software Management Protocol

Control Objective: To standardize the software approval, installation, and removal processes, ensuring consistency and security across the organization's IT environment.

Implementation Steps:

2.1 **Policy Development**: Develop a comprehensive policy detailing the processes for software requests, evaluation, approval, installation, and decommissioning.

2.2 **Implementation and Enforcement**: Implement the policy across the organization and enforce compliance through technical and administrative measures.

2.3 **Continuous Education and Communication**: Educate employees about the policy and the importance of adhering to established software management procedures.

Expected Outcome: The organization adopts a consistent and secure approach to software management, significantly reducing the risks associated with unauthorized software and ensuring compliance with established security policies.

Control 3: Regular System Audits for Unauthorized Software Detection

Control Objective: To identify and remediate instances of unauthorized software installations, maintaining compliance with organizational security policies and standards.

Implementation Steps:

3.1 **Audit Planning and Scheduling**: Develop a schedule for regular audits of all systems to search for unauthorized software installations.

3.2 **Conducting Audits**: Perform the audits as scheduled, using automated tools where possible to ensure thorough and efficient examination of all systems.

3.3 **Remediation and Reporting**: Remove any unauthorized software found during audits and report the findings to relevant stakeholders, documenting the actions taken and lessons learned.

Expected Outcome: Regular system audits effectively identify and eliminate unauthorized software, reducing potential security vulnerabilities and ensuring continuous compliance with organizational policies.

PR.PS-06: Secure Software Development Practices Are Integrated, and Their Performance Is Monitored Throughout the Software Development Lifecycle

Integrating secure software development practices throughout the Software Development Life Cycle (SDLC) is crucial for building and maintaining robust, secure applications. This approach, often termed the Secure SDLC, ensures that security is not an afterthought but a primary concern at every development phase. By embedding security requirements into the planning and design phases, organizations can foresee and mitigate potential security issues before they manifest in the final product.

Integrating security requirements from the outset sets a solid foundation for the development process. When security considerations guide the planning and design stages, the resulting software architecture inherently minimizes security risks. This proactive approach contrasts sharply with traditional methods that bolt on security features after development, often leading to less secure and more costly outcomes.

Implementing secure coding standards and guidelines is a pivotal step in the Secure SDLC. Developers equipped with clear, concise, and consistent coding practices can significantly reduce vulnerabilities in the software they create. These standards should be based on industry best practices and continuously updated to address emerging threats and new programming paradigms.

Conducting regular security reviews and audits during the development process is essential for maintaining and enhancing the security posture of software projects. These reviews should be systematic and cover all aspects of the software, from the code to its dependencies. Regular audits help identify vulnerabilities early when they are typically more manageable and less expensive to fix.

Automated tools for code analysis and vulnerability scanning streamline identifying security issues, allowing developers to focus on solving complex problems rather than finding them. These tools can rapidly assess vast quantities of code to detect common vulnerabilities, such as injection flaws or buffer overflows. However, they complement rather than replace the nuanced judgment and problem-solving abilities of experienced security professionals.

Secure testing and validation procedures before software release are vital for the Secure SDLC. This stage involves rigorous testing strategies to uncover and address security vulnerabilities, including static and dynamic analysis. Organizations can avoid costly and damaging security breaches by thoroughly vetting software before it reaches production.

Monitoring and measuring security practices throughout the SDLC are imperative for continuous improvement. By tracking key performance indicators, such as the number of vulnerabilities identified and fixed or the time taken to remediate issues, organizations can gauge the effectiveness of their security efforts and refine their practices accordingly.

Continuous feedback loops are instrumental in refining security practices within the development process. Organizations can create a culture of continuous improvement by fostering an environment where feedback is encouraged and acted upon. This culture ensures that lessons learned from past projects inform future ones, leading to progressively more secure software.

Training and awareness programs for developers on secure coding practices address the human element of software security. Organizations can significantly reduce the number of vulnerabilities introduced during development by providing developers with the knowledge and tools they need to write secure code. Continuous education and awareness efforts ensure that security remains a priority for everyone involved in software creation.

Recommendations

- **Embed Security in Every Phase**: Ensure that security is integral to every phase of the software development lifecycle, from initial design to final release and beyond.
- **Establish Clear Secure Coding Standards**: Develop and enforce clear, secure coding standards and guidelines, and ensure that all developers are trained in these practices.
- **Perform Regular Security Reviews**: Implement a schedule for regular security reviews and audits throughout development to identify and remediate vulnerabilities early.
- **Leverage Automated Security Tools**: Utilize automated security tools to complement manual security reviews and provide ongoing feedback to developers.
- **Cultivate a Culture of Security Awareness**: Foster an organizational culture that prioritizes security, encouraging continuous learning and improvement among all team members involved in software development.

NIST 800-53 Controls

(Refer to our appendix for a comprehensive description of the NIST 800-53 Controls, essential for organizations adhering to strict regulatory standards.)

- SA-03,SA-08,SA-10,SA-11,SA-15,SA-17

Simplified Security Controls (SSC)

This section outlines our SSC, tailored for environments beyond the scope of NIST compliance. These controls offer streamlined, effective security solutions suitable for any organization seeking to enhance its cybersecurity defenses with clear, actionable measures.

Control 1: Secure SDLC Integration and Compliance Framework

Control Objective: To ensure security measures are integrated into every phase of the Software Development Life Cycle (SDLC), from initial planning to deployment, complying with established secure coding standards and regulatory requirements.

Implementation Steps:

1.1 **Develop and Implement Secure SDLC Guidelines**: Create comprehensive guidelines that integrate security practices into each stage of the SDLC, ensuring these practices comply with current security standards and regulations.

1.2 **Integrate Security Checkpoints**: Establish mandatory security checkpoints at critical stages within the SDLC, such as requirement gathering, design, development, and deployment, to evaluate compliance with the secure SDLC guidelines.

1.3 **Compliance Audits and Reviews**: Regularly conduct audits and reviews to ensure ongoing compliance with the secure SDLC guidelines, immediately addressing deviations or gaps.

Expected Outcome: Implementation of a secure SDLC integration and compliance framework ensures that all software development processes are aligned with best security practices and regulatory standards, minimizing vulnerabilities and enhancing the overall security posture of developed applications.

Control 2: Continuous Security Training for Development Teams

Control Objective: To equip development teams with the necessary knowledge and skills to implement secure coding practices, thereby reducing the introduction of vulnerabilities during the software development process.

Implementation Steps:

2.1 **Develop a Security Training Program**: Create a comprehensive security training program for development teams, covering current secure coding practices, emerging threats, and case studies on common vulnerabilities.

2.2 **Implement Regular Training Sessions**: Schedule and conduct regular training sessions, workshops, and security awareness activities for all development personnel.

2.3 **Monitor and Evaluate Training Effectiveness**: Assess the effectiveness of the training program through regular testing, feedback, and adjustments based on current security trends and the evolving threat landscape.

Expected Outcome: Continuous security training for development teams increases awareness and understanding of secure coding practices, significantly reducing the number of security vulnerabilities introduced in new software and enhancing the overall security culture within the organization.

Control 3: Adoption of Automated Security Tools in the Development Pipeline

Control Objective: To streamline the detection and remediation of security vulnerabilities within the development pipeline by adopting automated security testing tools.

Implementation Steps:

3.1 **Tool Selection and Integration**: Identify, select, and integrate automated security testing tools, such as Static Application Security Testing (SAST) and Dynamic Application Security Testing (DAST), into the development and deployment pipeline.

3.2 **Configure and Customize Tools**: Configure the selected tools to match the organization's development environment and security requirements, ensuring comprehensive coverage and accurate vulnerability detection.

3.3 **Continuous Monitoring and Improvement**: Regularly monitor the effectiveness of automated tools, updating configurations as necessary and ensuring developers are trained in responding to and mitigating identified vulnerabilities.

Expected Outcome: The adoption of automated security tools within the development pipeline enables early detection and remediation of vulnerabilities, significantly improving the security of software products and reducing the time and resources required for manual security reviews.

18

Technology Infrastructure Resilience (PR.IR)

> True resilience in the digital domain is not just about erecting barriers; it's about creating an ecosystem where security, efficiency, and continuity coalesce, preparing us to thrive amidst adversities.

In safeguarding digital environments, a multifaceted approach is essential. Starting with network security and access control, organizations employ firewalls, intrusion detection, and prevention systems while enforcing strict authentication protocols to mitigate unauthorized access. Additionally, they focus on protecting technology assets from environmental hazards through physical security, environmental controls, and rigorous testing. Organizational resilience extends beyond preventing threats, incorporating redundant systems, crisis management teams, and continuous improvement practices to ensure business continuity under adverse conditions. Finally, maintaining operational efficiency relies on adequate resource capacity and availability management, where strategic capacity planning, scalable infrastructure, and effective load balancing ensure that growing business demands are met without compromising service quality.

PR.IR-01: Networks and Environments Are Protected from Unauthorized Logical Access and Usage

The significance of robust network security and access control in the digital era cannot be overstated. As businesses become increasingly reliant on technology, the potential for unauthorized access and cyber threats escalates, highlighting the necessity for stringent security measures. The foundation of network security begins with understanding the myriad of threats that networks face, from external attacks like phishing and malware to internal threats such as unauthorized access or data leaks. A solid network security strategy protects organizational assets and secures the data and privacy of clients and employees. It is critical to recognize that network security is not a one-time task but a continuous assessment, implementation, and improvement process.

Implementing firewalls, intrusion detection systems (IDS), and intrusion prevention systems (IPS) constitutes the frontline defense against unauthorized access. Firewalls serve as the gatekeepers of the network, controlling incoming and outgoing traffic based on predetermined security rules. Meanwhile, IDS and IPS operate to detect and prevent malicious activities within the network. IDS monitors the network for suspicious activities and alerts the security team, while IPS takes active measures to block these threats. These tools, when combined, create a multilayered defense system that significantly reduces the risk of cyberattacks and unauthorized access.

A Comprehensive Guide to the NIST Cybersecurity Framework 2.0: Strategies, Implementation, and Best Practice, First Edition. Jason Edwards.
© 2025 John Wiley & Sons Ltd. Published 2025 by John Wiley & Sons Ltd.

Network segmentation and isolation are additional pivotal strategies in safeguarding networks from unauthorized access. By dividing the network into smaller, manageable segments, organizations can limit the spread of malicious activities and reduce the attack surface. Isolation of critical systems and sensitive data from the rest of the network further enhances security, ensuring that even if a breach occurs, the damage is contained and does not spread across the entire network. These practices not only bolster security but also improve network performance and management.

Strong authentication and authorization mechanisms ensure that only authorized individuals can access network resources. Techniques such as multifactor authentication (MFA), biometrics, and robust password policies enhance security by adding layers of verification before granting access. Authorization ensures that individuals have access only to the resources necessary for their role, implementing the principle of least privilege. Regular updates and reviews of these mechanisms are crucial to adapting to emerging threats and changing organizational structures.

Regularly reviewing and updating access control policies and procedures are imperative for maintaining a secure network environment. This involves evaluating the effectiveness of existing policies, identifying new risks, and adjusting measures accordingly. Documenting all changes and ensuring they align with industry best practices and regulatory requirements is also essential. Regular reviews help identify potential vulnerabilities before they can be exploited and ensure access control measures remain effective and relevant.

Encryption is critical in protecting sensitive data in transit and at rest. Encryption ensures that even if data is intercepted or accessed without authorization, it remains incomprehensible and secure by converting data into a coded format that can only be deciphered with the correct key. Implementing strong encryption standards for data in transit, such as Secure Sockets Layer (SSL) or Transport Layer Security (TLS), and for data at rest ensures that sensitive information remains confidential and intact against unauthorized access.

Conducting regular security assessments and penetration testing is crucial for identifying vulnerabilities in the network and evaluating the effectiveness of existing security measures. These assessments provide a realistic view of the organization's defense capabilities and reveal weaknesses that attackers could exploit. Penetration testing, in particular, simulates cyberattacks to test the network's resilience and identify areas for improvement. Regular assessments ensure security measures are always one step ahead of potential threats.

Monitoring and logging network access and activities provide visibility into the network's security posture. This involves tracking who accesses the network, what resources they use, and detecting any unusual patterns that could indicate a security breach. Effective monitoring and logging not only aid in the detection of threats but also in investigating and remedying security incidents. Organizations need a comprehensive logging strategy that includes storing, analyzing, and regularly reviewing logs for any signs of unauthorized access or suspicious activities.

Recommendations

- **Regularly Update Security Protocols**: Continuously review and update network security protocols, ensuring they align with the latest cybersecurity trends and threats. This includes updating firewalls, IDS, and IPS to combat new vulnerabilities.
- **Strengthen Access Controls**: Implement stringent access controls and authentication measures, such as MFA and role-based access controls, to minimize the risk of unauthorized access.
- **Conduct Frequent Security Training**: Organize regular training sessions for employees to raise awareness about network security, the latest cyber threats, and best practices in preventing unauthorized access.

- **Perform Regular Security Audits**: Schedule routine security assessments, including penetration testing, to identify and rectify vulnerabilities within the network infrastructure.
- **Establish a Response Plan**: Develop and maintain a comprehensive incident response plan that includes procedures for detecting, reporting, and responding to network security breaches.

NIST 800-53 Controls

(Refer to our appendix for a comprehensive description of the NIST 800-53 Controls, essential for organizations adhering to strict regulatory standards.)

- AC-03,AC-04,SC-04,SC-05,SC-07

Simplified Security Controls (SSC)

This section outlines our simplified security controls (SSC), tailored for environments beyond the scope of NIST compliance. These controls offer streamlined, effective security solutions suitable for any organization seeking to enhance its cybersecurity defenses with clear, actionable measures.

Control 1: Enhanced Network Access Control Systems

Control Objective: To strengthen the organization's defense against unauthorized access by implementing robust network access control systems, ensuring that only authorized users and devices can access the network and its resources.

Implementation Steps:

1.1 **System Selection and Configuration**: Select and implement advanced network access control systems that support capabilities such as device assessment, user identity verification, and role-based access control. Configure the systems to align with organizational security policies and access requirements.

1.2 **User and Device Profiling**: Develop profiles for all users and devices accessing the network, detailing permitted access levels and times. Implement continuous monitoring to ensure compliance with established profiles and detect anomalies.

1.3 **Incident Response Integration**: Integrate network access control systems with the organization's incident response platform to enable immediate action on detected unauthorized access attempts, such as blocking access or alerting security personnel.

Expected Outcome: Implementing enhanced network access control systems will result in a more secure network environment. Unauthorized access attempts should be significantly reduced, with immediate detection and response capabilities mitigating potential security breaches.

Control 2: Comprehensive Encryption Deployment

Control Objective: To ensure the confidentiality and integrity of sensitive data in transit and at rest through the comprehensive deployment of encryption technologies across the organization's networks and systems.

Implementation Steps:

2.1 **Encryption Standards Selection**: Adopt industry-standard encryption protocols, such as AES for data at rest and TLS for data in transit, ensuring they meet current security best practices and regulatory requirements.

2.2 **Implementation Across Platforms**: Apply encryption to all sensitive data at rest, including databases and storage devices, and ensure encrypted data transmission across internal and external networks.

2.3 **Key Management and Rotation**: Establish secure encryption key management procedures, including regular rotation of keys and secure storage of key management infrastructure, to prevent unauthorized access and ensure data recovery capabilities.

Expected Outcome: The comprehensive deployment of encryption should ensure that all sensitive data within the organization is unreadable and tamper-proof to unauthorized individuals. This will help maintain the confidentiality and integrity of critical information, reducing the risk of data breaches and compliance violations.

Control 3: Regular Security Training and Awareness Programs

Control Objective: To cultivate a security-aware culture within the organization where employees are equipped with the knowledge and skills to identify, prevent, and respond to cybersecurity threats, particularly network access and data security.

Implementation Steps:

3.1 **Program Development**: Develop a comprehensive security training program that includes modules on network security, access controls, encryption, and incident response. Tailor the training to different roles within the organization for relevance and effectiveness.

3.2 **Mandatory Participation and Regular Updates**: All employees must complete security training upon hire and at regular intervals thereafter. Regularly update the training content to reflect the latest cybersecurity threats and defense strategies.

3.3 **Simulations and Assessments**: Conduct regular security simulations, such as phishing tests, to assess employee awareness and provide feedback. Use assessment results to identify areas for improvement in the training program.

Expected Outcome: Implementing regular security training and awareness programs is expected to significantly reduce the likelihood of security incidents caused by human error. Employees will become proactive participants in the organization's cybersecurity efforts, more effectively identifying and mitigating potential security risks.

PR.IR-02: The Organization's Technology Assets Are Protected from Environmental Threats

Environmental threats to technology assets encompass a range of potential risks, from natural disasters like floods and earthquakes to human-made issues such as fire or heating, ventilation, and air conditioning (HVAC) failure. These threats can disrupt business operations, cause data loss, and damage physical infrastructure, leading to significant financial and reputational harm. Understanding these threats is the first step in protecting organizational assets. It is essential to identify which environmental factors could impact the organization and to what extent, allowing for the development of targeted strategies to mitigate these risks.

The implementation of physical security measures is critical in protecting against environmental threats. Surveillance systems, secure access controls, and physical barriers can prevent unauthorized access to sensitive areas, especially where essential technology assets are stored. These measures should be complemented by security policies that limit access based on the principle of least privilege, ensuring that only authorized personnel can access sensitive areas. Regular audits and updates of these physical security measures ensure they remain effective and aligned with the current threat landscape.

Deploying environmental controls such as HVAC and fire suppression systems is crucial in safeguarding technology assets from environmental threats. Properly functioning HVAC systems prevent overheating and humidity issues that can lead to equipment failure or degradation. Fire suppression systems, designed explicitly for technological environments like data centers, can quickly extinguish fires without damaging equipment. These systems must be chosen carefully to ensure they meet the specific needs of the technology assets they are designed to protect.

Environmental protection systems are regularly tested and maintained to ensure their reliability and effectiveness. This involves conducting routine inspections, testing system components, and replacing parts that are not functioning correctly. Regular maintenance helps to identify potential issues before they result in system failure, thereby preventing downtime and damage to technology assets. It is also essential to keep records of maintenance activities to track the health and status of environmental protection systems over time.

Developing disaster recovery and business continuity plans is essential for responding effectively to environmental threats. These plans should outline the steps to be taken before, during, and after an environmental event to ensure quick recovery and minimal disruption to business operations. They should include clear communication strategies, roles and responsibilities, and detailed recovery procedures. Regular training and drills should be conducted to ensure all employees are familiar with these plans and can execute them effectively in an emergency.

Assessment of environmental risks and impact analysis are critical components of a comprehensive environmental protection strategy. Organizations should conduct regular risk assessments to identify potential environmental threats and evaluate their impact on business operations. This includes analyzing the likelihood of different types of ecological events and assessing the vulnerability of technology assets. Impact analysis helps to prioritize the assets and processes that are most critical to the organization, guiding the allocation of resources to protect against environmental threats.

Using redundancy and failover mechanisms for critical systems ensures continuous operation in the event of environmental threats. Redundant systems and data backups can be activated immediately if primary systems fail due to environmental factors, minimizing downtime and data loss. Failover mechanisms, such as uninterruptible power supplies and generator backups, provide an alternative power source to keep critical systems running during power outages. Implementing these mechanisms requires careful planning and regular testing to ensure they work effectively when needed.

Employee training on response procedures for environmental threats is vital for ensuring a quick and coordinated response to incidents. Employees should be aware of the potential environmental risks to the organization and know the procedures to follow in the event of an incident. This training should cover how to safely evacuate the premises, report incidents, and initiate disaster recovery and business continuity plans. Regular updates and drills will help keep this knowledge fresh and ensure employees are prepared to respond effectively to environmental threats.

Recommendations

- **Enhance Physical Security Infrastructure**: Regularly review and upgrade physical security measures, such as surveillance cameras and access controls, to safeguard against unauthorized entry and environmental hazards.
- **Implement Regular System Checks**: Schedule frequent inspections and maintenance of environmental controls like HVAC and fire suppression systems to ensure they function optimally.
- **Conduct Comprehensive Risk Assessments**: Perform regular environmental risk assessments to identify and mitigate potential threats to technology assets, adjusting strategies based on the findings as necessary.

- **Develop and Drill Recovery Plans**: Create and regularly update disaster recovery and business continuity plans, ensuring all employees are trained and prepared to implement these plans effectively during an environmental incident.
- **Invest in Redundant Systems**: Establish redundancy and failover mechanisms for critical infrastructure to maintain operations during environmental disruptions, testing these systems regularly to confirm their effectiveness.

NIST 800-53 Controls

(Refer to our appendix for a comprehensive description of the NIST 800-53 Controls, essential for organizations adhering to strict regulatory standards.)

- CP-02,PE-09,PE-10,PE-11,PE-12,PE-13,PE-14,PE-15,PE-18,PE-23

Simplified Security Controls (SSC)

This section outlines our SSC, tailored for environments beyond the scope of NIST compliance. These controls offer streamlined, effective security solutions suitable for any organization seeking to enhance its cybersecurity defenses with clear, actionable measures.

Control 1: Enhanced Environmental Threat Protection Measures

Control Objective: To safeguard the organization's technology assets from environmental threats through comprehensive physical and environmental controls.

Implementation Steps:

1.1 **Installation of Advanced Security Systems**: Implement advanced physical security measures, including surveillance cameras, secure access controls, and IDS, to prevent unauthorized access and monitor environmental conditions.

1.2 **Deployment of Robust Environmental Controls**: Install and maintain environmental controls such as HVAC and fire suppression systems to protect technology assets from environmental risks.

1.3 **Regular Systems Review and Maintenance**: Schedule regular inspections and maintenance of physical and environmental control systems to ensure they remain effective and operational.

Expected Outcome: The organization's technology assets will be better protected against environmental threats, reducing the risk of damage and ensuring continuous operation. Enhanced security measures and environmental controls will contribute to a safer and more secure technology environment.

Control 2: Comprehensive Environmental Risk Assessment and Management

Control Objective: To systematically assess and manage environmental risks to the organization's technology assets, ensuring appropriate measures are in place to mitigate these risks.

Implementation Steps:

2.1 **Conduct an Environmental Risk Assessment**: Conduct a detailed assessment to identify potential environmental risks to technology assets, including natural disasters, fire, and water damage.

2.2 **Develop and Implement Mitigation Strategies**: Based on the risk assessment, develop and implement strategies to mitigate identified risks, such as relocating assets away from flood-prone areas or reinforcing structures against earthquakes.

2.3 **Regular Review and Update of Risk Assessment and Mitigation Strategies**: Regularly review and update the environmental risk assessment and mitigation strategies to reflect changes in the organization's environment and operations.

Expected Outcome: The organization will clearly understand the environmental risks to its technology assets and will have implemented effective measures to mitigate these risks, reducing the likelihood and impact of environmental incidents.

Control 3: Disaster Recovery and Business Continuity Integration

Control Objective: To integrate disaster recovery and business continuity planning with environmental threat protection strategies, ensuring the organization can quickly recover from any disruptions caused by environmental incidents.

Implementation Steps:

3.1 **Development of Integrated Plans**: Develop integrated disaster recovery and business continuity plans that address potential environmental threats and outline specific recovery procedures.

3.2 **Regular Testing and Updating of Plans**: Conduct regular drills to test the effectiveness of the disaster recovery and business continuity plans, making necessary adjustments based on test outcomes.

3.3 **Employee Training and Awareness**: Train employees on their roles and responsibilities in implementing disaster recovery and business continuity plans, particularly in response to environmental incidents.

Expected Outcome: The organization will have robust and effective disaster recovery and business continuity plans specifically designed to address environmental threats, ensuring a quick and efficient response to incidents and minimizing downtime and disruption.

PR.IR-03: Mechanisms Are Implemented to Achieve Resilience Requirements in Normal and Adverse Situations

Organizational resilience is fundamental in today's rapidly evolving cyber landscape. It encompasses an organization's ability to anticipate, prepare for, respond to, and recover from adverse situations while maintaining continuous business operations. Developing and implementing a resilience strategy covering various adverse scenarios form the backbone of a robust cybersecurity framework. This strategy should be comprehensive, addressing not just technological aspects but also human factors and business processes.

Implementing redundant systems and data backup solutions is crucial in maintaining business continuity. Redundancy ensures that if one component fails, another can immediately take its place without impacting operations. In contrast, effective backup solutions ensure data can be recovered swiftly and accurately following a data loss incident. These measures must be accompanied by the establishment of dedicated incident response and crisis management teams, which are tasked with responding to incidents promptly and organized, minimizing their impact.

Regular testing and exercises of resilience plans and mechanisms are essential for validating their effectiveness and identifying areas for improvement. These drills should simulate various realistic scenarios, challenging the technical infrastructure and the personnel involved in crisis response. By doing so, organizations can ensure that their response will be swift, coordinated, and effective when a real crisis occurs.

Integration of resilience planning with business continuity management is another critical element. This ensures that all aspects of the organization's operations are considered in the resilience strategy, facilitating a seamless transition between normal and emergency operations. This integration also ensures that the recovery strategies align with the organization's overall business objectives and operational requirements.

Monitoring and adapting to emerging threats and vulnerabilities is an ongoing process. As the threat landscape evolves, so must the organization's resilience strategies. This requires a proactive approach to cybersecurity, continuously scanning for new threats and adjusting strategies accordingly. By staying ahead of threats, organizations can ensure they are always prepared for what may come rather than merely reacting after the fact.

Employee training on roles and responsibilities in resilience efforts is vital. Every organization member should understand their role in maintaining and restoring operations, from the executive to the frontline staff. This training should be regularly updated to reflect the current threat environment and organizational policies. Organizations can foster a culture of resilience by ensuring that all employees are knowledgeable and prepared.

A continuous improvement process for resilience planning and execution ensures that strategies remain effective and relevant. This involves regularly reviewing and updating the resilience plan, incorporating lessons learned from drills and actual incidents, and staying informed about new threats and recovery techniques. By committing to continuous improvement, organizations can adapt to internal and external changes, maintaining their resilience over time.

Recommendations

- **Develop Comprehensive Resilience Plans**: Create detailed and actionable resilience plans that cover a range of adverse scenarios, ensuring all aspects of the organization are prepared for potential disruptions.
- **Enhance Data Backup and Redundancy**: Regularly review and update data backup and system redundancy strategies to ensure they can meet current and future organizational needs.
- **Conduct Regular Resilience Training and Drills**: Implement training programs and regular drills for all employees to ensure they understand their roles in maintaining organizational resilience.
- **Integrate Resilience with Business Continuity**: Seamlessly integrate resilience planning with overall business continuity management to ensure a unified approach to organizational preparedness.
- **Monitor and Adapt to Emerging Threats**: Continuously monitor the cybersecurity landscape for new threats and adapt resilience strategies to ensure they protect the organization effectively.

NIST 800-53 Controls

(Refer to our appendix for a comprehensive description of the NIST 800-53 Controls, essential for organizations adhering to strict regulatory standards.)

- CP,IR,SA-08,SC-06,SC-24,SC-36,SC-39,SI-13

Simplified Security Controls (SSC)

This section outlines our SSC, tailored for environments beyond the scope of NIST compliance. These controls offer streamlined, effective security solutions suitable for any organization seeking to enhance its cybersecurity defenses with clear, actionable measures.

Control 1: Resilience Strategy Development and Implementation

Control Objective: To create a comprehensive resilience strategy that enables the organization to maintain critical functions and quickly recover in various adverse situations.

Implementation Steps:

1.1 **Strategy Development**: Assemble a multidisciplinary team to develop a resilience strategy that addresses various adverse scenarios, from cyberattacks to natural disasters.

1.2 **Strategy Integration**: Integrate the resilience strategy with existing business continuity and disaster recovery plans, ensuring a cohesive approach to organizational resilience.

1.3 **Policy and Procedure Formulation**: Develop specific policies and procedures that support the resilience strategy, detailing roles, responsibilities, and actions for different scenarios.

Expected Outcome: The organization will have a clear, comprehensive resilience strategy integrated with other continuity efforts, ensuring preparedness and unified response to various incidents, thereby minimizing the impact on critical functions and services.

Control 2: Resilience Testing and Improvement Program

Control Objective: To ensure the effectiveness of the resilience strategy and its related mechanisms through regular testing, evaluation, and refinement.

Implementation Steps:

2.1 **Testing Schedule**: Establish a regular schedule for testing resilience plans and mechanisms, including tabletop exercises, simulations, and live drills.

2.2 **Performance Evaluation**: After each test, evaluate performance against predefined criteria to identify strengths and weaknesses in the resilience strategy and its execution.

2.3 **Continuous Improvement**: Use the insights from testing and evaluation to improve the resilience plans, policies, and procedures continuously.

Expected Outcome: Through regular, structured testing and improvement, the organization will enhance its resilience capabilities, ensuring that strategies, plans, and mechanisms are effective and up-to-date, improving overall preparedness and response to adverse situations.

Control 3: Organizational Resilience Training and Awareness

Control Objective: To ensure that all employees understand their roles and responsibilities within the organization's resilience efforts, enhancing the collective ability to respond to and recover from adverse events.

Implementation Steps:

3.1 **Training Program Development**: Develop a comprehensive training program that covers the organization's resilience strategies, employee roles and responsibilities, and response procedures for different scenarios.

3.2 **Mandatory Participation**: Require all employees to participate in resilience training, with specialized sessions for teams with critical roles in resilience efforts.

3.3 **Regular Updates and Refresher Sessions**: Update the training program regularly to reflect changes in the resilience strategy and conduct refresher sessions to ensure ongoing awareness and preparedness.

Expected Outcome: Employees across the organization will be well-informed about their roles in resilience efforts, leading to more effective and coordinated responses to incidents and faster recovery, thereby enhancing overall resilience.

PR.IR-04: Adequate Resource Capacity to Ensure Availability Is Maintained

Understanding the intricacies of capacity planning and availability management is fundamental for ensuring that an organization's technological infrastructure can handle current and future demands. This concept involves assessing present and prospective resource requirements to maintain seamless operations. As digital transformations accelerate, organizations must adapt by implementing scalable infrastructure and cloud solutions, facilitating expansion and flexibility while optimizing costs.

Monitoring resource usage and performance metrics provides vital insights into the health and efficiency of IT systems. This ongoing surveillance aids in identifying potential bottlenecks or underutilized resources, allowing for timely adjustments. Deploying advanced load balancing and failover techniques further ensures that services remain uninterrupted, even during unexpected surges in demand or infrastructure failures, thus preserving user experience and operational continuity.

Regular reviews of capacity plans against business growth and needs are essential. This alignment ensures that IT capabilities evolve within the organization, avoiding over- or under-provisioning resources. By regularly assessing these plans, companies can stay ahead of growth curves, ensuring they are neither squandering resources nor caught unprepared for demand increases.

Optimization of resource allocation for efficiency and cost-effectiveness is an ongoing challenge. Organizations can achieve more with less by employing strategies such as virtualization, containerization, and cloud services, reducing physical infrastructure costs while increasing flexibility and scalability. This conserves financial resources and aligns IT expenditures more closely with actual usage patterns.

Planning for peak load management and conducting regular stress testing is critical for understanding how systems will perform under extreme conditions. These activities help identify vulnerabilities before they cause problems, allowing for the refinement of architectures and processes to handle high traffic volumes or data processing peaks, thus avoiding potential service outages or degradations.

Employee training on capacity management tools and techniques equips staff with the knowledge and skills to manage and respond to capacity-related issues effectively. This empowers teams to make informed decisions and take proactive measures, enhancing the overall resilience and efficiency of the organization's IT infrastructure.

Finally, the continuous improvement process for capacity planning and execution ensures that strategies remain relevant and effective. By embracing a cycle of assessment, adjustment, and review, organizations can adapt their capacity planning and management practices to meet changing needs and technologies, thereby maintaining optimal performance and service availability.

Recommendations

- **Implement an Integrated Capacity Management Tool**: Utilize comprehensive capacity management software that provides real-time insights into resource utilization and performance, enabling proactive adjustments.
- **Establish a Capacity Planning Committee**: Form a dedicated group responsible for overseeing capacity planning and management, ensuring alignment with business objectives and growth.

- **Regularly Conduct Scenario-based Planning**: Engage in regular scenario planning exercises to understand potential impacts on capacity and prepare appropriate responses.
- **Leverage Cloud Scalability**: Maximize the use of cloud technologies to provide scalable and flexible resource solutions that can adapt quickly to changing requirements.
- **Foster a Culture of Efficiency**: Encourage a company-wide ethos that prioritizes efficiency and continuous resource usage and capacity planning improvement.

NIST 800-53 Controls

(Refer to our appendix for a comprehensive description of the NIST 800-53 Controls, essential for organizations adhering to strict regulatory standards.)

- CP-06,CP-07,CP-08,PM-03,PM-09

Simplified Security Controls (SSC)

This section outlines our SSC, tailored for environments beyond the scope of NIST compliance. These controls offer streamlined, effective security solutions suitable for any organization seeking to enhance its cybersecurity defenses with clear, actionable measures.

Control 1: Dynamic Resource Assessment and Forecasting

Control Objective: To ensure the organization has sufficient resource capacity to meet current and future demands, thereby maintaining availability and performance standards.

Implementation Steps:

1.1 **Continuous Assessment**: Implement a continuous assessment process to monitor current resource utilization and performance metrics across all critical systems.

1.2 **Predictive Analysis**: Utilize predictive analytics tools to forecast future resource requirements based on historical data, business growth trends, and upcoming projects.

1.3 **Adjustment Procedures**: Establish procedures for timely resource adjustment, including scaling up or reallocating resources to meet predicted demands without impacting system performance.

Expected Outcome: The organization will maintain optimal performance and availability by continuously assessing and forecasting resource needs. This proactive approach prevents system overload and efficiently allocates resources to meet current and future demands.

Control 2: Scalable Infrastructure Implementation

Control Objective: To provide a flexible and scalable infrastructure that can adapt to changing business needs and workloads, ensuring uninterrupted service and performance.

Implementation Steps:

2.1 **Scalability Evaluation**: Evaluate existing infrastructure for scalability and identify areas for improvement or replacement with more scalable solutions, such as cloud services or modular hardware.

2.2 **Implementation of Scalable Solutions**: Implement scalable infrastructure solutions, focusing on areas critical to business operations and future growth, ensuring they can be easily adjusted without significant downtime.

2.3 **Regular Scalability Testing**: Regular tests ensure the infrastructure can handle increased loads and seamlessly scale up as needed, adjusting configurations based on test results.

Expected Outcome: The organization's infrastructure will be able to adapt rapidly to increased demands, maintaining service availability and performance even under varying loads. This ensures business continuity and supports growth without compromising service quality.

Control 3: Comprehensive Load Management and Stress Testing

Control Objective: To ensure the organization's systems can handle peak loads and continue operating under stress, preventing service disruptions and performance degradation.

Implementation Steps:

3.1 **Load Balancing Strategy Development**: Develop and implement a load balancing strategy that distributes workloads evenly across servers and resources, preventing overload and ensuring optimal performance.

3.2 **Regular Stress Testing**: Schedule regular stress tests to simulate peak loads and identify potential bottlenecks in the system, providing insights into how the infrastructure performs under extreme conditions.

3.3 **Optimization and Remediation**: Based on stress test results, optimize the infrastructure and remediate any identified issues to ensure the system can handle peak loads without failure or significant performance loss.

Expected Outcome: The organization will have a robust system capable of managing peak loads and stressful conditions, ensuring continuous availability and performance. Regular testing and optimization will prepare the system for real-world challenges, reducing the risk of downtime and service interruptions.

19

Continuous Monitoring (DE.CM)

In the vast and intricate digital security landscape, the meticulous monitoring of networks, environments, personnel, and resources emerges as a cornerstone of the defense, weaving a multilayered safeguard that enhances the organization's resilience and responsiveness to the ever-evolving cybersecurity challenges.

In the modern digital age, monitoring networks, physical environments, personnel activities, external service providers, and computing resources have become indispensable for maintaining organizational security. Implementing robust detection systems like intrusion detection systems (IDS) and employing real-time traffic and anomaly detection techniques are crucial for identifying potentially adverse events in networks and network services. Similarly, physical security monitoring incorporates surveillance, environmental sensors, and regular audits to protect an organization's physical assets. Monitoring personnel activity and technology usage, including implementing user and entity behavior analytics (UEBA) and data loss prevention (DLP) technologies, ensures compliance and safeguards against internal threats. Furthermore, overseeing external service provider activities through clear contractual requirements and regular security assessments helps manage third-party risks. Finally, comprehensive monitoring of computing hardware and software through endpoint detection and response (EDR) solutions, application performance monitoring (APM) tools, and regular security assessments ensures the integrity and availability of IT assets. These strategies form a multilayered defense, significantly enhancing an organization's ability to respond to and mitigate security incidents.

Author's Note: DE.CM-04, DE.CM-05, DE.CM-07, and DE.CM-08 were deprecated from the CSF 2 and are not included in this chapter.

DE.CM-01: Networks and Network Services Are Monitored to Find Potentially Adverse Events

Network monitoring is critical for defending an organization's information systems from malicious activities and unauthorized access. Effective network monitoring involves observing and analyzing network traffic to detect suspicious activities that could indicate a security breach or other adverse events. This practice helps identify and mitigate potential threats before they escalate into serious incidents. Organizations can swiftly detect irregularities and respond accordingly by maintaining vigilance over network traffic.

A Comprehensive Guide to the NIST Cybersecurity Framework 2.0: Strategies, Implementation, and Best Practice, First Edition. Jason Edwards.
© 2025 John Wiley & Sons Ltd. Published 2025 by John Wiley & Sons Ltd.

Deploying IDS and network analysis tools is fundamental to establishing a robust network monitoring environment. IDS are designed to automatically detect and alert potential intrusions by comparing network traffic with known threat signatures. These systems are crucial in identifying malicious activities such as unauthorized access, malware distribution, and other security threats. Network analysis tools complement IDS by providing a more detailed view of network traffic, enabling security professionals to conduct in-depth analyses and identify patterns indicative of potential security issues. Together, these tools form the backbone of an effective network monitoring strategy, enabling organizations to detect and respond to threats in real time.

Implementing real-time traffic and anomaly detection techniques further enhances network security. These techniques involve analyzing network traffic to identify deviations from established norms, which may indicate a security incident. Anomaly detection is particularly effective in identifying zero-day attacks and other novel threats that traditional signature-based detection methods may not recognize. By continuously monitoring network traffic, organizations can detect unusual patterns, such as spikes in data transfer or uncharacteristic access attempts, enabling them to respond quickly to potential threats.

Establishing baselines for normal network behavior is essential for effective anomaly detection. Baselines provide a reference point against which current network activity can be compared, making it easier to identify unusual behavior. This involves collecting and analyzing data to understand typical traffic patterns, including peak usage times, common service requests, and normal data flows. Having well-defined baselines allows organizations to swiftly detect deviations, which could indicate a security incident or other issues requiring attention.

Regular updates and tuning of monitoring tools are necessary to adapt to new threats and changes in the network environment. As cyber threats evolve, monitoring tools must be updated to recognize new attack patterns and vulnerabilities. Additionally, as organizations grow and change, their network traffic patterns may develop, necessitating baselines and detection parameters adjustments. Regularly updating and tuning these tools ensures they effectively detect current threats and reduce the likelihood of false positives, which can divert resources from genuine threats.

Integration of network monitoring with incident response processes is vital to ensure a coordinated and effective response to detected threats. When a potential security incident is identified, the incident response team should be alerted immediately to assess the situation and take necessary actions. This integration enables swift containment and mitigation of threats, minimizing potential damage. Effective communication between the network monitoring and incident response teams ensures that all relevant information is shared promptly, facilitating a well-informed response to incidents.

Training for network administrators on advanced monitoring techniques is crucial for maintaining an effective network security posture. Network administrators must have the latest skills and knowledge to detect and respond effectively as cyber threats become more sophisticated. This training should cover the latest trends in cyber threats, advanced detection techniques, and best practices for network security. Organizations can enhance their ability to detect and respond to security incidents by ensuring that network administrators are well-trained.

Artificial intelligence (AI) and machine learning (ML) for threat detection represent a significant advancement in network monitoring capabilities. AI and ML algorithms can analyze vast network data to identify patterns and anomalies that may indicate a security threat. These technologies can learn from historical data, improving their detection capabilities. By incorporating AI and ML into network monitoring, organizations can enhance their ability to detect sophisticated threats more quickly and accurately, strengthening their overall cybersecurity posture.

Documentation and reporting of identified network anomalies are critical components of effective network monitoring. Detailed records of detected incidents, including their nature, response, and outcome, are essential for analyzing trends, improving security measures, and complying with regulatory requirements. Regular reporting on network security status to relevant stakeholders, including management and regulatory bodies, ensures transparency and accountability in the organization's cybersecurity efforts.

Recommendations

- **Refine Anomaly Detection Techniques**: Continuously refine real-time traffic and anomaly detection methods to improve accuracy. This involves adjusting detection parameters based on evolving threat landscapes and changes in network behavior.
- **Enhance Network Visibility with IDS and Analysis Tools**: Prioritize the integration of IDS and comprehensive network analysis tools to form a robust foundation for network monitoring. These technologies are critical for identifying unauthorized access, malware distribution, and other security threats. Their deployment enables real-time detection and response, significantly strengthening your organization's ability to combat cyber threats.
- **Establish and Regularly Update Network Behavior Baselines**: To enhance anomaly detection capabilities, develop clear baselines for normal network activity. Regularly analyzing traffic patterns, peak usage times, and typical data flows is essential for setting these benchmarks. As your organization evolves, continually adjust and refine these baselines to maintain accuracy in threat detection.
- **Integrate AI and ML Technologies for Advanced Threat Detection**: Adopt AI and ML technologies to enhance your network monitoring capabilities significantly. These tools can analyze vast network data to detect complex patterns and anomalies indicative of security threats, offering a proactive approach to cybersecurity.
- **Foster Transparency through Comprehensive Documentation and Reporting**: Maintain detailed records of all detected network anomalies and security incidents, including their investigation, response, and resolution. Regularly share updates on network security status with key stakeholders to ensure organizational transparency and compliance with regulatory standards.

NIST 800-53 Controls

(Refer to our appendix for a comprehensive description of the NIST 800-53 Controls, essential for organizations adhering to strict regulatory standards.)

- AC-02,AU-12,CA-07,CM-03,SC-05,SC-07,SI-04

Simplified Security Controls (SSC)

This section outlines our simplified security controls (SSC), tailored for environments beyond the scope of NIST compliance. These controls offer streamlined, effective security solutions suitable for any organization seeking to enhance its cybersecurity defenses with clear, actionable measures.

Control 1: Advanced Network Monitoring Integration

Control Objective: To enhance network security by integrating advanced monitoring tools and techniques, ensuring the timely detection of potentially adverse events within networks and network services.

Implementation Steps:

1.1 **Deployment and Configuration**: Implement state-of-the-art IDS and network analysis tools tailored to the organization's network architecture and threat landscape. Configure these tools to monitor critical network traffic and activities continuously.

1.2 **Anomaly Detection Setup**: Employ real-time traffic and anomaly detection techniques, utilizing AI and ML to analyze network behavior. Establish normal network activity baselines to identify deviations that may indicate security incidents.

1.3 **Continuous Improvement Process**: Regularly update and tune the network monitoring tools to adapt to new threats and technological changes. Establish a routine for reviewing the effectiveness of monitoring strategies and incorporating feedback for continuous improvement.

Expected Outcome: Integrating advanced network monitoring tools and techniques will lead to earlier detection of security threats, reducing potential damage. The organization will achieve a more dynamic and adaptive security posture, capable of responding swiftly to evolving cyber threats.

Control 2: Network Security Awareness and Response Training

Control Objective: To ensure network administrators and relevant personnel are fully trained in advanced monitoring techniques and incident response procedures, enhancing the organization's ability to identify and mitigate network-related security events.

Implementation Steps:

2.1 **Development of Training Programs**: Create comprehensive training programs focusing on advanced network monitoring techniques, IDS and network analysis tools, and the latest threat detection methodologies.

2.2 **Regular Training Sessions**: Conduct regular, mandatory training sessions for network administrators and security personnel. Include practical exercises, simulations, and the latest cybersecurity trends and threats.

2.3 **Incident Response Drills**: Integrate network monitoring with incident response processes. Conduct regular drills to practice detecting, analyzing, and responding to simulated network threats, ensuring procedures are effective and well-understood.

Expected Outcome: Network administrators and security personnel will be well-versed in using advanced monitoring tools and responding to network threats. This control will improve the capacity to prevent, detect, and respond to adverse events, strengthening network security.

Control 3: Comprehensive Documentation and Reporting Mechanism

Control Objective: To establish a robust framework for documenting and reporting network anomalies and security incidents, ensuring that all detected events are accurately recorded, analyzed, and addressed.

Implementation Steps:

3.1 **Documentation Standards**: Develop and implement standards for documenting all detected network anomalies and security incidents. This should include the detection time, nature of the anomaly, affected systems, and steps taken in response.

3.2 **Reporting System Setup**: Set up a centralized reporting system for network security events. Ensure that it is accessible to relevant personnel and facilitates the easy reporting and tracking of incidents.

3.3 **Review and Analysis Procedures**: Establish regular review and analysis sessions to examine reported anomalies and incidents. Utilize findings to refine monitoring strategies and response procedures, identifying trends and areas for improvement.

Expected Outcome: Establishing a comprehensive documentation and reporting mechanism will ensure that all network security events are recorded and analyzed, leading to informed decision-making and continuous improvement in network security practices. This will enhance the organization's ability to respond effectively to and learn from security incidents.

DE.CM-02: The Physical Environment Is Monitored to Find Potentially Adverse Events

Physical security monitoring plays a vital role in the holistic approach to cybersecurity, safeguarding the tangible assets and environments where critical data and systems reside. This practice involves continuously observing and assessing physical spaces to detect and mitigate potential security breaches or environmental hazards that could compromise an organization's information assets. Effective physical security monitoring encompasses various tools and strategies to protect against unauthorized access, damage, and other threats to physical resources.

Surveillance cameras and access control systems form the backbone of modern physical security monitoring. Surveillance cameras serve as both a deterrent to unauthorized individuals and a means of recording potential security breaches. On the other hand, access control systems ensure that only authorized personnel can enter sensitive areas, thereby preventing unauthorized access. These systems work in tandem to monitor and control the environment, ensuring that any irregular activity is detected and addressed promptly. Implementing these systems requires careful planning and regular maintenance to ensure they safeguard the physical premises effectively.

Implementing environmental sensors such as smoke detectors, water leak detectors, and temperature monitors is essential for identifying environmental threats that could damage physical assets or disrupt operations. These sensors provide real-time alerts to potential hazards, allowing swift action to mitigate risks. For instance, temperature sensors can prevent overheating in server rooms, which could lead to equipment failure and data loss. By integrating these sensors into the physical security monitoring framework, organizations can protect against various environmental risks.

Regular security audits of the physical infrastructure are necessary to identify vulnerabilities and ensure the effectiveness of security measures. These audits should assess the condition and effectiveness of surveillance systems, access controls, and environmental sensors. They should also evaluate the security of doors, locks, windows, and other physical barriers. Regular audits help organizations identify areas for improvement and ensure that their physical security measures remain current with the latest threats and best practices.

Integrating physical security systems with the security operations center (SOC) enhances the organization's security posture. This integration allows real-time monitoring and coordinated response to cyber and physical threats. The SOC team can analyze data from physical security devices alongside cyber threat intelligence, providing a comprehensive view of the organization's security status. This holistic approach ensures that all security aspects are covered, enabling a more effective and efficient response to incidents.

Training on physical security best practices and emergency procedures is critical for all staff members. Employees should be aware of the security measures in place, their role in maintaining security, and the correct actions to take in case of a security breach or environmental hazard. Regular training and drills can help ensure staff members are prepared to respond effectively in emergencies, reducing the risk of damage or loss.

Monitoring and testing physical security devices and systems are essential to ensure they function correctly and provide the intended protection. Regular testing helps identify any malfunctions or weaknesses that unauthorized individuals could exploit. It also ensures that environmental sensors are accurately detecting hazards. Continuous monitoring and periodic testing help maintain the integrity of physical security measures, ensuring they always provide effective protection.

Procedures for responding to physical security incidents should be clearly defined and communicated to all relevant staff. These procedures should outline the steps for detecting a security breach or environmental hazard, including who to contact and how to contain the situation. A well-defined response plan helps minimize the impact of security incidents and ensures a swift and coordinated response.

Documentation of physical security events and investigations is crucial for legal compliance and improving security measures. Detailed records help analyze the causes of incidents and develop strategies to prevent future occurrences. Documentation also provides evidence for legal proceedings and helps assess the effectiveness of current security measures.

Recommendations

- **Enhance Surveillance and Access Control**: Upgrade and maintain surveillance cameras and access control systems to ensure comprehensive monitoring and control of physical spaces. Consider advanced technologies like biometric scanners for higher security areas.
- **Expand Environmental Monitoring**: Implement a wide range of environmental sensors to detect and alert on various threats such as smoke, water, and extreme temperatures. Integrate these sensors with the overall security system for real-time threat detection.
- **Conduct Regular Physical Security Audits**: Schedule and perform regular audits of all physical security measures to identify vulnerabilities and ensure the effectiveness of existing protocols. Use the findings to strengthen security and mitigate identified risks.
- **Integrate Physical and Cybersecurity Efforts**: Ensure physical security systems are integrated with the organization's SOC. This allows for a unified threat detection and response approach, enhancing the overall security posture.
- **Develop Comprehensive Training and Response Plans**: Train employees regularly on physical security best practices and emergency procedures. Develop clear, actionable response plans for different physical security incidents, ensuring a swift and coordinated reaction to threats.

NIST 800-53 Controls

(Refer to our appendix for a comprehensive description of the NIST 800-53 Controls, essential for organizations adhering to strict regulatory standards.)

- CA-07,PE-03,PE-06,PE-20

Simplified Security Controls (SSC)

This section outlines our SSC, tailored for environments beyond the scope of NIST compliance. These controls offer streamlined, effective security solutions suitable for any organization seeking to enhance its cybersecurity defenses with clear, actionable measures.

Control 1: Enhanced Surveillance and Access Control Integration

Control Objective: To secure the physical premises by implementing advanced surveillance cameras and access control systems, ensuring only authorized individuals access sensitive areas and monitor all activities for potentially adverse events.

Implementation Steps:

1.1 **System Installation and Configuration**: Install high-definition surveillance cameras and electronic access control systems at all entry points and sensitive areas. Configure systems to ensure optimal coverage and control, including time-based access restrictions and multifactor authentication for sensitive zones.

1.2 **Regular System Testing and Updates**: Conduct regular tests to ensure all physical security devices function correctly. Update firmware and software to address vulnerabilities and improve functionality.

1.3 **Integration with SOC**: Integrate surveillance and access control systems with the SOC for real-time monitoring and incident management. Establish protocols for immediate response to unauthorized access attempts or suspicious activities detected by these systems.

Expected Outcome: By enhancing surveillance and access control integration, the organization will achieve tighter security over physical assets, reducing the risk of unauthorized access and enabling real-time response to security incidents. This will ensure a secure and monitored environment, contributing to the overall safety of personnel and the protection of assets.

Control 2: Comprehensive Environmental Monitoring Implementation

Control Objective: To protect physical assets from environmental risks by implementing a comprehensive environmental monitoring system that detects smoke, water leakage, and extreme temperature changes.

Implementation Steps:

2.1 **Sensor Deployment**: Install environmental sensors throughout the facility, focusing on high-risk areas such as server rooms, electrical rooms, and storage areas. Ensure sensors can detect smoke, water leaks, and significant temperature fluctuations.

2.2 **Alert Configuration and Emergency Protocols**: Configure sensors to send immediate alerts to the SOC and relevant personnel upon detecting adverse environmental conditions. Develop and document emergency response protocols for different types of environmental incidents.

2.3 **Regular Maintenance and Testing**: Schedule regular maintenance checks and system tests to ensure all environmental sensors are operational and accurately reporting conditions. Address any issues promptly to maintain continuous monitoring.

Expected Outcome: Implementing a comprehensive environmental monitoring system will safeguard the organization's physical assets from environmental hazards, minimizing the risk of damage and ensuring business continuity. Prompt detection and response to environmental threats will prevent significant losses and maintain operational integrity.

Control 3: Physical Security Training and Incident Response Drills

Control Objective: To ensure that all employees understand physical security best practices and are prepared to respond effectively to physical security incidents.

Implementation Steps:

3.1 **Develop and Conduct Training Programs**: Create detailed training programs covering physical security awareness, best practices, and emergency procedures. Conduct these training sessions regularly for all employees, with specialized sessions for security personnel.

3.2 **Incident Response Drills**: Plan and execute regular drills simulating various physical security incidents involving all relevant staff. Use these drills to practice emergency procedures and improve the organization's incident response capabilities.

3.3 **Feedback and Continuous Improvement**: Gather feedback from participants after each drill to identify areas for improvement. Update training programs and emergency procedures based on this feedback to enhance physical security and preparedness.

Expected Outcome: Through comprehensive training and regular drills, employees will be better prepared to identify and respond to physical security threats. This proactive approach will enhance the overall security culture within the organization, reduce the likelihood of security breaches, and ensure a swift and coordinated response to any incidents that do occur.

DE.CM-03: Personnel Activity and Technology Usage Are Monitored to Find Potentially Adverse Events

Monitoring personnel activity and technology usage is paramount in identifying and mitigating internal threats and ensuring all users adhere to established security policies—the focus on tracking underlines the importance of transparency and accountability within an organization's cybersecurity framework. By understanding normal user behavior, organizations can more effectively spot anomalies that may signify a security threat, such as data breaches or insider threats.

Deploying UEBA solutions enhances security monitoring. UEBA systems analyze the behavior of users and entities within the network, employing advanced analytics to detect activities that deviate from established patterns. This approach enables the identification of suspicious activities that could indicate malicious intent or policy noncompliance. By leveraging ML and statistical analysis, UEBA tools provide a dynamic and adaptive approach to security, capable of evolving with the ever-changing landscape of cyber threats and user behavior patterns.

Establishing policies for the acceptable use of technology resources is crucial in setting clear expectations for employee behavior. These policies should outline permitted and prohibited activities, ensuring all users understand the organization's rules regarding technology usage. By establishing clear guidelines, organizations can foster an environment of compliance and security consciousness among their staff. These policies also serve as a foundation for monitoring activities, providing a benchmark against which user behavior can be measured.

Access logs and audit trails are critical for monitoring user activities and detecting potential security incidents. These logs provide a chronological record of system access and activities, offering valuable insights into user behavior and system usage. By regularly reviewing and analyzing these logs, security teams can identify unusual or unauthorized activities, such as access to sensitive information outside of normal working hours or repeated login failures, which may indicate an attempted security breach.

Implementing DLP technologies is essential for safeguarding sensitive information from unauthorized access or exfiltration. DLP systems monitor and control data transfer, preventing sensitive information from leaving the network without proper authorization. By tracking the movement of critical data across an organization's network, DLP solutions help prevent data breaches and ensure that sensitive information remains secure.

Integration of monitoring tools with incident response plans is necessary to ensure a swift and effective reaction to detected anomalies. When a security threat is identified through monitoring, the organization must have a well-defined procedure for responding to the incident. This integration allows for immediate action, minimizing the potential damage caused by security incidents. Effective collaboration between monitoring and incident response teams enhances the organization's ability to manage and mitigate threats.

Training on secure and responsible use of IT resources is vital in fostering a culture of cybersecurity awareness among employees. Regular training sessions should inform personnel about the risks associated with improper technology usage, the importance of adhering to security policies, and the role of monitoring in maintaining organizational security. Educating employees about their part in the cybersecurity ecosystem can lead to more conscientious behavior and reduced risk of security incidents.

Regular review and analysis of monitoring data for anomaly detection are indispensable for maintaining an effective security posture. Continuous evaluation of monitoring data enables organizations to adapt to new threats and identify emerging malicious behavior patterns. This proactive approach ensures that security measures remain effective, protecting the organization from internal and external threats.

Ensuring privacy and compliance in monitoring practices is essential to maintaining trust and adherence to legal standards. Monitoring should be conducted transparently and in accordance with applicable laws and regulations to protect individual privacy rights. Organizations must balance the need for security with the requirement to respect employee privacy, ensuring that monitoring practices are ethical and lawful.

Recommendations

- **Refine and Communicate Monitoring Policies**: Develop comprehensive policies regarding personnel activity and technology usage monitoring. Communicate these policies effectively to all employees, ensuring they understand the purpose and extent of monitoring.
- **Enhance Anomaly Detection Capabilities**: Invest in advanced UEBA solutions and DLP technologies to detect unusual activities and safeguard sensitive information. Regularly update these tools to adapt to new threats and changes in user behavior.
- **Strengthen Incident Response Integration**: Ensure monitoring systems fully integrate with the organization's incident response framework. This alignment allows for quicker detection, reporting, and remediation of security incidents.
- **Expand Employee Training Programs**: Implement ongoing training and awareness programs focused on secure technology usage, monitoring, and compliance with organizational policies.
- **Audit and Review Monitoring Practices**: Regularly review and audit monitoring practices to ensure they remain effective, compliant with laws and regulations, and respectful of privacy. Use audit findings to improve monitoring strategies and technologies.

NIST 800-53 Controls

(Refer to our appendix for a comprehensive description of the NIST 800-53 Controls, essential for organizations adhering to strict regulatory standards.)

- AC-02,AU-12,AU-13,CA-07,CM-10,CM-11

Simplified Security Controls (SSC)

This section outlines our SSC, tailored for environments beyond the scope of NIST compliance. These controls offer streamlined, effective security solutions suitable for any organization seeking to enhance its cybersecurity defenses with clear, actionable measures.

Control 1: Implementation of Comprehensive User Activity Monitoring System

Control Objective: To ensure the security of technology resources by monitoring and analyzing user activities and behaviors, thereby detecting and mitigating potential security threats and policy violations.

Implementation Steps:

1.1 **Deploy UEBA Systems**: Select and implement a UEBA solution that aligns with the organization's specific needs and integrates seamlessly with existing security infrastructure to monitor user activities and detect anomalies.

1.2 **Establish Monitoring Policies**: Develop comprehensive policies for acceptable use of technology resources and communicate these to all employees. Ensure that these policies cover all aspects of technology usage and specify what constitutes acceptable and unacceptable behavior.

1.3 **Continuous Monitoring and Alerting**: Set up the UEBA system to continuously monitor user activities across the organization's network and systems and configure alerting mechanisms for any detected anomalies that could indicate security threats or policy breaches.

Expected Outcome: By implementing a comprehensive user activity monitoring system, the organization can detect and respond to insider threats, unauthorized access, and other security risks more effectively. This will improve overall security posture and compliance with established policies and regulations.

Control 2: Data Loss Prevention and Integration Strategy

Control Objective: To prevent unauthorized access and leakage of sensitive information while integrating data protection measures with broader incident response plans.

Implementation Steps:

2.1 **DLP Technology Implementation**: Deploy DLP technologies across the organization's network, endpoints, and data storage systems to monitor, detect, and prevent unauthorized access and transfer of sensitive information.

2.2 **Policy Development and Enforcement**: Develop and enforce data protection policies that define sensitive data, outline how it should be handled, and specify the consequences of policy violations. Integrate these policies with the DLP technologies to ensure automated enforcement.

2.3 **Integration with Incident Response**: Ensure that DLP systems are integrated with the organization's incident response plans. This allows immediate action when potential data breaches are detected, including automated alerts to the appropriate personnel.

Expected Outcome: The implementation of DLP technologies and integration with incident response plans will significantly reduce the risk of sensitive data being exposed or stolen. The organization will be able to promptly respond to potential data breaches, minimizing the impact on business operations and reputation.

Control 3: Regular Security Awareness and Compliance Training

Control Objective: To promote a culture of security awareness and ensure all employees understand and comply with the organization's technology usage policies and data protection practices.

Implementation Steps:

3.1 **Develop Training Program**: Create an ongoing security awareness training program that includes modules on acceptable use of technology, data protection best practices, and how to recognize and respond to security threats.

3.2 **Mandatory Training Sessions**: Make security awareness training mandatory for all employees, with regular sessions scheduled to address new threats, refresh knowledge, and ensure ongoing compliance.

3.3 **Monitoring and Enforcement**: Monitor compliance with training requirements and technology usage policies and enforce consequences for noncompliance, such as additional training or disciplinary action.

Expected Outcome: Regular security awareness and compliance training will enhance employees' understanding of security risks and the importance of following established policies, reducing incidents caused by human error, and creating an overall stronger security culture within the organization.

DE.CM-06: External Service Provider Activities and Services Are Monitored to Find Potentially Adverse Events

Monitoring external service provider activities and services is a crucial component of a comprehensive cybersecurity strategy. As organizations increasingly rely on third parties for critical services and infrastructure, the potential for security vulnerabilities introduced by these providers has grown significantly. Understanding and implementing effective monitoring practices is essential for maintaining the integrity, availability, and confidentiality of data and systems managed by external parties.

Establishing clear security and monitoring requirements in contract contracts is the first step in ensuring external services align with an organization's security posture. These agreements should detail the expected security standards, monitoring capabilities, and incident response expectations. By formalizing these requirements, organizations can set a benchmark for security practices and facilitate accountability. Clear contractual obligations also provide a legal basis for enforcement should a provider fail to meet the agreed-upon standards, helping to mitigate potential risks associated with third-party services.

Third-party security assessments and audits are valuable methods for verifying the security posture of external service providers. These evaluations should be conducted regularly, including comprehensive reviews of the provider's infrastructure, policies, and procedures. The assessments can help identify vulnerabilities and noncompliance issues that could pose a risk to the organization. Additionally, audits can provide insights into the effectiveness of the provider's security practices and their ability to maintain the confidentiality, integrity, and availability of the managed data and services.

Implementation of shared security and incident reporting tools facilitates collaboration and transparency between organizations and their service providers. By using common platforms for monitoring and reporting, both parties can have real-time access to security information, enhancing the ability to detect and respond to incidents quickly. Shared tools also help streamline communication and coordination, improving the overall efficiency of security operations and incident management.

Regular reviews of service provider security postures and compliance reports are essential for maintaining an ongoing understanding of the external service environment. These reviews should assess the provider's adherence to contractual security requirements and industry best practices. By continuously evaluating the security performance of service providers, organizations can identify and address potential security gaps before they lead to serious incidents.

Coordination of incident response efforts with external providers is critical when dealing with security incidents that impact shared services or data. A well-defined and practiced incident response plan that includes roles and responsibilities for the organization and its service providers ensures a cohesive and effective approach to managing and mitigating incidents. Coordination helps prevent duplication of efforts, reduces response times, and ensures all parties are aligned in their response strategies.

Training on managing and securing third-party integrations is essential for employees involved in the procurement, management, and oversight of external services. Staff should be aware of the potential risks associated with third-party services and understand the security measures that should be in place to protect the organization. Training should cover best practices for integrating and securing third-party solutions, including managing access controls and monitoring data flows.

Monitoring access controls and data flow between providers and the organization is crucial for preventing unauthorized access and data exfiltration. Organizations must implement and maintain appropriate access control measures, restricting provider access to what is necessary for service delivery. Additionally, monitoring data flow to and from service providers can help detect and prevent the unauthorized transfer of sensitive information.

Documentation of service provider performance and security issues is key in managing third-party relationships. Keeping detailed records of service delivery, security incidents, and resolution outcomes helps organizations evaluate provider performance and make informed decisions about continued partnerships. Documentation also supports compliance efforts and can provide valuable evidence in the event of a dispute or legal action.

Recommendations

- **Clarify Security Expectations and Requirements**: Ensure all contracts with external service providers clearly outline security expectations, monitoring requirements, and incident response protocols. This establishes a solid foundation for security and accountability.
- **Conduct Regular Security Assessments**: Schedule and conduct regular security assessments and audits of external service providers to ensure continuous compliance and identify potential security gaps.
- **Implement Collaborative Tools and Processes**: Utilize shared security and incident reporting tools to enhance communication and real-time monitoring of security postures between your organization and external providers.
- **Enhance Coordination for Incident Response**: Develop and regularly update coordinated incident response plans that include clear roles and responsibilities for your organization and external service providers.
- **Educate Staff on Third-Party Risk Management**: Provide training for staff involved in managing and overseeing external service providers, emphasizing the importance of security in third-party integrations and the management of access controls.

NIST 800-53 Controls

(Refer to our appendix for a comprehensive description of the NIST 800-53 Controls, essential for organizations adhering to strict regulatory standards.)

- CA-07,PS-07,SA-04,SA-09,SI-04

Simplified Security Controls (SSC)

This section outlines our SSC, tailored for environments beyond the scope of NIST compliance. These controls offer streamlined, effective security solutions suitable for any organization seeking to enhance its cybersecurity defenses with clear, actionable measures.

Control 1: Contractual Security Requirements and Compliance Verification

Control Objective: To ensure external service providers adhere to the organization's security standards and comply with contractual obligations for protecting sensitive information and systems.

Implementation Steps:

1.1 **Define Security Clauses**: Develop comprehensive security clauses for inclusion in all contracts with external service providers. These clauses should cover data protection, access controls, incident reporting, and compliance with industry standards.

1.2 **Regular Compliance Assessments**: Schedule regular assessments and audits to verify compliance with the security clauses. Use third-party security assessments to provide an unbiased evaluation of the provider's security posture.

1.3 **Remediation and Penalty Provisions**: Include clear remediation steps and penalties for noncompliance within the contract. Ensure that providers know the consequences of failing to meet security requirements.

Expected Outcome: Implementing strict contract security clauses and regularly verifying compliance will ensure that external service providers meet the organization's security standards. This control will minimize the risk of data breaches and other security incidents from third-party services, contributing to a more secure overall operational environment.

Control 2: Secure Integration and Information Sharing Framework

Control Objective: To establish a secure and effective framework for sharing information and integrating systems with external service providers, ensuring the confidentiality, integrity, and availability of data.

Implementation Steps:

2.1 **Integration Security Guidelines**: Develop and implement security guidelines for integrating systems and sharing information with external providers. This includes defining secure communication channels, encryption standards, and data handling procedures.

2.2 **Shared Security Tools**: Implement shared security and incident reporting tools that enable real-time information sharing and collaboration between the organization and external providers. Ensure that these tools are secure and comply with organizational security policies.

2.3 **Access Control and Monitoring**: Continuously monitor data flow and access controls between the organization and external providers. Regularly review and adjust permissions to ensure that providers only have access to necessary information and systems.

Expected Outcome: Creating a secure integration and information-sharing framework will protect sensitive data shared with or accessed by external service providers. This will help prevent unauthorized access and data breaches, ensuring that data remains confidential and integral while facilitating necessary collaboration.

Control 3: Comprehensive External Provider Security Management Program

Control Objective: To systematically monitor the security performance and risk associated with external service providers, ensuring consistent adherence to the organization's security standards.

Implementation Steps:

3.1 **Security Management Policy for External Providers**: Develop a comprehensive policy that outlines the management, monitoring, and review of external service providers' security practices. This policy should include criteria for selecting providers, ongoing monitoring, and handling security incidents.

3.2 **Continuous Monitoring and Evaluation**: Implement a continuous monitoring program to assess the security posture of external providers. This includes regular reviews of security reports, audits, and compliance documents provided by the service providers.

3.3 **Training and Awareness**: Conduct training sessions for employees managing and securing third-party integrations. Educate them on the risks associated with external services and the importance of maintaining security and compliance.

Expected Outcome: Establishing a comprehensive external provider security management program will enable the organization to manage the risks associated with outsourcing effectively. By continuously monitoring and evaluating the security practices of external providers, the organization can ensure that third-party services do not compromise its security posture.

DE.CM-09: Computing Hardware and Software, Runtime Environments, and Their Data Are Monitored to Find Potentially Adverse Events

In cybersecurity, it is indispensable to monitor computing hardware, software, runtime environments, and their data for potentially adverse events. This continuous oversight is fundamental to identifying, understanding, and mitigating threats that could compromise organizational assets' integrity, availability, and confidentiality. Effective system and application monitoring can differentiate between a minor security incident and a catastrophic data breach.

Deploying EDR solutions is crucial in modern cybersecurity defenses. EDR platforms provide comprehensive monitoring and analysis of endpoint data, enabling organizations to quickly detect, investigate, and respond to potential threats. These solutions are designed to identify suspicious activities that might indicate a cybersecurity threat, such as malware infection or unauthorized data exfiltration. By equipping endpoints with EDR solutions, organizations can enhance their visibility into potentially malicious activities and improve their overall security posture.

APM tools are also essential in the monitoring landscape. APM solutions offer real-time visibility into the performance and health of applications, enabling organizations to detect and diagnose complex application issues quickly. While primarily focused on performance optimization, APM tools can also contribute to security by identifying unusual application behavior that may indicate a security issue. By integrating APM with security processes, organizations can ensure that performance issues do not become security vulnerabilities.

Implementation of configuration and vulnerability scanning is another cornerstone of effective system monitoring. Regular scans of hardware and software configurations can identify unauthorized changes and misconfigurations that could expose systems to risk. Vulnerability scanning goes hand-in-hand with configuration scanning by identifying known vulnerabilities within systems and applications, allowing for timely remediation. These practices are essential for maintaining a secure and compliant computing environment.

Monitoring for unauthorized changes and activities within systems is vital for maintaining security. This involves tracking and analyzing file changes, configurations, and user behaviors to identify actions deviating from established norms. Unauthorized changes can indicate a security compromise and may lead to data breaches or system outages if not promptly addressed. Effective change monitoring can help organizations maintain control over their IT environments and respond quickly to unauthorized modifications.

Integrating system logs with centralized security information and event management (SIEM) solutions enhances an organization's ability to detect and respond to security incidents. SIEM systems aggregate and analyze log data from the organization's IT infrastructure, providing a unified view of security events. This integration allows for real-time analysis and correlation of security data, improving the detection of complex threats and facilitating a rapid response to incidents.

Establishing baselines for system and application behavior is essential for effective monitoring. Baselines provide a reference point against which current activity can be compared, making identifying deviations that may signify a security issue easier. Organizations can more accurately detect anomalies and potential threats by understanding what normal behavior looks like. Regularly updating these baselines is crucial to account for changes in the IT environment and evolving business practices.

Training on secure configuration and maintenance of IT assets is crucial for ensuring all personnel understand their role in cybersecurity. Employees should be educated on best practices for system configuration, the importance of regular updates, and how to recognize signs of a security breach. Empowering employees with this knowledge helps create a security culture and reinforces the importance of vigilance in everyday operations.

Regular security assessments of runtime environments help ensure they remain secure against evolving threats. These assessments should evaluate the security of operating systems, applications, and middleware to identify vulnerabilities and misconfigurations. By conducting these assessments routinely, organizations can stay ahead of potential threats and ensure that their runtime environments support secure operations.

Recommendations

- **Strengthen Endpoint Security**: Implement advanced EDR solutions on all endpoints to provide comprehensive monitoring, detection, and response capabilities. Regularly update and configure these tools to adapt to new cybersecurity threats.
- **Leverage APM for Security Insights**: Integrate APM tools with security processes to detect unusual application behavior that could indicate security issues. Use APM data to complement overall security monitoring.
- **Enhance Scanning Practices**: Conduct regular configuration and vulnerability scans to identify and remediate potential security weaknesses. Prioritize the remediation of identified vulnerabilities based on their severity and potential impact.
- **Implement Rigorous Change Monitoring**: Establish robust processes for monitoring unauthorized system changes and activities. Automated tools track and alert unauthorized modifications to critical files and configurations.
- **Integrate and Optimize SIEM Solutions**: Ensure system logs are fully integrated with a centralized SIEM solution for real-time security monitoring and incident response. Regularly review and tune SIEM rules to improve accuracy and reduce false positives.
- **Educate and Train Staff**: Provide ongoing employee training on secure system configuration, maintenance practices, and threat recognition. Encourage a culture of security awareness and responsibility throughout the organization.

NIST 800-53 Controls

(Refer to our appendix for a comprehensive description of the NIST 800-53 Controls, essential for organizations adhering to strict regulatory standards.)

- AC-04,AC-09,AU-12,CA-07,CM-03,CM-06,CM-10,CM-11,SC-34,SC-35,SI-04,SI-07

Simplified Security Controls (SSC)

This section outlines our SSC, tailored for environments beyond the scope of NIST compliance. These controls offer streamlined, effective security solutions suitable for any organization seeking to enhance its cybersecurity defenses with clear, actionable measures.

Control 1: Enhanced Endpoint Protection with EDR Implementation

Control Objective: To safeguard computing devices against advanced threats by deploying comprehensive EDR solutions, ensuring continuous monitoring and rapid response to security incidents.

Implementation Steps:

1.1 **Solution Selection and Deployment**: Evaluate and select an EDR solution that best fits the organizational needs, considering factors such as integration capabilities, threat detection efficiency, and response features. Deploy it across all endpoints within the organization.

1.2 **Configuration and Customization**: Configure the EDR solution according to organizational security policies and threat intelligence, ensuring it is optimized for detecting the most relevant threats. Customize the solution to minimize false positives while maintaining high detection rates.

1.3 **Continuous Monitoring and Incident Response**: Establish protocols for continuous monitoring of endpoint activities and ensure that the EDR solution is integrated with the organization's incident response plan for immediate action on detected threats.

Expected Outcome: Implementing an enhanced EDR solution will lead to more effective detection and mitigation of threats at the endpoint level, significantly reducing the organization's exposure to malware and other malicious activities, thereby strengthening its overall security posture.

Control 2: System and Application Behavior Baseline Establishment

Control Objective: Identify and respond to unusual activities and potential threats by establishing and maintaining baselines for normal system and application behavior.

Implementation Steps:

2.1 **Baseline Development**: Collect and analyze data over a defined period to establish normal operational baselines for systems and applications, focusing on regular activity patterns and performance metrics.

2.2 **Anomaly Detection Configuration**: Configure monitoring tools, including APM and SIEM solutions, to continuously compare current activity against established baselines, alerting on significant deviations indicative of potential security issues or system failures.

2.3 **Regular Baseline Review and Update**: Periodically review and update the baselines to reflect changes in the operational environment and emerging threat landscapes, ensuring anomaly detection efforts' continuous relevance and effectiveness.

Expected Outcome: Establishing clear baselines for system and application behavior will enable the early detection of unusual activities, facilitating prompt investigation and response. This proactive approach will help in maintaining system integrity and operational continuity.

Control 3: Comprehensive Security Configuration and Vulnerability Management Program

Control Objective: To minimize the attack surface and protect against known vulnerabilities by implementing a robust security configuration and vulnerability management program.

Implementation Steps:

3.1 **Configuration Standards and Vulnerability Scanning**: Develop and implement security configuration standards for all IT assets. Schedule regular vulnerability scanning to identify and assess vulnerabilities within the organization's systems and software.

3.2 **Patch Management and Remediation**: Establish a systematic patch management process to promptly apply critical updates and patches. Include procedures for prioritizing and remediating identified vulnerabilities based on their severity and potential impact.

3.3 **Continuous Improvement and Staff Training**: Foster a culture of continuous improvement through regular reviews of configuration and vulnerability management practices. Provide training for IT staff on secure configuration, maintenance practices, and the latest cybersecurity trends and threats.

Expected Outcome: Implementing a comprehensive security configuration and vulnerability management program will significantly reduce vulnerabilities within the organization's IT environment. This control will lead to a more secure operational state, decreasing the likelihood of successful cyberattacks and enhancing overall security resilience.

20

Adverse Event Analysis (DE.AE)

> The essence of a formidable cybersecurity response lies not merely in the strength of defenses but in the depth of analysis and agility of action, transforming information into insight and adversity into advantage, thereby crafting a resilient bastion in the digital age.

In the evolving cybersecurity landscape, the analytical depth and operational agility of an organization's response to adverse events are critical. Organizations can significantly enhance their security posture by establishing procedures for the meticulous assessment and categorization of incidents, leveraging forensic tools for in-depth analysis, and integrating these findings into incident response planning. This comprehensive approach extends to correlating information across multiple sources, enriching event analysis with cyber threat intelligence (CTI), and understanding incidents' impact and scope through quantitative and qualitative methods. Essential to this process is the effective dissemination of information to authorized personnel and systems, ensuring a coordinated response to identified threats. Moreover, the declaration of incidents is streamlined by developing clear criteria supported by training, communication protocols, and decision-support tools. Together, these practices foster a culture of continuous improvement and adaptation to the ever-changing cybersecurity threat landscape, underpinning the resilience of organizations in the face of potential disruptions.

Author's Note: DE.AE-01 and DE.AE-05 were deprecated from the CSF 2 and are not included in this chapter.

DE.AE-02: Potentially Adverse Events Are Analyzed to Better Understand Associated Activities

Analyzing potentially adverse cybersecurity events is critical for understanding the activities associated with these incidents. Such analysis not only aids in the immediate response but also contributes to the long-term cybersecurity posture of an organization. This comprehensive examination encompasses the initial identification of incidents, the detailed forensic analysis, and the integration of findings into future preventive measures.

The initial step in analyzing adverse events involves establishing standardized procedures for assessing and categorizing incidents as they are identified. This process ensures that each event is evaluated according to its severity, potential impact, and the immediate steps required for containment and mitigation. By categorizing events, organizations can prioritize their response efforts

A Comprehensive Guide to the NIST Cybersecurity Framework 2.0: Strategies, Implementation, and Best Practice, First Edition. Jason Edwards.
© 2025 John Wiley & Sons Ltd. Published 2025 by John Wiley & Sons Ltd.

effectively, focusing resources where they are needed most urgently. This structured approach is fundamental in managing the cybersecurity threats organizations face today.

Employing forensic tools and techniques for an in-depth analysis of adverse events is critical to the investigative process. Forensic analysis allows organizations to uncover the root cause of an incident, understand the tactics, techniques, and procedures (TTPs) used by attackers, and gather evidence for potential legal proceedings. Using sophisticated forensic tools enables a deeper dive into the incident, providing insights invaluable for strengthening cybersecurity defenses and aiding recovery.

Identifying patterns and techniques in adverse events is essential for understanding the threat landscape and preventing future incidents. By analyzing the modus operandi of attackers, organizations can anticipate potential vulnerabilities and implement targeted defenses. This analysis extends beyond single incidents, encompassing broader trends and tactics employed across the cybersecurity domain. Recognizing these patterns allows for proactively adjusting security measures, enhancing resilience against evolving threats.

Integrating event analysis with incident response planning is a strategic step toward a comprehensive cybersecurity framework. Insights gained from analyzing adverse events inform the development and refinement of incident response plans, ensuring they are effective against observed threats. This integration facilitates a dynamic approach to cybersecurity, where plans are continuously updated based on real-world experiences and emerging trends. Such a strategy not only improves immediate response capabilities but also contributes to the long-term security posture of the organization.

Documenting findings and recommendations for future prevention is a critical outcome of the analysis process. Comprehensive reports detailing the incident, analysis, and lessons learned are valuable resources for ongoing security efforts. These documents support implementing improved practices, policies, and technologies to prevent the recurrence of similar events. Additionally, they provide a historical record of cybersecurity incidents, offering insights that can inform future strategies and training programs.

Training staff on advanced analysis and investigation techniques is pivotal to enhancing the analytical capabilities of an organization. By equipping employees with the skills to conduct thorough investigations, organizations can ensure a more effective response to cybersecurity incidents. This training should cover the latest forensic tools, analysis methodologies, and threat intelligence resources, preparing staff to face emerging cybersecurity challenges.

Continuous improvement of analysis methods based on lessons learned ensures that organizations remain at the forefront of cybersecurity. This iterative process involves regularly reviewing and updating analysis techniques, incorporating new tools and technologies, and adapting to the changing threat landscape. By committing to continuous improvement, organizations can enhance their analytical capabilities, better understand the nature of cyber threats, and strengthen their overall cybersecurity posture.

Sharing analysis outcomes with relevant stakeholders for awareness and action is an essential practice. Disseminating findings across the organization and with external partners where appropriate fosters a culture of cybersecurity awareness and collaboration. Information sharing supports a unified approach to cybersecurity, leveraging collective knowledge and resources to combat threats more effectively.

Recommendations

- **Develop a Structured Incident Categorization System**: Implement a comprehensive system for assessing and categorizing cybersecurity incidents. This will facilitate the prioritization of response efforts and ensure a consistent approach to incident management.

- **Enhance Forensic Analysis Capabilities**: Invest in advanced forensic tools and techniques to deepen the investigation of adverse events. This will aid in uncovering the root cause of incidents and gathering critical evidence for mitigating future threats.
- **Implement Pattern Recognition Training**: Provide specialized training for staff in identifying patterns and techniques used in cyberattacks. This knowledge will enable the proactive defense against evolving cybersecurity threats.
- **Integrate Analysis Findings into Response Planning**: Regularly update incident response plans based on insights gained from event analysis. This ensures that response strategies are aligned with the latest threat intelligence and best practices.
- **Promote Continuous Learning and Improvement**: Establish mechanisms for continuously reviewing and enhancing analysis methods. Encourage a culture of learning that values sharing knowledge and experiences, both within the organization and with external stakeholders.

NIST 800-53 Controls

(Refer to our appendix for a comprehensive description of the NIST 800-53 Controls, essential for organizations adhering to strict regulatory standards.)

- AU-06,CA-07,IR-04,SI-04

Simplified Security Controls (SSC)

This section outlines our simplified security controls (SSC), tailored for environments beyond the scope of NIST compliance. These controls offer streamlined, effective security solutions suitable for any organization seeking to enhance its cybersecurity defenses with clear, actionable measures.

Control 1: Development of Comprehensive Forensic Analysis Protocols

Control Objective: To establish standardized procedures for forensic analysis of potentially adverse cybersecurity events, ensuring a systematic and thorough investigation into such incidents to uncover root causes and mitigate future risks effectively.

Implementation Steps:

1.1 **Protocol Development:** Draft comprehensive forensic analysis protocols that detail the steps for conducting in-depth investigations of cybersecurity incidents, including the initial assessment, evidence collection, analysis phases, and reporting.

1.2 **Forensic Tool Selection:** Identify and procure a suite of forensic tools and technologies that align with the organization's IT infrastructure and the types of cybersecurity threats it faces, ensuring these tools are integrated into the forensic analysis protocols.

1.3 **Protocol Training and Dissemination:** Train cybersecurity personnel in applying the forensic analysis protocols and ensure that these protocols are readily accessible to all relevant staff, promoting a standardized approach to incident analysis.

Expected Outcome: Establishing comprehensive forensic analysis protocols and integrating appropriate tools will enhance the organization's capacity to conduct detailed investigations into cybersecurity incidents. This will lead to a better understanding of threat actors, techniques, and vulnerabilities, informing more effective prevention and response strategies.

Control 2: Integration of Advanced Analysis Training into Cybersecurity Education Programs

Control Objective: To equip cybersecurity staff with advanced analysis and investigation skills through targeted training programs, enhancing the organization's analytical capabilities and improving its response to cybersecurity incidents.

Implementation Steps:

2.1 **Identify Training Needs:** Assess the current capability of the cybersecurity team to identify gaps in knowledge and skills related to advanced analysis and investigation techniques.

2.2 **Develop or Procure Training Programs:** Based on identified needs, develop in-house training modules or procure external training programs that cover advanced cybersecurity analysis techniques, including forensic tools, pattern recognition, and incident analysis.

2.3 **Implement Regular Training Schedules:** Establish a regular schedule for conducting these training programs, ensuring that all cybersecurity staff can continually enhance their skills.

Expected Outcome: By prioritizing advanced analysis training, the organization will foster a team of cybersecurity professionals adept at conducting complex investigations and analyses. This will lead to improved identification, understanding, and mitigation of cybersecurity threats, strengthening the overall security posture.

Control 3: Continuous Improvement Program for Cybersecurity Analysis Methods

Control Objective: To implement a continuous improvement program that regularly evaluates and enhances the methods and tools used for cybersecurity analysis based on lessons learned from past incidents and evolving threat landscapes.

Implementation Steps:

3.1 **Establish a Continuous Improvement Framework:** Create a framework for regularly assessing current cybersecurity analysis methods, including feedback mechanisms from incident analyses, trend monitoring, and external benchmarking.

3.2 **Conduct Periodic Reviews:** Schedule and conduct periodic reviews of analysis methods and tools to identify areas for improvement, incorporating feedback from cybersecurity staff, incident reports, and industry best practices.

3.3 **Implement Enhancements:** Based on review outcomes, implement enhancements to analysis methods and tools, including the acquisition of new technologies, refinement of procedures, and additional staff training as required.

Expected Outcome: Establishing a continuous improvement program for cybersecurity analysis methods ensures that the organization remains at the cutting edge of cybersecurity defense capabilities. This approach enables proactive adjustments to analysis practices, keeping pace with the dynamic nature of cyber threats and technological advancements, thereby enhancing the effectiveness of incident analysis and response.

DE.AE-03: Information Is Correlated from Multiple Sources

Correlating information from multiple sources is a cornerstone of effective cybersecurity defense, enabling organizations to detect complex threats that single-source analysis might miss. This process involves gathering, comparing, and analyzing data across various platforms and technologies to identify potential security incidents. Integrating disparate data sources enhances the ability to detect sophisticated cyber threats, providing a comprehensive view of an organization's security posture.

Implementing centralized logging and security information and event management (SIEM) systems is foundational to effective multisource information correlation. SIEM systems aggregate data from various sources within an organization's IT environment, including network devices, servers, and security appliances, enabling the centralized monitoring and analysis of security alerts. This data consolidation facilitates the rapid detection of potential threats by providing a holistic view of the organization's security landscape. Centralized logging simplifies the security monitoring process and enhances the ability to track and analyze trends over time.

The complex technique for correlating events across diverse data sources requires sophisticated analytical tools and methodologies. Effective correlation involves more than simply collecting data; it requires identifying relationships between seemingly unrelated events. This might include pattern recognition, anomaly detection, and the application of artificial intelligence and machine learning algorithms to sift through vast quantities of data and highlight potential security incidents. Such techniques enable security analysts to discern patterns indicative of cyberattacks, even when those attacks are designed to evade traditional detection methods.

Establishing contextual relationships between disparate security alerts is crucial for interpreting the significance of correlated information. Contextual analysis involves evaluating security alerts within the broader framework of an organization's operations, threat environment, and known vulnerabilities. This approach helps differentiate between false positives and genuine threats, ensuring that security teams focus their efforts where they are most needed. Analysts can make informed decisions about prioritizing and responding to incidents by understanding the context in which security alerts occur.

Using threat intelligence platforms (TIPs) to enrich correlation adds another layer of depth to the analysis of cybersecurity data. TIPs collect and analyze information on emerging threats from various external sources, providing insights into cyber adversaries' TTPs. Integrating this external threat intelligence with internal correlation efforts enhances the organization's ability to detect and respond to attacks, providing a more comprehensive understanding of the threat landscape.

Automating correlation rules and responses for efficiency is essential in managing the volume and velocity of security data generated by modern IT environments. Automation allows for the real-time analysis of security events and the instant implementation of predefined response actions to identified threats. This speeds up the detection and mitigation of attacks and reduces the workload on security teams, enabling them to focus on more strategic tasks. Automation in correlation processes ensures a swift and consistent response to detected threats, minimizing potential damage.

Training on interpretation and action based on correlated data is vital for security personnel. Even the most advanced correlation tools and processes require skilled analysts to interpret the results and decide on the appropriate action. Ongoing training programs should cover the latest techniques in data analysis, threat identification, and incident response, ensuring that security teams are well-equipped to leverage correlated information effectively. This training enhances the analytical capabilities of security staff, enabling them to make informed decisions rapidly in the face of complex threats.

Regularly reviewing and updating correlation rules and algorithms ensures that information correlation processes remain effective. Cyber threats constantly evolve, requiring adaptive and dynamic correlation strategies to detect new and sophisticated attacks. Regularly updating correlation rules and algorithms in response to emerging threats and changing IT environments ensures that correlation efforts are aligned with the current threat landscape. This continuous improvement process is critical for maintaining an effective cybersecurity defense posture.

Recommendations

- **Implement a Robust SIEM System**: Adopt a comprehensive SIEM system that aggregates data across your IT environment, enabling effective multisource correlation and centralized security monitoring.

- **Leverage Advanced Correlation Techniques**: Utilize advanced data analytics, including machine learning and AI, to identify complex patterns and relationships across diverse data sets, enhancing threat detection capabilities.
- **Integrate External Threat Intelligence**: Incorporate external threat intelligence into your correlation processes to provide additional context and enhance the detection of sophisticated cyber threats.
- **Automate Correlation and Response Processes**: Implement automation in your correlation rules and responses to improve efficiency and ensure a rapid reaction to detected threats, minimizing potential damage.
- **Invest in Continuous Training and Development**: Ensure your security team is well-versed in the latest correlation techniques and technologies through ongoing training, enhancing their ability to interpret and act on correlated data effectively.

NIST 800-53 Controls

(Refer to our appendix for a comprehensive description of the NIST 800-53 Controls, essential for organizations adhering to strict regulatory standards.)

- AU-06,CA-07,PM-16,IR-04,IR-05,IR-08,SI-04,PM-1

Simplified Security Controls (SSC)

This section outlines our SSC, tailored for environments beyond the scope of NIST compliance. These controls offer streamlined, effective security solutions suitable for any organization seeking to enhance its cybersecurity defenses with clear, actionable measures.

Control 1: Establishment of an Integrated SIEM and Centralized Logging Infrastructure

Control Objective: To implement a robust SIEM system combined with centralized logging capabilities, ensuring that data from various sources can be aggregated, analyzed, and correlated efficiently to identify potential security incidents.

Implementation Steps:

1.1 **Assess Requirements and Select SIEM Solution:** Conduct an assessment of the organization's needs, considering the complexity of its IT environment, to select an SIEM solution that offers comprehensive logging, monitoring, and event correlation capabilities.

1.2 **Centralize Log Management:** Integrate logs from all critical systems, applications, and security devices into the SIEM system, ensuring comprehensive data capture for analysis and correlation.

1.3 **Configure and Optimize SIEM Correlation Rules:** Customize the SIEM system's correlation rules to reflect the organization's specific security concerns and operational realities, optimizing its ability to detect relevant security events across disparate data sources.

Expected Outcome: Implementing an SIEM system with centralized logging capabilities will enable the organization to effectively aggregate, monitor, and analyze data across its digital infrastructure. This integrated approach enhances the detection of complex security threats through efficient correlation, supporting timely and informed response actions.

Control 2: Development and Application of Advanced Correlation Techniques

Control Objective: To refine the organization's capability to correlate events across diverse data sources by developing and applying advanced correlation techniques, thereby improving the detection of sophisticated security threats.

Implementation Steps:

2.1 **Identify Correlation Needs:** Analyze previous security incidents and current threat intelligence to identify patterns and relationships that require advanced correlation techniques.

2.2 **Develop Advanced Correlation Algorithms:** Create or adapt correlation algorithms to analyze data from multiple sources, identify subtle patterns, and highlight potential security incidents based on the organization's specific threat landscape.

2.3 **Implement and Monitor Correlation Techniques:** Deploy these advanced correlation techniques within the SIEM system and other analytical tools, continuously monitoring their effectiveness in identifying potential threats.

Expected Outcome: By developing and applying advanced correlation techniques, the organization will enhance its ability to identify and respond to complex security threats that may not be detectable through standard correlation methods. This leads to a more nuanced understanding of the organization's security posture and more effective incident prevention and response strategies.

Control 3: Comprehensive Training Program on Correlated Data Analysis and Response

Control Objective: To ensure that cybersecurity personnel are proficient in analyzing correlated data from multiple sources and can take appropriate action based on this analysis, thereby improving the organization's incident response capabilities.

Implementation Steps:

3.1 **Design Tailored Training Modules:** Create training programs that cover the specific tools, technologies, and processes used by the organization for data correlation, including the interpretation of correlation outputs and decision-making based on correlated data.

3.2 **Conduct Regular Training Sessions:** Schedule and conduct regular training sessions for all relevant cybersecurity staff, ensuring they are up to date with the latest correlation techniques and incident response protocols.

3.3 **Assess and Update Training Content:** Regularly assess the effectiveness of the training program and update its content to reflect new correlation techniques, tools, and evolving security threats.

Expected Outcome: A comprehensive training program on correlated data analysis and response ensures cybersecurity personnel can interpret complex data correlations and take informed action. This enhances the organization's incident response speed and effectiveness, reducing the potential impact of security threats.

DE.AE-04: The Estimated Impact and Scope of Adverse Events Are Understood

Understanding adverse events' estimated impact and scope is fundamental to effective cybersecurity management. This process involves assessing the potential damage and reach of security incidents, enabling organizations to prioritize responses and allocate resources efficiently. Such assessments are critical for immediate incident response and long-term risk management and mitigation strategies. By accurately evaluating the impact and scope of adverse events, organizations can better understand the vulnerabilities in their systems and the potential consequences of security breaches.

A structured framework for evaluating the severity and reach of incidents is essential for consistent and effective impact assessment. This framework should incorporate qualitative and quantitative methods to gauge the potential consequences of security events. Data sensitivity, system criticality, and potential financial or reputational damage should be considered. Establishing clear criteria for severity levels aids in quickly classifying incidents and allocating response resources according to the potential impact.

Using quantitative and qualitative methods for impact analysis provides a comprehensive understanding of an incident's consequences. Quantitative analysis might include metrics such as the number of affected users, downtime duration, and financial losses. In contrast, qualitative analysis assesses the less tangible impacts, such as damage to reputation or customer trust. Combining these approaches offers a nuanced view of incidents, facilitating informed decision-making and strategic planning.

Collaboration with business units to understand operational impact is crucial for accurate impact assessment. By engaging with various departments, cybersecurity teams can gain insights into the specific operational contexts and potential effects of security incidents on different areas of the organization. This cross-functional approach ensures that assessments consider the full range of consequences, from disruption of services to regulatory compliance issues, enhancing the organization's overall response strategy.

Techniques for scope containment and mitigation during analysis are vital components of the response process. While assessing the impact, it is equally important to implement measures to limit the spread and severity of incidents. This might involve isolating affected systems, applying temporary security controls, or activating contingency plans. Effective scope containment not only aids in accurately assessing impact but also minimizes damage during the analysis phase.

Documenting and communicating the impact assessment findings is essential in the incident response process. Detailed reports should be prepared, outlining the nature of the incident, the estimated impact, and recommended response actions. These findings must be communicated promptly to relevant stakeholders, including senior management, affected business units, and, where necessary, external parties such as customers or regulatory bodies. Clear and transparent communication helps ensure a coordinated and effective response to security incidents.

Incorporating impact assessments into risk management practices enables organizations to enhance their cybersecurity posture. By analyzing the impacts of past incidents, organizations can identify patterns, vulnerabilities, and areas for improvement in their security frameworks. This information is invaluable for updating risk management strategies, prioritizing security investments, and developing more resilient systems and processes.

Training on rapid impact analysis and decision-making empowers cybersecurity teams and decision-makers to act swiftly and confidently during incidents. Such training should cover the methodologies for impact assessment, tools and technologies for data analysis, and strategies for effective communication and decision-making under pressure. Equipping personnel with these skills ensures organizations can respond to security threats effectively, minimizing impact and recovering more quickly.

Updating impact assessment strategies based on new threats is crucial in the ever-evolving cybersecurity landscape. As new types of attacks emerge and organizational contexts change, impact assessment methodologies must be revisited and refined. This proactive approach ensures that organizations remain prepared to assess and respond to incidents, regardless of their nature or scale.

Recommendations

- **Develop a Comprehensive Impact Assessment Framework**: Establish a structured framework that includes qualitative and quantitative methods for evaluating the impact of cybersecurity incidents. This ensures consistent and thorough assessments.

- **Foster Cross-Functional Collaboration**: Engage with various business units to understand the operational implications of security incidents. This collaborative approach ensures that impact assessments consider the full potential consequences.
- **Implement Scope Containment Measures**: Apply techniques for limiting the spread and severity of incidents during the impact analysis phase. Effective containment strategies are essential for accurate impact assessment and mitigation.
- **Enhance Documentation and Communication Processes**: Ensure that findings from impact assessments are thoroughly documented and communicated to all relevant stakeholders. Clear communication facilitates informed decision-making and coordinated response efforts.
- **Prioritize Continuous Training and Strategy Updates**: Invest in ongoing training for rapid impact analysis and decision-making. Regularly update impact assessment strategies to reflect emerging threats and changes in the organizational context.

NIST 800-53 Controls

(Refer to our appendix for a comprehensive description of the NIST 800-53 Controls, essential for organizations adhering to strict regulatory standards.)

- PM-09,PM-11,PM-18,PM-28,PM-30

Simplified Security Controls (SSC)

This section outlines our SSC, tailored for environments beyond the scope of NIST compliance. These controls offer streamlined, effective security solutions suitable for any organization seeking to enhance its cybersecurity defenses with clear, actionable measures.

Control 1: Development of an Impact and Scope Assessment Framework

Control Objective: To establish a comprehensive framework for evaluating the severity and reach of cybersecurity incidents, ensuring that the potential impact on the organization's operations and assets is accurately assessed.

Implementation Steps:

1.1 **Framework Design:** Create a detailed framework outlining the methodologies and criteria for assessing the impact and scope of adverse cybersecurity events, incorporating quantitative and qualitative analysis techniques.

1.2 **Stakeholder Collaboration:** Engage with key stakeholders across various business units to tailor the impact assessment criteria to the specific operational realities of the organization, ensuring that all potential impacts are considered.

1.3 **Framework Implementation and Training:** Implement the framework across the organization and train relevant staff on its application, ensuring they can effectively conduct impact assessments.

Expected Outcome: Establishing an impact and scope assessment framework enables the organization to evaluate the potential consequences of cybersecurity incidents systematically. This facilitates informed decision-making regarding response strategies and resource allocation, ultimately minimizing adverse events' operational and financial impact.

Control 2: Integration of Impact Assessments with Organizational Risk Management

Control Objective: To integrate the findings from impact and scope assessments directly into the organization's risk management practices, enhancing the organization's ability to manage and mitigate cybersecurity risks.

Implementation Steps:

2.1 **Assessment Integration:** Establish processes for incorporating the results of impact and scope assessments into the organization's overall risk management framework, ensuring that these assessments inform risk prioritization and mitigation strategies.

2.2 **Risk Management Collaboration:** Facilitate regular collaboration sessions between the cybersecurity and risk management teams to discuss impact assessment outcomes and integrate these insights into the risk management process.

2.3 **Continuous Risk Management Enhancement:** Use insights gained from impact assessments to continuously refine and enhance risk management practices, including developing new mitigation strategies and adjusting existing controls.

Expected Outcome: Integrating impact and scope assessment findings with organizational risk management practices ensures that cybersecurity risks are comprehensively understood and effectively managed. This leads to a more resilient organizational posture against cyber threats, with a proactive risk mitigation and management approach.

Control 3: Continuous Training Program on Rapid Impact Analysis and Decision-Making

Control Objective: To develop and maintain a continuous training program focused on rapid impact analysis and decision-making, enabling cybersecurity personnel to assess and respond to incidents with enhanced speed and effectiveness.

Implementation Steps:

3.1 **Program Development:** Develop a training program that covers the principles of rapid impact analysis, including quantitative and qualitative methods and decision-making processes under pressure.

3.2 **Scenario-Based Training:** Incorporate realistic scenarios and simulations into the training program to provide hands-on experience with impact analysis and making critical decisions during cybersecurity incidents.

3.3 **Ongoing Training and Skill Refreshment:** Schedule regular training sessions and updates to the training content to reflect the latest threats, technologies, and best practices in impact analysis and incident response.

Expected Outcome: Implementing a continuous training program on rapid impact analysis and decision-making equips cybersecurity personnel with the skills and knowledge to swiftly assess the impact of incidents and make informed decisions. This leads to more effective and timely responses to cybersecurity incidents, reducing their potential impact on the organization.

DE.AE-06: Information on Adverse Events Is Provided to Authorized Staff and Tools

The dissemination of information regarding adverse events to authorized personnel and systems is a critical component of an effective cybersecurity response strategy. This process ensures that those responsible for managing and mitigating cybersecurity threats are informed promptly, enabling a swift and coordinated response. Proper information-sharing mechanisms enhance the organization's resilience against cyber threats by ensuring that vital information reaches the right people and tools at the right time.

Establishing secure and timely information dissemination protocols is foundational to effective communication during cybersecurity events. These protocols should specify the methods and channels for sharing information, the categorization of information based on sensitivity, and the prioritization of events based on their impact. Ensuring the security of the dissemination process is paramount to prevent sensitive information from being intercepted or misused. On the other hand, timeliness ensures that responses to cybersecurity incidents are initiated without undue delay, minimizing potential damage.

The selection and use of communication tools for event reporting play a significant role in the effectiveness of information dissemination. Tools should be chosen based on their reliability, security, and ease of use, ensuring that they facilitate rather than hinder communication. These may include encrypted messaging systems, secure email platforms, and specialized incident management software. As required, the tools should support the rapid distribution of information to all relevant parties, including security personnel, IT staff, and management.

Defining roles and responsibilities for receiving and acting on information is critical to ensure that cybersecurity events are managed efficiently. This involves identifying which individuals or teams are responsible for various aspects of the response process, from initial assessment to containment, eradication, and recovery. Clearly defined roles and responsibilities prevent overlapping gaps in the response process, ensuring a coordinated effort. Additionally, they help in establishing accountability and improving the overall effectiveness of the cybersecurity incident response plan.

Integrating event information with security and management tools enhances the organization's ability to respond to and manage adverse events. By feeding real-time data about security incidents into SIEM systems, TIPs, and other relevant tools, organizations can automate aspects of the response process, improve situational awareness, and facilitate decision-making. This integration allows for a more dynamic and informed response to cybersecurity threats, leveraging the capabilities of both human and technological resources.

Procedures for feedback and follow-up on shared information are essential for continuous improvement in handling cybersecurity events. These procedures should facilitate evaluating the incident response, including communication effectiveness and stakeholder coordination. Feedback mechanisms can include debriefings, incident reports, and lessons-learned sessions. Follow-up actions might involve adjusting response strategies, updating protocols, and conducting additional training based on the insights gained from analyzing the handling of past events.

Training on effective communication during security events is vital for ensuring that all relevant personnel are prepared to share and act on information efficiently. This training should cover the protocols for information dissemination, the use of communication tools, and the roles and responsibilities in the response process. It should also emphasize the importance of security and timeliness in communication, equipping staff with the skills necessary to effectively manage information flow during cybersecurity incidents.

Compliance with legal and regulatory requirements for information sharing is a key consideration for organizations. Regulations may dictate specific requirements for reporting cybersecurity incidents to external entities, such as government agencies or industry regulators. Organizations must ensure their information dissemination practices align with these requirements, avoiding potential legal and regulatory penalties. Additionally, compliance helps maintain trust and confidence among customers, partners, and other stakeholders.

Recommendations

- **Establish Comprehensive Communication Protocols**: Develop secure and efficient protocols for disseminating information regarding adverse events. These protocols should ensure that information is shared with authorized personnel and systems in a timely manner.

- **Leverage Secure Communication Tools**: Select and implement communication tools that are secure and reliable and facilitate easy sharing of information. Ensure these tools are integrated into the organization's cybersecurity incident response plan.
- **Define Clear Roles and Responsibilities**: Clarify the roles and responsibilities of all stakeholders involved in the information sharing and incident response process. This clarity will support efficient and coordinated action during cybersecurity events.
- **Integrate Information Sharing with Security Tools**: Enhance the organization's response capabilities by integrating event information with security and management tools. This integration should support automated responses and improved decision-making.
- **Foster a Culture of Continuous Improvement**: Implement procedures for feedback and follow-up on shared information to continuously improve the organization's approach to managing cybersecurity events. This includes regular training and updates to communication protocols and tools.

NIST 800-53 Controls

(Refer to our appendix for a comprehensive description of the NIST 800-53 Controls, essential for organizations adhering to strict regulatory standards.)

- IR-04,PM-15,PM-16,RA-04,RA-10

Simplified Security Controls (SSC)

This section outlines our SSC, tailored for environments beyond the scope of NIST compliance. These controls offer streamlined, effective security solutions suitable for any organization seeking to enhance its cybersecurity defenses with clear, actionable measures.

Control 1: Secure and Efficient Information Dissemination System

Control Objective: To establish a system that ensures the secure and timely dissemination of information regarding adverse events to authorized staff and integrated tools, enhancing the organization's ability to respond effectively to cybersecurity incidents.

Implementation Steps:

1.1 **Protocol Development:** Develop and document clear protocols for disseminating information related to adverse events, specifying secure channels, encryption standards, and authentication processes to protect sensitive information.

1.2 **Tool Selection and Integration:** Select and implement communication tools that comply with security standards and integrate seamlessly with the organization's cybersecurity and management tools, ensuring efficient information flow.

1.3 **Regular System Testing and Audits:** Conduct regular tests and audits of the information dissemination system to ensure its effectiveness and security, adjusting protocols and tools based on audit outcomes as needed.

Expected Outcome: Implementing a secure and efficient information dissemination system will ensure that critical information regarding adverse events is shared promptly and securely with authorized personnel and systems. This facilitates a swift and coordinated response to incidents, minimizing their impact on the organization.

Control 2: Comprehensive Training Program on Incident Communication

Control Objective: To develop and maintain a comprehensive training program for all relevant personnel on effective communication during security events, ensuring clear and efficient information sharing in times of crisis.

Implementation Steps:

2.1 **Training Program Development:** Create a detailed training program that covers the protocols for information dissemination, communication tool use, and handling sensitive information during adverse events.

2.2 **Simulation Exercises:** Incorporate simulation exercises that mimic real-world security events into the training program, allowing personnel to practice and refine their communication skills in a controlled environment.

2.3 **Continuous Training and Updates:** Ensure the training program is an ongoing initiative with regular updates to reflect new communication tools, protocols, and emerging threats, keeping personnel well-prepared for effective communication during incidents.

Expected Outcome: A comprehensive training program on incident communication will equip personnel with the necessary skills and knowledge to share information efficiently and securely during adverse events. This ensures that the organization can respond quickly and cohesively to mitigate the impact of cybersecurity incidents.

Control 3: Legal and Regulatory Compliance Review for Information Sharing

Control Objective: To ensure that the organization's practices for sharing information on adverse events comply with all applicable legal and regulatory requirements, minimizing legal risks and maintaining stakeholder trust.

Implementation Steps:

3.1 **Compliance Assessment:** Conduct a thorough assessment of current information-sharing practices to identify gaps in compliance with relevant laws, regulations, and industry standards.

3.2 **Protocol Adjustment:** Adjust information dissemination protocols and practices as needed to ensure full compliance with legal and regulatory requirements, involving legal counsel to ensure accuracy and completeness.

3.3 **Regular Compliance Updates and Training:** Implement a process for regularly updating the organization's information-sharing protocols to reflect changes in legal and regulatory landscapes and provide updated training to all relevant personnel.

Expected Outcome: By ensuring that information-sharing practices comply with legal and regulatory requirements, the organization minimizes legal risks and maintains the trust of stakeholders, including customers, partners, and regulators. This control reinforces the organization's commitment to responsible and secure information handling during adverse cybersecurity events.

DE.AE-07: Cyber Threat Intelligence and Other Contextual Information Are Integrated into the Analysis

Integrating CTI into the analysis of cybersecurity incidents elevates the understanding and contextualization of threats facing an organization. This approach enables cybersecurity professionals to not just react to incidents as they occur but also to anticipate potential threats and vulnerabilities, crafting more informed and proactive defense strategies. By weaving CTI into the fabric of cybersecurity operations, organizations can achieve a more dynamic and adaptive security posture.

Establishing processes for collecting and evaluating CTI is the foundation for effective integration. These processes should encompass a variety of sources, including industry reports, threat intelligence feeds, law enforcement notifications, and information sharing and analysis centers

(ISACs). Evaluating the credibility, relevance, and timeliness of the intelligence gathered is crucial for its usefulness. Organizations must develop criteria for assessing intelligence sources, ensuring that the CTI incorporated into their security operations is reliable and pertinent to their specific threat landscape.

Techniques for incorporating CTI into event analysis for enhanced context involve aligning intelligence with observed cybersecurity events. This can be achieved by mapping incoming intelligence against indicators of compromise within the organization's network, thereby identifying potential threats before they manifest into breaches. Such techniques require the integration of CTI with SIEM systems, allowing for the automatic correlation of intelligence with internal security data and enhancing the speed and efficiency of the analysis.

CTI feeds and platforms for real-time threat information are indispensable in today's fast-paced cybersecurity environment. These feeds provide organizations with timely updates on emerging threats, TTPs used by adversaries, and vulnerabilities being exploited in the wild. Real-time access to CTI allows security teams to quickly adapt their defense mechanisms to counteract evolving threats, significantly reducing the window of opportunity for attackers.

Training on applying CTI in security operations is essential for ensuring that security teams are equipped to utilize intelligence in their daily activities effectively. This training should cover the processes for collecting and evaluating CTI, techniques for integrating intelligence into security operations, and methods for applying CTI to enhance incident analysis and response strategies. Empowering security personnel with the knowledge and skills to leverage CTI can significantly improve an organization's ability to anticipate and mitigate cyber threats.

Assessing the reliability and relevance of CTI sources is a continuous necessity. The cybersecurity threat landscape is dynamic, constantly emerging new actors, techniques, and vulnerabilities. As such, the value of CTI sources can change over time, necessitating regular reassessment to ensure that the intelligence integrated into security operations remains current and applicable. This assessment should consider the source's history of accuracy, breadth of coverage, and timeliness of information.

Documentation of how CTI influenced analysis outcomes provides valuable insights for future incident response and threat-hunting efforts. Keeping detailed records of intelligence applications not only aids in refining the CTI integration process but also helps demonstrate the value of intelligence to stakeholders. Documenting the role of CTI in identifying, mitigating, or preventing cyber incidents can support continuous improvement in security operations and justify investment in intelligence capabilities.

The integration of CTI with incident response and prevention strategies ensures that intelligence is actionable, guiding the development of tactics and measures to counter identified threats. This involves the reactive application of CTI to incidents as they occur and the proactive adjustment of security controls and policies based on intelligence about emerging threats. The dynamic nature of cyber threats requires that this integration be an ongoing process, with CTI continually informing and enhancing organizational security practices.

Recommendations

- **Develop Robust CTI Collection and Evaluation Processes**: Establish comprehensive mechanisms for acquiring and assessing CTI, ensuring the use of diverse, credible sources.
- **Enhance SIEM Systems with CTI Integration**: Upgrade SIEM systems to incorporate CTI feeds, enabling real-time threat detection and analysis.

- **Invest in Continuous Training for Security Teams**: Provide ongoing training for security personnel on the effective application of CTI in cybersecurity operations, focusing on the latest techniques and tools.
- **Regularly Assess CTI Source Quality**: Implement a routine evaluation process for CTI sources to ensure their relevance and reliability, adjusting intelligence collection strategies as needed.
- **Document CTI Impact on Security Operations**: Maintain detailed records of how CTI contributes to incident analysis, response, and prevention efforts, leveraging these insights for continuous process improvement.

NIST 800-53 Controls

(Refer to our appendix for a comprehensive description of the NIST 800-53 Controls, essential for organizations adhering to strict regulatory standards.)

- PM-16,RA-03,RA-10

Simplified Security Controls (SSC)

This section outlines our SSC, tailored for environments beyond the scope of NIST compliance. These controls offer streamlined, effective security solutions suitable for any organization seeking to enhance its cybersecurity defenses with clear, actionable measures.

Control 1: Centralized Cyber Threat Intelligence (CTI) Management System

Control Objective: To create a centralized system for managing and distributing CTI across the organization, ensuring that all relevant personnel access up-to-date threat information for enhanced situational awareness and decision-making.

Implementation Steps:

1.1 **System Development:** Develop and implement a centralized CTI management system that aggregates intelligence from various feeds and platforms, providing a unified source of threat information.

1.2 **Access Control and Distribution:** Define access controls to ensure that CTI is shared with authorized personnel and establish mechanisms for efficiently distributing intelligence across the organization.

1.3 **Integration with Security Tools:** Integrate the CTI management system with existing security and management tools, enabling automated threat indicator sharing and response actions based on current intelligence.

Expected Outcome: Implementing a centralized CTI management system will ensure that threat intelligence is consistently and efficiently distributed throughout the organization, enabling timely and informed responses to emerging threats and enhancing overall security posture.

Control 2: CTI Source Evaluation and Quality Assurance Program

Control Objective: To establish a program for evaluating the reliability and relevance of CTI sources, ensuring that the organization bases its security decisions on accurate and timely intelligence.

Implementation Steps:

2.1 **Evaluation Criteria Development:** Develop a set of criteria for assessing the quality, reliability, and relevance of CTI sources, considering factors such as source credibility, timeliness of information, and applicability to the organization's threat landscape.

2.2 **Regular Source Assessment:** Conduct regular evaluations of current CTI sources against these criteria, identifying areas for improvement and potential new intelligence sources.

2.3 **Feedback Loop Creation:** Create a feedback loop involving CTI users and security analysts to continually assess the usefulness of the intelligence received and make adjustments to source selection and evaluation criteria as needed.

Expected Outcome: A robust CTI source evaluation and quality assurance program will ensure that the organization utilizes high-quality and relevant threat intelligence, enhancing its ability to detect, analyze, and respond to cyber threats more effectively.

Control 3: Continuous CTI Training and Awareness Program

Control Objective: To continuously educate and train security personnel on the latest CTI application techniques, tools, and best practices, ensuring that the workforce remains adept at leveraging CTI for security operations enhancement.

Implementation Steps:

3.1 **Program Establishment:** Establish a continuous training and awareness program focused on applying CTI in security operations, including the interpretation of intelligence, integration with security tools, and operational decision-making based on CTI.

3.2 **Customized Training Modules:** Develop training modules customized to the organization's technology stack, threat profile, and operational practices, ensuring relevance and applicability of the training content.

3.3 **Regular Training Updates and Exercises:** Update training content regularly to reflect the latest CTI trends, tools, and methodologies and incorporate practical exercises and simulations to reinforce learning and application skills.

Expected Outcome: Implementing a continuous CTI training and awareness program will ensure that security personnel are proficient in leveraging CTI to enhance security operations, leading to more effective detection, analysis, and mitigation of cyber threats across the organization.

DE.AE-08: Incidents Are Declared When Adverse Events Meet the Defined Incident Criteria

Declaring cybersecurity incidents is a pivotal process within the broader organizational security posture management framework. This process begins with establishing clear, actionable criteria defining what constitutes an incident. These criteria are essential for differentiating between events that can be routinely managed and those that necessitate formal declaration and response.

Developing clear criteria for incident declaration based on severity, impact, and scope is fundamental to an effective cybersecurity strategy. These criteria should be meticulously designed to encompass many potential scenarios, from data breaches and malware infections to unauthorized access and system disruptions. The severity of an event, its impact on operations and data integrity, and the scope of its potential or actual damage are critical factors in determining whether an adverse event should be escalated to an incident.

Procedures for rapid assessment and declaration of incidents are crucial for minimizing the impact of cybersecurity threats. Once potential incidents are identified, a swift and systematic evaluation must be conducted to determine if they meet the predefined declaration criteria. This rapid assessment process ensures that incidents are declared in a timely manner, facilitating immediate activation of response protocols to contain and mitigate threats.

Training on the incident declaration process and criteria is essential for all personnel involved in cybersecurity operations. This training ensures that team members can accurately identify potential incidents, understand the assessment process, and know the steps required to declare an incident. Empowering employees with this knowledge fosters a culture of vigilance and preparedness, significantly enhancing the organization's ability to respond to cybersecurity threats.

Establishing communication channels for incident reporting and escalation is key to the incident declaration process. These channels must be secure, reliable, and accessible, enabling the prompt sharing of information about potential incidents. Clear protocols for reporting and escalation ensure that information flows efficiently from the point of detection to decision-makers, enabling swift incident declaration and response actions.

Documentation and record-keeping of declared incidents are vital for several reasons. This practice aids post-incident analysis, regulatory compliance, and continuous improvement of security protocols. Maintaining detailed records of incidents, including their nature, assessment, declaration, and response actions, provides valuable insights that can be leveraged to strengthen future security measures.

Using decision support tools for incident assessment and declaration can significantly enhance the effectiveness and efficiency of these processes. These tools can automate aspects of the assessment process, provide recommendations based on predefined criteria, and facilitate the rapid declaration of incidents. By leveraging technology, organizations can ensure a consistent and objective approach to incident declaration, minimizing the potential for human error.

Regularly reviewing and updating incident criteria based on the evolving threat landscape is essential to maintaining an effective incident declaration process. As cybersecurity threats evolve, so must the criteria used to define and declare incidents. This iterative process ensures that organizations remain prepared to identify and respond to an ever-changing array of security challenges.

Integrating incident declaration with the broader incident response framework is critical for a cohesive security posture. The declaration process should seamlessly transition into incident response protocols, ensuring that response actions are initiated immediately once an incident is declared. This integration facilitates a coordinated, efficient approach to managing cybersecurity incidents, from detection to resolution.

Recommendations

- **Refine Incident Declaration Criteria**: Continuously develop and refine criteria for incident declaration to ensure they remain aligned with the evolving cybersecurity threat landscape and organizational priorities.
- **Enhance Training Programs**: Implement comprehensive training for all relevant personnel on the incident declaration process, including assessing and escalating potential incidents based on defined criteria.
- **Strengthen Communication Protocols**: Establish clear, secure channels for reporting and escalating incidents to ensure rapid assessment and declaration processes.
- **Leverage Technology**: Invest in decision support tools that automate and facilitate the incident assessment and declaration process, enhancing accuracy and efficiency.
- **Conduct Regular Reviews**: Periodically review and update incident declaration criteria and processes to reflect changes in the threat environment and organizational structure or operations, ensuring an agile and responsive cybersecurity strategy.

NIST 800-53 Controls

(Refer to our appendix for a comprehensive description of the NIST 800-53 Controls, essential for organizations adhering to strict regulatory standards.)

- IR-04,IR-08

Simplified Security Controls (SSC)

This section outlines our SSC, tailored for environments beyond the scope of NIST compliance. These controls offer streamlined, effective security solutions suitable for any organization seeking to enhance its cybersecurity defenses with clear, actionable measures.

Control 1: Formalization of Incident Declaration Criteria

Control Objective: To formalize and standardize the criteria for incident declaration based on the severity, impact, and scope of adverse events, ensuring a consistent approach to recognizing and escalating incidents across the organization.

Implementation Steps:

1.1 **Criteria Development:** Collaborate with cybersecurity, IT, business units, and risk management teams to develop comprehensive criteria for incident declaration, considering data sensitivity, system criticality, and potential operational impact.

1.2 **Criteria Documentation:** Document the incident declaration criteria in an easily accessible format, ensuring clarity on the thresholds for severity, impact, and scope that warrant an incident declaration.

1.3 **Criteria Dissemination and Training:** Disseminate the documented criteria across the organization and conduct training sessions for all relevant personnel to familiarize them with the incident declaration process and criteria.

Expected Outcome: The formalization of incident declaration criteria ensures that all personnel are equipped with a clear understanding of when and how to escalate adverse events as incidents. This leads to a more efficient and uniform response to cybersecurity threats, enhancing the organization's overall security posture.

Control 2: Implementation of Decision Support Systems for Incident Assessment

Control Objective: To leverage decision support tools and systems in the incident assessment process, facilitating the rapid and accurate determination of whether an adverse event meets the defined criteria for incident declaration.

Implementation Steps:

2.1 **Tool Selection:** Identify and select decision support tools to assess adverse events against the formalized incident declaration criteria, considering the integration capabilities with existing security infrastructure.

2.2 **Tool Configuration and Integration:** Configure the selected tools to reflect the organization's specific incident declaration criteria and integrate them with existing security systems to automate the initial assessment process.

2.3 **User Training and Tool Testing:** Train relevant personnel on using decision support tools and conduct regular testing to ensure accuracy and effectiveness in supporting incident assessment and declaration.

Expected Outcome: Implementing decision support systems streamlines the incident assessment process, enabling rapid and accurate declarations. This control reduces the time to respond to incidents, minimizing potential damage and enhancing the organization's responsiveness to cybersecurity threats.

Control 3: Continuous Improvement Process for Incident Declaration Practices

Control Objective: To establish a continuous improvement process for reviewing and updating the organization's incident declaration criteria and processes, ensuring they remain relevant in the face of an evolving threat landscape.

Implementation Steps:

3.1 **Regular Review Schedule:** Set a regular schedule for reviewing the incident declaration criteria and processes involving stakeholders from cybersecurity, IT, business units, and risk management.

3.2 **Incident Analysis and Feedback Integration:** Analyze declared incidents and collect feedback from involved personnel to identify areas for improvement in the declaration criteria and process.

3.3 **Criteria and Process Updates:** Update the incident declaration criteria and processes based on the outcomes of the reviews and incident analyses, ensuring the organization's approach remains aligned with the latest threat intelligence and business objectives.

Expected Outcome: A continuous improvement process for incident declaration ensures that the organization can adapt its criteria and processes to the changing threat landscape and operational needs. This control fosters an agile, responsive incident declaration mechanism, contributing to more robust and effective cybersecurity incident management.

21

Incident Management (RS.MA)

> Mastering incident response is akin to conducting an orchestra of expertise, where precision in execution and harmony in collaboration set the stage for a resilient defense, turning the complexity of multi-party coordination into a symphony of cybersecurity readiness.

In cybersecurity, responding to incidents with precision and collaboration is paramount. Executing an incident response plan with relevant third parties once an incident is declared is critical for safeguarding organizational assets. This comprehensive approach involves identifying and engaging with key third-party stakeholders, activating detailed incident response protocols, and employing robust coordination mechanisms to ensure seamless collaboration. Communication protocols are vital in maintaining timely information exchange among all parties involved. In contrast, clearly defined roles and responsibilities ensure that each team member and third-party contributor understands their part in the response effort. Through meticulous documentation, post-incident reviews, and the continuous refinement of coordination processes based on lessons learned, organizations can strengthen their third-party collaborations, enhancing their overall incident response capabilities.

RS.MA-01: The Incident Response Plan Is Executed in Coordination with Relevant Third Parties Once an Incident Is Declared

When an organization faces a cybersecurity incident, executing a well-coordinated incident response plan in collaboration with relevant third parties is critical. This coordinated approach ensures that the response is swift and effective and leverages the specialized capabilities of each participant. The inception of this process is the immediate activation of the incident response plan upon declaring an incident. This plan outlines the necessary steps to mitigate the incident's impact, including identifying and engaging third-party partners.

Engaging with relevant third parties is a nuanced process that requires careful planning and understanding of each party's capabilities and resources. These third parties may include external cybersecurity firms, law enforcement agencies, and other cyber threat intelligence community entities. Their involvement can provide additional expertise, tools, and resources that are pivotal in managing the incident. Identifying which third parties to engage with and under what circumstances should be a predefined aspect of the incident response plan, ensuring a prompt activation and collaboration process.

A Comprehensive Guide to the NIST Cybersecurity Framework 2.0: Strategies, Implementation, and Best Practice, First Edition. Jason Edwards.
© 2025 John Wiley & Sons Ltd. Published 2025 by John Wiley & Sons Ltd.

The activation of the incident response plan sets into motion a series of predefined steps designed to manage and mitigate the incident. This phase is crucial and must be executed with precision and speed. Activation includes mobilizing the internal incident response team, engaging with identified third parties, and implementing containment and remediation measures. This process's efficiency dramatically depends on establishing clear protocols and communication channels with all involved parties.

Coordination mechanisms are pivotal in the seamless collaboration between an organization and its third-party partners. These mechanisms include shared communication platforms, regular update calls, and joint decision-making protocols. Establishing these mechanisms before an incident occurs is imperative to ensure all parties can work together effectively during a crisis. They facilitate a unified approach to incident response, leveraging the strengths and capabilities of each participant.

Effective communication is the backbone of successful coordination during incident response. It ensures that all parties are kept informed of the incident's status, actions taken, and any changes in strategy. Communication protocols should outline how information is shared, the frequency of updates, and the escalation paths for critical decisions. This clarity in communication helps prevent misunderstandings and ensures a cohesive response effort.

Defining the roles and responsibilities of internal teams and third parties is critical to the incident response plan. This clarity prevents overlaps and gaps in the response process, ensuring each party knows their tasks and responsibilities. Roles can range from technical analysis and containment to communication with stakeholders and legal advice. Clearly defined roles and responsibilities facilitate a more organized and effective response to incidents.

Documentation throughout the incident response process is vital. It records the actions taken, decisions made, and information shared. This documentation is crucial for post-incident reviews, legal proceedings, and compliance. It also serves as a valuable resource for learning and improving the incident response process and coordination with third parties.

Post-incident reviews are an opportunity to assess the effectiveness of third-party coordination and the overall response to the incident. These reviews should examine what worked well, what didn't, and why. The insights gained from these reviews are essential for continuous improvement, enabling organizations to refine their incident response plans, coordination processes, and partnerships with third parties.

Recommendations

- **Establish Predefined Engagement Protocols**: Create clear protocols for engaging with third parties, including the conditions under which they are activated and the communication channels used. This ensures a rapid and coordinated response to incidents.
- **Develop a Joint Incident Response Exercise Program**: Regularly conduct joint incident response exercises with third-party partners to test and improve coordination mechanisms and communication protocols. This practice helps identify potential issues and areas for improvement in a controlled environment.
- **Implement a Shared Communication Platform**: Utilize a shared communication platform for real-time information sharing and coordination with third parties. This platform should be secure, accessible, and familiar to all parties involved.
- **Define and Regularly Review Roles and Responsibilities**: Clearly define and document all internal and external parties involved in incident response roles and responsibilities. Regularly review and update these roles to reflect capabilities or organizational structure changes.

- **Establish a Continuous Improvement Framework**: Adopt a framework for continuous improvement based on lessons learned from incidents and post-incident reviews. This framework should include mechanisms for updating the incident response plan and coordination protocols with third parties.

NIST 800-53 Controls

(Refer to our appendix for a comprehensive description of the NIST 800-53 Controls, essential for organizations adhering to strict regulatory standards.)

- IR-06,IR-07,IR-08,SR-03,SR-08

Simplified Security Controls (SSC)

This section outlines our Simplified Security Controls (SSC), tailored for environments beyond the scope of NIST compliance. These controls offer streamlined, effective security solutions suitable for any organization seeking to enhance its cybersecurity defenses with clear, actionable measures.

Control 1: Development of a Third-Party Incident Response Coordination Protocol

Control Objective: To establish a comprehensive protocol that outlines the procedures for effective coordination between the organization and relevant third parties during incident response efforts, ensuring a unified and efficient response to cybersecurity incidents.

Implementation Steps:

1.1 **Identify and Classify Third Parties**: Catalog relevant third parties based on their roles, responsibilities, and level of involvement in the organization's incident response processes. This classification helps understand the criticality and nature of collaboration required with each party.

1.2 **Develop Coordination Protocol**: Create a detailed coordination protocol document that includes communication channels, escalation procedures, shared tools and platforms, and roles and responsibilities of both internal teams and third parties. This protocol should be developed with inputs from all relevant stakeholders to ensure comprehensiveness and feasibility.

1.3 **Conduct Joint Coordination Drills**: Regularly conduct joint incident response exercises with third parties to test and refine the coordination protocol. These drills should simulate various cybersecurity incidents to ensure the protocol covers multiple scenarios.

Expected Outcome: Developing and implementing a third-party incident response coordination protocol ensures that the organization and its relevant third parties are well-prepared to respond collaboratively to cybersecurity incidents. This preparation will lead to quicker incident resolution times, minimized impacts, and strengthened relationships with key third-party partners, enhancing the organization's cybersecurity resilience.

Control 2: Establishment of a Real-Time Incident Communication Hub

Control Objective: To create a centralized, real-time communication platform that facilitates efficient and secure information sharing among internal response teams and relevant third parties during a cybersecurity incident.

Implementation Steps:

2.1 **Select a Secure Communication Platform**: Choose a secure, reliable, and accessible communication platform that supports real-time information sharing and collaboration.

The platform should be capable of handling classified information and be accessible to all relevant stakeholders.

2.2 **Define Access Controls and Protocols**: Establish strict access controls and usage protocols for the communication platform, ensuring only authorized personnel can access and share information. This includes defining what information can be shared, with whom, and under what circumstances.

2.3 **Train Participants on Platform Use**: Provide comprehensive training for all potential users of the communication platform, including internal teams and third-party partners. Training should cover operational procedures, security practices, and emergency communication protocols.

Expected Outcome: Establishing a real-time incident communication hub significantly enhances the efficiency and effectiveness of communication during cybersecurity incidents. This centralized approach ensures that all relevant parties are informed and can collaborate effectively, leading to faster incident resolution and reduced impact on the organization.

Control 3: Implementation of a Continuous Third-Party Coordination Improvement Program

Control Objective: To continuously improve the effectiveness of third-party coordination in incident response through regular reviews, feedback, and updates to protocols and processes.

Implementation Steps:

3.1 **Establish a Review and Feedback Mechanism**: Create mechanisms for gathering and analyzing feedback from internal teams and third parties regarding the coordination effectiveness during and after incident response. This could include after-action reviews, surveys, and debriefing sessions.

3.2 **Analyze Performance and Identify Improvement Areas**: Regularly analyze incident response outcomes and feedback to identify areas for improvement in third-party coordination. Focus on communication efficiency, role clarity, and the effectiveness of shared tools and protocols.

3.3 **Update Coordination Protocols and Processes**: Make targeted updates to coordination protocols and processes based on the analysis. This includes refining communication channels, adjusting roles and responsibilities, and enhancing shared incident response tools and platforms.

Expected Outcome: Implementing a continuous third-party coordination improvement program ensures that the organization and its third parties regularly refine their collaboration approach based on actual incident response experiences. This ongoing improvement leads to more effective and efficient incident response, strengthening the organization's cybersecurity posture and relationships with critical third parties.

RS.MA-02: Incident Reports Are Triaged and Validated

The initial response to a cybersecurity incident report is critical in determining the subsequent direction of incident management efforts. This response starts with the triage and validation of incident reports—a process that efficiently categorizes and assesses the severity and authenticity of an incident. Effective triage ensures that resources are allocated appropriately, focusing efforts on incidents that pose the most significant risk to the organization.

Establishing triage criteria is the first step in creating an efficient incident response process. These criteria should be clear, measurable, and adaptable, enabling rapid assessment of incidents based on their potential impact, urgency, and complexity. Developing these criteria requires a deep understanding of the organization's operational environment, assets, and possible threats. By applying these criteria, organizations can prioritize incidents, ensuring that responses are proportionate to the threat level.

Rapidly validating incident reports is essential to confirm their authenticity and prevent the misallocation of resources to false alarms. This process involves verifying the report's details and assessing the source's credibility. Techniques for validation may include cross-referencing incident details with known threats, analyzing logs, and conducting preliminary investigations. The goal is to quickly distinguish between genuine threats and erroneous reports, allowing for a focused response to actual incidents.

Once an incident has been triaged, the allocation of resources is tailored based on the triage outcome. High-priority incidents may warrant the mobilization of specialized response teams, while lower-priority incidents might be addressed through standard operational procedures. This strategic allocation of resources ensures that the organization's response capabilities are optimized, balancing the need to address current incidents with the maintenance of normal operations.

Implementing automated tools can significantly enhance the efficiency of the incident triage process. These tools can rapidly analyze incident reports, apply triage criteria, and prioritize incidents for further action. Automation also helps consistently apply triage criteria, reducing the potential for human error and bias. However, the use of automated tools should be complemented by human oversight to address complexities and nuances that automated systems may overlook.

Training on triage procedures and decision-making is crucial for ensuring that personnel can handle the complexities of incident triage effectively. This training should cover triage criteria, the operation of automated triage tools, and the decision-making process for incident prioritization. By developing a skilled and knowledgeable incident response team, organizations can improve their initial response to cybersecurity incidents.

Documentation of triage decisions and the rationale behind them is important for transparency and accountability. This documentation records actions taken and supports post-incident analysis, enabling organizations to review and assess their response to incidents. It also provides valuable data for training purposes and for refining triage criteria and procedures over time.

Continuous monitoring and adjustment of triage criteria ensure that the incident response process remains effective and responsive to evolving threats. As new types of incidents emerge and organizational priorities change, triage criteria may need to be updated to reflect these developments. Additionally, a feedback mechanism for improving incident reporting and triage can help organizations refine their approach to incident management, incorporating lessons learned from past incidents.

Recommendations

- **Refine Triage Criteria Regularly**: Continuously monitor and update triage criteria to ensure they remain effective against evolving cybersecurity threats. This involves reviewing incident outcomes, threat intelligence, and organizational changes that may affect prioritization.
- **Invest in Automated Triage Tools**: Implement and integrate advanced automated tools for incident triage to enhance efficiency and consistency. Ensure these tools are regularly updated and configured to reflect current triage criteria and organizational priorities.
- **Conduct Regular Training on Triage Procedures**: Provide ongoing training for all incident response team members on the latest triage procedures, decision-making processes, and automated tools. This ensures a skilled and prepared response team.

- **Establish a Robust Documentation Process**: Implement a comprehensive documentation strategy for all triage decisions, including the rationale for prioritization and resource allocation. This aids in transparency, accountability, and continuous improvement of the triage process.
- **Create a Feedback Loop for Incident Reporting and Triage**: Develop mechanisms for gathering and analyzing stakeholders' feedback on the incident reporting and triage process. Use this feedback to make iterative improvements, enhancing the effectiveness and efficiency of incident response efforts.

NIST 800-53 Controls

(Refer to our appendix for a comprehensive description of the NIST 800-53 Controls, essential for organizations adhering to strict regulatory standards.)

- IR-04,IR-05,IR-06

Simplified Security Controls (SSC)

This section outlines our SSC, tailored for environments beyond the scope of NIST compliance. These controls offer streamlined, effective security solutions suitable for any organization seeking to enhance its cybersecurity defenses with clear, actionable measures.

Control 1: Development of Dynamic Triage Criteria for Incident Assessment

Control Objective: To establish a set of dynamic, clearly defined criteria for the initial assessment and triage of incident reports, enabling rapid and accurate categorization of incidents based on severity, urgency, and impact.

Implementation Steps:

1.1 **Define Triage Criteria**: Develop comprehensive triage criteria that include indicators of severity, urgency, and potential impact on the organization. These criteria should be flexible enough to accommodate various incident types.

1.2 **Implement Criteria Review Process**: Establish a regular review process for the triage criteria involving key stakeholders from IT security, business operations, and risk management. This process ensures the criteria remain relevant and effective in identifying the most critical incidents.

1.3 **Train Response Teams on Criteria Application**: Conduct training sessions for all incident response team members on effectively applying the triage criteria. Include scenario-based exercises to enhance understanding and application in real-world situations.

Expected Outcome: Developing and implementing dynamic triage criteria leads to a more efficient and effective incident response process. By accurately categorizing incidents, the organization can prioritize responses based on actual risk, ensuring that critical resources are allocated to the most severe incidents, thereby minimizing potential damage and disruption.

Control 2: Integration of Automated Incident Triage Tools

Control Objective: To leverage automated tools for the efficient and consistent triage of incident reports, enhancing the speed and accuracy of the initial incident assessment process.

Implementation Steps:

2.1 **Select Automated Triage Tools**: Identify and select appropriate automated triage tools to analyze incident reports against the established triage criteria and categorize them accordingly. Consider tools with AI and machine learning capabilities for enhanced accuracy over time.

2.2 **Integrate Tools with Incident Reporting Systems**: Seamlessly integrate the selected automated triage tools with existing incident reporting systems to ensure all reports are automatically assessed and categorized upon receipt.

2.3 **Monitor and Adjust Tool Settings**: Regularly monitor the performance of the automated triage tools and adjust their settings as necessary to maintain high accuracy and effectiveness. Use feedback from incident response teams to fine-tune the tools.

Expected Outcome: Integrating automated incident triage tools streamlines the initial assessment of incident reports, reducing the time to categorize and prioritize incidents. This automation supports a quicker response to critical incidents and a more efficient allocation of incident response resources.

Control 3: Establishment of an Incident Triage Feedback Loop

Control Objective: To continuously improve the incident triage process by establishing a feedback loop that collects, analyzes, and incorporates feedback from incident response teams and stakeholders.

Implementation Steps:

3.1 **Develop Feedback Collection Mechanisms**: Implement mechanisms for collecting detailed feedback on the triage process from incident response teams, including surveys, debrief sessions, and incident review meetings.

3.2 **Analyze Feedback for Insights**: Regularly analyze the collected feedback to identify trends, challenges, and opportunities for improvement in the triage process.

3.3 **Implement Improvements**: Based on feedback analysis, make targeted improvements to the triage criteria, processes, and tools. This may include revising triage criteria, enhancing training programs, or adjusting automated tool settings.

Expected Outcome: Establishing an incident triage feedback loop ensures that the triage process is continuously refined and improved based on experience and stakeholder feedback. This iterative approach leads to more accurate incident categorization, more effective resource allocation, and a more resilient incident response capability.

RS.MA-03: Incidents Are Categorized and Prioritized

A systematic approach to categorizing and prioritizing cybersecurity incidents is essential for effective incident response and management. This process begins with a comprehensive understanding of the different types of incidents affecting an organization, ranging from data breaches and malware infections to insider threats and denial of service attacks. Accurate incident categorization and prioritization enable organizations to respond effectively, ensuring that resources are allocated efficiently and the most critical incidents are addressed promptly.

Developing a standardized incident categorization scheme is a foundational step in this process. Such a scheme should be comprehensive, covering all the organization's potential incidents. It must also be adaptable, allowing for the inclusion of new types of incidents as they emerge. This standardization ensures that all incidents are classified consistently, facilitating effective communication and coordination within the incident response team and with external partners.

Criteria for prioritizing incidents are based on their impact on the organization's operations, reputation, legal obligations, and the urgency of the response required. High-priority incidents may involve significant data loss, affect critical systems, or pose immediate regulatory compliance issues. Establishing clear criteria for prioritization helps ensure that responses are proportionate to the severity of the incident and that resources are allocated where they are needed most.

Automated systems for categorization and prioritization can significantly enhance the efficiency and consistency of these processes. These systems can quickly analyze incident reports, classify them according to the established scheme, and prioritize them based on predefined criteria. Automation also helps handle high volumes of incidents, ensuring that each is addressed promptly.

Integrating the categorization and prioritization processes with response efforts is critical for a seamless and effective incident management strategy. This integration ensures that the response is tailored to the specific type of incident and its severity level. It also facilitates the allocation of appropriate resources and the implementation of the most effective response strategies.

Training on the importance of accurate incident categorization is vital for all incident response team members. Such training should cover the categorization scheme, the criteria for prioritization, and the use of automated tools. This ensures all team members are prepared to apply these processes consistently and effectively.

Regular review of the categorization and prioritization processes is necessary to ensure that they remain effective and relevant. This review should consider changes in the threat landscape, new types of incidents emerge, and feedback from incident response activities. Adjustments may be required to reflect these changes and to incorporate lessons learned from past incidents.

Adjustments to prioritization criteria based on emerging threats are essential to maintaining an effective incident response capability. As new threats emerge, their potential impact and the urgency of response may differ from those of known threats. Updating the prioritization criteria to reflect these changes ensures that the organization remains prepared to respond effectively to the evolving threat landscape.

Recommendations

- **Implement a Dynamic Incident Categorization Framework**: Establish a comprehensive and adaptable incident categorization framework that can accommodate the evolving nature of cybersecurity threats. Ensure it is widely understood across the organization.
- **Leverage Advanced Automation Tools**: Invest in advanced automation tools for the rapid and consistent categorization and prioritization of incidents. This facilitates a swift response to critical incidents and efficient resource allocation.
- **Integrate Categorization with Incident Response**: Seamlessly integrate incident categorization and prioritization processes with the broader incident response strategy. Tailor response efforts to the specific characteristics and severity of incidents.
- **Conduct Regular Training and Simulations**: Conduct ongoing training and regular simulation exercises for all incident response personnel. Focus on the importance of accurate incident categorization and applying prioritization criteria.
- **Review and Adjust Processes Periodically**: Review and adjust the incident categorization and prioritization processes. Consider emerging threats and lessons learned from past incidents to ensure continued relevance and effectiveness.

NIST 800-53 Controls

(Refer to our appendix for a comprehensive description of the NIST 800-53 Controls, essential for organizations adhering to strict regulatory standards.)

- IR-04,IR-05,IR-06

Simplified Security Controls (SSC)

This section outlines our SSC, tailored for environments beyond the scope of NIST compliance. These controls offer streamlined, effective security solutions suitable for any organization seeking to enhance its cybersecurity defenses with clear, actionable measures.

Control 1: Implementation of a Comprehensive Incident Categorization Framework

Control Objective: To develop and implement a standardized incident categorization scheme that accurately classifies cybersecurity incidents based on their nature, severity, and impact, facilitating effective prioritization and response.

Implementation Steps:

1.1 **Develop Categorization Scheme**: Create a comprehensive incident categorization scheme that covers various cybersecurity incidents, incorporating industry standards and specific organizational needs. This scheme should clearly define categories and subcategories of incidents.

1.2 **Establish Categorization Guidelines**: Draft detailed guidelines for applying the categorization scheme to incidents, ensuring consistency and accuracy in classification. These guidelines should include examples and scenarios for clarity.

1.3 **Train Incident Response Teams**: Provide extensive training for all members of the incident response teams on the categorization scheme and guidelines. Regularly update training materials to reflect any changes in the categorization scheme or emerging threats.

Expected Outcome: Implementing a comprehensive incident categorization framework enables the organization to quickly and accurately classify incidents, improving the efficiency and effectiveness of the incident response process. This systematic approach ensures that resources are appropriately allocated to incidents based on severity and impact, enhancing the organization's overall cybersecurity posture.

Control 2: Deployment of Automated Categorization and Prioritization Tools

Control Objective: To leverage technology by deploying automated systems that assist in categorizing and prioritizing incidents, reducing human error and speeding up the response time to critical incidents.

Implementation Steps:

2.1 **Select and Deploy Tools**: Identify and deploy automated tools that use artificial intelligence and machine learning to categorize and prioritize incidents based on predefined criteria and historical data.

2.2 **Integrate Tools with Existing Systems**: Ensure seamless integration of the automated tools with the organization's existing incident management and reporting systems for efficient workflow and data exchange.

2.3 **Monitor and Refine Tool Performance**: Regularly monitor the performance of the automated tools to ensure they are categorizing and prioritizing incidents accurately. Refine tool configurations and criteria based on feedback and evolving cybersecurity landscapes.

Expected Outcome: The deployment of automated categorization and prioritization tools streamlines the incident response process by ensuring rapid and accurate classification of incidents. This leads to more timely and focused response efforts, particularly for high-priority incidents, thereby minimizing potential damage and disruption.

Control 3: Regular Review and Update of Categorization and Prioritization Criteria

Control Objective: To ensure that the incident categorization and prioritization process remains effective and relevant in the face of evolving cybersecurity threats and organizational changes.

Implementation Steps:

3.1 **Establish a Review Schedule**: Set a regular schedule for reviewing and updating the categorization and prioritization criteria involving key cybersecurity, IT, and business unit stakeholders.

3.2 **Collect and Analyze Feedback**: Systematically collect feedback on the categorization and prioritization process from incident response teams, IT staff, and other relevant stakeholders. Analyze this feedback to identify trends, challenges, and opportunities for improvement.

3.3 **Implement Updates**: Based on the review and analysis, make targeted updates to the categorization and prioritization criteria. Communicate changes clearly to all stakeholders and update training materials accordingly.

Expected Outcome: Regular review and update of categorization and prioritization criteria ensure that the organization's incident response process adapts to changing threats and business requirements. This continuous improvement cycle fosters a resilient cybersecurity posture capable of effectively managing incidents with an informed, strategic approach to resource allocation and response prioritization.

RS.MA-04: Incidents Are Escalated or Elevated as Needed

A robust incident response framework includes clear processes for escalating and elevating cybersecurity incidents. This ensures that incidents are addressed by the appropriate level of expertise and authority within an organization. Escalation and elevation are critical components that allow for the dynamic allocation of resources and attention based on an incident's evolving nature and severity. Understanding the nuances of these processes is crucial for effective incident management.

Escalation thresholds and incident elevation criteria are foundational elements guiding the decision-making process. These criteria should be clearly defined based on the severity of the incident, its potential impact on the organization, and the required expertise for effective resolution. Establishing these thresholds ensures that incidents are escalated promptly and orderly, facilitating a swift and appropriate response.

Procedures for rapid escalation to appropriate response teams are essential for effective incident management. These procedures should outline the steps to be taken when an incident meets predefined escalation criteria, including identifying key decision-makers and activating specialized response teams. Rapid escalation ensures that incidents are promptly addressed, minimizing potential damage and disruption.

Communication channels for escalating or elevating incidents are vital in ensuring information flows efficiently to all relevant parties. These channels should be secure, reliable, and well-established to facilitate clear and timely communication. Effective communication supports the decision-making process and enables the coordination of response efforts across different levels of the organization.

The role of management in incident escalation decisions is to provide leadership, oversight, and strategic direction. Management should set escalation criteria, approve escalation procedures, and participate in significant escalation decisions. Their involvement ensures that incident response efforts are aligned with the organization's overall risk management strategy and business objectives.

Integration of escalation processes with overall incident management ensures a cohesive and coordinated response to cybersecurity incidents. This integration includes aligning escalation procedures with other aspects of incident response, such as incident detection, analysis, and recovery efforts. A well-integrated approach enhances the organization's ability to manage incidents effectively, from initial detection to resolution.

Training on escalation protocols and decision points is crucial for all incident response team members and key stakeholders throughout the organization. This training should cover the criteria for escalation, the steps involved in the escalation process, and the roles and responsibilities of all parties involved. Adequate training ensures everyone understands how and when to escalate an incident, supporting effective and efficient management.

Documentation and tracking of escalation and elevation actions are essential for accountability, analysis, and continuous improvement. This documentation should include the decision-making process, actions taken, and the rationale for escalation or elevation decisions. Tracking these actions facilitates post-incident reviews and audits, enabling organizations to learn from incidents and refine their escalation processes.

Reviewing and updating escalation procedures based on incident feedback is necessary to ensure these processes remain effective and relevant. This review should consider feedback from incident response teams, management, and external partners. Updates may be required to address new threats, incorporate lessons learned from past incidents, and refine escalation criteria and procedures.

Recommendations

- **Establish Clear Escalation Thresholds**: Define and communicate clear thresholds for escalating and elevating incidents based on severity, impact, and the expertise required for resolution. This ensures timely and appropriate responses to incidents.
- **Enhance Communication Channels**: Strengthen communication channels for escalation and elevation to ensure that information is shared efficiently and securely among all relevant parties. Effective communication supports quick decision-making and coordinated response efforts.
- **Empower Management Involvement**: Encourage active management involvement in the escalation process to provide strategic oversight and ensure that incident response efforts are aligned with organizational priorities and risk management strategies.
- **Integrate Escalation with Incident Management**: Seamlessly integrate escalation processes with overall incident management strategies, ensuring a coordinated and efficient response across all stages of incident handling.
- **Prioritize Continuous Training and Improvement**: Invest in regular training for all relevant personnel on escalation protocols and conduct periodic reviews of escalation procedures to incorporate feedback and adapt to evolving cybersecurity landscapes.

NIST 800-53 Controls

(Refer to our appendix for a comprehensive description of the NIST 800-53 Controls, essential for organizations adhering to strict regulatory standards.)

- IR-04,IR-05,IR-06,IR-07

Simplified Security Controls (SSC)

This section outlines our SSC, tailored for environments beyond the scope of NIST compliance. These controls offer streamlined, effective security solutions suitable for any organization seeking to enhance its cybersecurity defenses with clear, actionable measures.

Control 1: Formalization of Escalation Thresholds and Criteria

Control Objective: To establish clear and formalized thresholds and criteria for incident escalation, ensuring consistent decision-making and timely escalation of incidents to appropriate response teams or management levels.

Implementation Steps:

1.1 **Define Escalation Thresholds**: Collaborate with key cybersecurity, IT, and business stakeholders to define specific thresholds for incident escalation, considering factors such as severity, impact, and urgency.

1.2 **Document Criteria and Procedures**: Document the escalation criteria and procedures, including specific scenarios and examples to guide decision-making. Ensure this documentation is accessible to all relevant personnel.

1.3 **Regular Review and Update**: Set a schedule for the regular review and update of escalation thresholds and criteria to ensure they remain relevant and effective in the face of evolving cybersecurity threats and business needs.

Expected Outcome: Formalizing escalation thresholds and criteria leads to more consistent and efficient incident escalation processes. It ensures incidents are escalated promptly and appropriately, facilitating rapid response actions and minimizing potential damage.

Control 2: Establishment of a Centralized Incident Escalation Hub

Control Objective: To create a centralized communication and coordination hub for incident escalation, streamlining the process of escalating incidents and ensuring effective communication between all relevant parties.

Implementation Steps:

2.1 **Develop a Centralized Hub**: Implement a centralized incident escalation hub using technology platforms that support secure communication and information sharing among incident response teams, management, and other relevant stakeholders.

2.2 **Integrate with Existing Systems**: Ensure the escalation hub fully integrates with existing incident management and reporting systems for seamless operation and information flow.

2.3 **Train Stakeholders**: Provide comprehensive training for all stakeholders on how to use the escalation hub effectively, including protocols for escalating incidents and communicating during an incident.

Expected Outcome: Establishing a centralized incident escalation hub enhances the organization's ability to quickly and effectively escalate and manage incidents. It ensures clear, secure communication among all parties, facilitating more coordinated and efficient incident response efforts.

Control 3: Continuous Improvement Program for Escalation Processes

Control Objective: To continuously improve incident escalation processes through regular reviews, feedback collection, and updates, ensuring the escalation system remains effective.

Implementation Steps:

3.1 **Implement Feedback Mechanisms**: Set up mechanisms for collecting feedback on the incident escalation process from all involved parties, including after-action reviews and incident debriefings.

3.2 **Analyze Feedback and Identify Improvements**: Regularly analyze the feedback to identify trends, challenges, and opportunities for improvement in the escalation process.
3.3 **Update Processes and Training**: Based on feedback analysis, make targeted updates to escalation processes, criteria, and training programs. Communicate changes clearly to all stakeholders and provide updated training as necessary.

Expected Outcome: Implementing a continuous improvement program for escalation processes ensures that the organization's approach to incident escalation constantly evolves and improves. This leads to more effective and efficient handling of incidents, better coordination among response teams, and a more robust overall cybersecurity posture.

RS.MA-05: The Criteria for Initiating Incident Recovery Are Applied

In the incident response lifecycle, transitioning to the recovery phase is a critical step that signifies the shift from immediate response and containment to restoring normal operations. Establishing clear criteria for initiating incident recovery is essential for a structured and effective transition. These criteria serve as benchmarks to assess whether an incident has been sufficiently contained and if it is appropriate to begin recovery efforts. A well-defined transition phase ensures that recovery efforts are timely, organized, and effective, minimizing the impact on business operations.

The definition of clear criteria for transitioning to the incident recovery phase involves several considerations, including the extent of damage, the success of containment measures, and the readiness of recovery resources. These criteria should be specific, measurable, and aligned with the organization's risk management objectives. They are crucial in determining the appropriate timing for initiating recovery efforts, ensuring that such efforts are neither premature nor delayed.

Assessing the incident's impact and the effectiveness of resolution efforts is a pivotal step in determining recovery readiness. This assessment should evaluate the situation's stability, the integrity of systems and data, and the resources required for recovery. Only when an incident is deemed under control and the risk of further damage has been mitigated should the recovery phase be initiated. This approach ensures that recovery efforts are focused and efficient, leading to a quicker restoration of normal operations.

The procedures for formally declaring the initiation of the recovery phase are critical for a coordinated response. This declaration should be formal, involve key decision-makers, and follow a predefined protocol. It serves as a clear signal to all stakeholders that the focus is shifting from containment to recovery, enabling teams to activate recovery plans and allocate resources accordingly.

Coordination with business continuity planning (BCP) is essential for effective recovery efforts. The incident recovery phase should be integrated with the organization's broader business continuity plans, ensuring recovery efforts are consistent with overall business recovery objectives. This integration facilitates a more resilient and unified response to incidents, supporting the seamless restoration of critical functions and minimizing business disruption.

The communication plan for notifying stakeholders of the recovery initiation is integral to effective incident management. This plan should outline the channels, messages, and timing for communications, ensuring that all relevant parties, including employees, customers, and partners, are appropriately informed. Clear and timely communication helps manage expectations, reduce uncertainty, and maintain trust throughout recovery.

The role of recovery teams in executing recovery plans is to implement the specific actions required to restore systems, data, and services to their normal state. These teams should comprise individuals with the expertise and authority to carry out recovery tasks efficiently. The effectiveness of the recovery teams is contingent upon their clear understanding of the recovery criteria, plans, and roles within the recovery process.

Documentation of the application of recovery criteria and actions taken during the recovery phase is crucial for accountability, learning, and compliance. This documentation should provide a detailed record of the decision-making process, the rationale for initiating recovery, and the steps taken to restore normal operations. It serves as a valuable resource for post-incident reviews, enabling organizations to evaluate the effectiveness of their recovery criteria and processes.

Post-recovery analysis to assess the criteria's effectiveness is essential in continuously improving incident management practices. This analysis should review the application of recovery criteria, the efficiency of the recovery process, and the outcomes achieved. Insights gained from this analysis can inform adjustments to recovery criteria and processes, enhancing the organization's resilience and preparedness for future incidents.

Recommendations

- **Develop Comprehensive Recovery Criteria**: Establish comprehensive, clear, and actionable criteria for initiating the recovery phase. These criteria should encompass technical, operational, and business considerations to ensure a holistic approach to recovery.
- **Integrate Recovery Planning with Business Continuity**: Ensure that incident recovery planning is seamlessly integrated with overall BCP. This alignment guarantees that recovery efforts support and are supported by broader business recovery strategies.
- **Formalize the Recovery Declaration Process**: Implement a formal process for declaring the initiation of the recovery phase, involving key stakeholders and decision-makers. This formalization ensures clarity and coordination across the organization.
- **Enhance Stakeholder Communication**: Strengthen the communication plan for the recovery initiation phase to include timely, transparent, and clear communication with all stakeholders. Effective communication is crucial for maintaining confidence and managing expectations during recovery.
- **Conduct Thorough Post-Recovery Reviews**: Regularly conduct post-recovery reviews to assess the effectiveness of the recovery criteria and the overall recovery process. Use insights from these reviews to refine and improve recovery strategies continuously.

NIST 800-53 Controls

(Refer to our appendix for a comprehensive description of the NIST 800-53 Controls, essential for organizations adhering to strict regulatory standards.)

- IR-04,IR-08

Simplified Security Controls (SSC)

This section outlines our SSC, tailored for environments beyond the scope of NIST compliance. These controls offer streamlined, effective security solutions suitable for any organization seeking to enhance its cybersecurity defenses with clear, actionable measures.

Control 1: Establishment of Incident Recovery Readiness Criteria

Control Objective: To define clear and actionable criteria for determining when an incident has been sufficiently contained to begin recovery, ensuring a seamless transition from incident response to recovery.

Implementation Steps:

1.1 **Develop Readiness Criteria**: Collaborate with cybersecurity, business continuity, and risk management stakeholders to develop comprehensive criteria for recovery readiness. These criteria should assess the situation's stability, the integrity of impacted systems, and the availability of resources for recovery.

1.2 **Formalize Recovery Initiation Protocol**: Document a formal protocol for initiating the recovery phase, incorporating the defined readiness criteria and outlining the decision-making process and authority for declaring the start of recovery.

1.3 **Integrate with Incident Response Plan**: Ensure the recovery readiness criteria and initiation protocol are fully integrated into the organization's broader incident response plan, creating a cohesive approach to incident management.

Expected Outcome: Establishing clear incident recovery readiness criteria and a formal initiation protocol ensures the organization can make informed decisions about transitioning to recovery efforts. This control leads to a more efficient recovery process, minimizing downtime and reducing the impact of incidents on operations.

Control 2: Coordination Mechanism Between Incident Response and Business Continuity Teams

Control Objective: To establish a formal coordination mechanism between incident response and BCP teams, ensuring recovery efforts align with business continuity requirements and objectives.

Implementation Steps:

2.1 **Define Coordination Roles and Responsibilities**: Clearly outline the roles and responsibilities of the incident response and BCP teams, highlighting areas of collaboration and decision-making authority during the recovery phase.

2.2 **Develop Joint Recovery Planning Sessions**: Schedule and conduct joint planning sessions between incident response and BCP teams to develop integrated recovery strategies that address both cybersecurity and business continuity concerns.

2.3 **Implement Cross-Training Programs**: Initiate cross-training programs to enhance mutual understanding and collaboration between incident response and BCP teams. This ensures that personnel are prepared to work together effectively during recovery efforts.

Expected Outcome: Establishing a formal coordination mechanism between incident response and BCP teams enhances the organization's ability to execute a cohesive and effective recovery strategy. This approach ensures that recovery efforts are technically sound and aligned with the overarching business continuity objectives, leading to a more resilient organizational posture.

Control 3: Recovery Phase Communication Plan

Control Objective: To develop a comprehensive communication plan that outlines the protocols for notifying internal and external stakeholders of the initiation of the recovery phase, ensuring clear and timely communication during critical transitions.

Implementation Steps:

3.1 **Identify Key Stakeholders**: Identify and list all internal and external stakeholders who must be informed about the initiation of the recovery phase, including employees, customers, partners, and regulatory bodies.

3.2 **Draft Communication Templates**: Prepare communication templates for various scenarios that convey the status of the incident, the initiation of recovery efforts, and what stakeholders can expect during the recovery phase.

3.3 **Establish Notification Protocols**: Define the protocols for how and when stakeholders will be notified, including the methods of communication (e.g., email, internal announcements, and press releases) and the timeline for dissemination.

Expected Outcome: Implementing a recovery phase communication plan ensures that all relevant stakeholders are promptly and effectively informed about the transition to recovery, minimizing confusion and uncertainty. This control fosters trust and transparency, supporting smoother recovery operations and helping to maintain stakeholder confidence during and after cybersecurity incidents.

22

Incident Analysis (RS.AN)

> Delving into the heart of cybersecurity incidents with forensic precision uncovers the symptoms and the systemic weaknesses, guiding us with clear-eyed insights to fortify our digital domains against the unseen battles yet to come.

In cybersecurity, the meticulous analysis of incidents to identify their root causes is paramount, integrating technical and nontechnical factors through collaboration with subject matter experts (SMEs). This comprehensive approach encompasses deploying forensic analysis techniques, leveraging log files, system monitors, and network traffic analyzers to gather evidence, and applying root cause analysis (RCA) frameworks to unearth underlying issues. Continuous documentation and the validation of findings through cross-verification techniques ensure transparency and accuracy in the analysis process. Moreover, developing corrective actions based on these discoveries and strategic recommendations to prevent similar incidents in the future underscores the importance of preserving the integrity and provenance of investigative records. This holistic methodology not only aids in accurately estimating and validating an incident's magnitude but also facilitates effective communication with stakeholders, thereby enhancing the organization's resilience against future cybersecurity threats.

Author's Note: RS.AN-01, RS.AN-02, RS.AN-04, and RS.AN-05 were deprecated from the CSF 2 and are not included in this chapter.

RS.AN-03: Analysis Is Performed to Establish What Has Taken Place During an Incident and the Root Cause of the Incident

The critical importance of advanced incident analysis in cybersecurity is undeniable, especially when thoroughly examining and understanding the events and root causes of security incidents. This in-depth approach aims to clarify the sequence of events and pinpoint the underlying causes, thereby significantly empowering organizations to improve their defense strategies. At the heart of this analysis is a grasp of incident dynamics and the basic principles of identifying root causes, which require a detailed and organized process.

The introduction of forensic analysis techniques marks a significant milestone in comprehending the dynamics of cybersecurity incidents. Forensic analysts are equipped to dissect and examine digital evidence meticulously by employing a suite of sophisticated tools and methodologies.

A Comprehensive Guide to the NIST Cybersecurity Framework 2.0: Strategies, Implementation, and Best Practice, First Edition. Jason Edwards.
© 2025 John Wiley & Sons Ltd. Published 2025 by John Wiley & Sons Ltd.

This analysis is pivotal in reconstructing the sequence of events, thereby illuminating the pathways through which the security breach occurred. Deployment of these techniques not only aids in identifying vulnerabilities exploited by adversaries but also furnishes critical insights into the tactics, techniques, and procedures employed, forming the bedrock for developing more resilient security postures.

In the arsenal of tools for incident analysis, log files, system monitors, and network traffic analyzers stand out for their ability to provide a granular view of an incident's landscape. These resources are indispensable for gathering evidence, offering a timeline of activities that can be scrutinized for anomalies or malicious activities. The meticulous examination of this data allows for a precise understanding of how the incident unfolded, highlighting potential entry points and the scope of the impact. This evidence-based approach is fundamental in piecing together the puzzle of complex cybersecurity incidents, paving the way for accurate root cause analysis.

Applying RCA frameworks introduces a structured methodology to identify underlying issues that precipitate security incidents. RCA is predicated on the notion that effective problem-solving extends beyond symptom alleviation, seeking to unearth the foundational weaknesses within systems or processes that adversaries exploit. By systematically peeling back the layers of an incident, RCA facilitates a deeper understanding of the vulnerabilities inherent in an organization's cybersecurity posture, thus guiding the development of more targeted and effective remediation strategies.

An effective incident analysis process transcends technical scrutiny boundaries, incorporating technical and nontechnical factors. This holistic approach acknowledges that human elements, procedural gaps, and policy shortcomings play significant roles in the genesis of cybersecurity incidents. By integrating these diverse factors into the analysis, organizations can cultivate a more comprehensive understanding of their security vulnerabilities, encompassing not only the technological dimensions but also the human and procedural aspects, often the linchpins of robust cybersecurity defenses.

Collaboration with SMEs is crucial for comprehensively analyzing cybersecurity incidents. SMEs bring specialized knowledge and insights that can significantly enhance understanding of complex technical issues and their implications for business operations. This collaborative approach ensures that the analysis is enriched with expert perspectives, facilitating a more nuanced and informed understanding of the incident. By leveraging the expertise of SMEs, organizations can ensure that their analysis and subsequent response strategies are grounded in deep technical knowledge and operational realities.

Continuous documentation throughout the analysis process is pivotal for maintaining transparency and accountability. This documentation serves as a detailed record of the investigative journey, cataloging the steps taken, the findings unearthed, and the rationale behind specific analytical conclusions. Such comprehensive record-keeping supports the integrity of the analysis process and provides a valuable resource for post-incident reviews, training, and future reference. This meticulous approach to documentation ensures that the insights gained from the analysis are preserved and accessible, facilitating continuous improvement in cybersecurity practices.

The validation of analysis findings through cross-verification techniques is a critical step in ensuring the accuracy and reliability of the conclusions drawn. This process involves the independent verification of evidence and hypotheses, seeking to eliminate biases and confirm the robustness of the analysis. Such validation is essential for establishing confidence in the findings, which informs the development of corrective actions. By rigorously testing the analysis against alternative interpretations and perspectives, organizations can solidify their understanding of the incident and refine their response strategies accordingly.

Recommendations

- **Foster Interdisciplinary Collaboration**: Cultivate a culture of interdisciplinary collaboration that brings together cybersecurity experts, forensic analysts, and SMEs. This collective expertise will enhance the quality and depth of incident analysis, enabling a more nuanced understanding of threats and vulnerabilities.
- **Invest in Advanced Analytical Tools**: Allocate resources to acquire and integrate advanced analytical tools into the cybersecurity arsenal. These tools will significantly augment the capacity to gather, analyze, and interpret complex data related to security incidents, facilitating more accurate and timely analysis.
- **Enhance Training Programs**: Develop comprehensive training programs on advanced incident analysis techniques, including forensic and root cause analysis. Such programs should equip cybersecurity professionals with the skills to effectively dissect and understand complex incidents.
- **Implement a Continuous Improvement Process**: Establish a continuous improvement process that leverages the insights from incident analysis to refine and enhance cybersecurity measures. This process should involve regularly reviewing security policies, procedures, and technologies to address identified vulnerabilities.
- **Promote a Culture of Documentation and Transparency**: Encourage a culture that prioritizes meticulous documentation and transparency throughout the incident analysis process. This approach will support accountability and ensure valuable insights are captured and shared, contributing to the collective cybersecurity knowledge base.

NIST 800-53 Controls

(Refer to our appendix for a comprehensive description of the NIST 800-53 Controls, essential for organizations adhering to strict regulatory standards.)

- AU-07,IR-04

Simplified Security Controls (SSC)

This section outlines our simplified security controls (SSC), tailored for environments beyond the scope of NIST compliance. These controls offer streamlined, effective security solutions suitable for any organization seeking to enhance its cybersecurity defenses with clear, actionable measures.

Control 1: Enhanced Incident Analysis and Response Training Program

Control Objective: To develop and implement a comprehensive training program to enhance the skills and knowledge of the cybersecurity incident response team, particularly in incident analysis, forensic techniques, and root cause analysis.

Implementation Steps:

1.1 **Identify Training Needs**: Conduct a skills gap analysis for the incident response team to identify specific areas where enhanced training is needed, focusing on incident analysis, forensic techniques, root cause analysis, and the integration of technical and nontechnical factors.

1.2 **Develop Customized Training Modules**: Based on the identified needs, develop a series of training modules or courses that cover the essentials of incident analysis, the deployment of forensic analysis techniques, the application of RCA frameworks, and effective collaboration strategies.

1.3 **Schedule Regular Training Sessions**: Implement a regular training schedule that allows for initial training and ongoing refresher courses to ensure the team's skills remain sharp and up-to-date with the latest cybersecurity threats and analysis techniques.

Expected Outcome: Implementing an enhanced incident analysis and response training program equips the team with advanced skills in identifying, analyzing, and mitigating cybersecurity incidents. This leads to more effective identification of root causes, quicker resolution of incidents, and improved prevention of future threats, strengthening the organization's overall cybersecurity resilience.

Control 2: Integration of Advanced Forensic Tools and Analysis Platforms

Control Objective: To incorporate advanced forensic analysis tools and platforms into the incident response process, facilitating more in-depth analysis of incident dynamics and evidence gathering.

Implementation Steps:

2.1 **Tool Selection**: Evaluate and select advanced forensic analysis tools and platforms that offer comprehensive capabilities for analyzing log files, system monitors, and network traffic and support the integration of technical and nontechnical factors in the analysis.

2.2 **Tool Integration**: Integrate the selected forensic tools into the existing cybersecurity incident response infrastructure, ensuring compatibility and seamless workflow between evidence gathering, analysis, and response stages.

2.3 **Training and Familiarization**: Provide specialized training for the incident response team on the effective use of the new forensic tools, including best practices for evidence gathering, analysis, and preserving data integrity and provenance.

Expected Outcome: Integrating advanced forensic tools and analysis platforms significantly enhances the incident response team's ability to gather detailed evidence and conduct thorough analyses of cybersecurity incidents. This leads to more accurate identification of root causes, more effective mitigation strategies, and a stronger defense against future incidents.

Control 3: Cross-Functional Incident Analysis Task Force

Control Objective: To establish a cross-functional task force that brings together SMEs from various departments for comprehensive incident analysis, ensuring a holistic understanding of incidents and their impacts on the organization.

Implementation Steps:

3.1 **Task Force Formation**: Form a cross-functional incident analysis task force that includes SMEs from IT security, network operations, legal, human resources, and business units. This diversity ensures a comprehensive analysis incorporating multiple perspectives.

3.2 **Define Roles and Responsibilities**: Clearly outline the roles and responsibilities of each task force member, ensuring that all aspects of incident analysis are covered, from technical analysis to the assessment of business impact and legal considerations.

3.3 **Establish Incident Analysis Protocols**: Develop and document specific protocols for incident analysis, including procedures for collaboration, information sharing, documentation, and the validation of findings through cross-verification techniques.

Expected Outcome: Creating a cross-functional incident analysis task force enables a more thorough and nuanced analysis of cybersecurity incidents. By leveraging the diverse expertise of task force members, the organization can achieve a deeper understanding of incidents, leading to more effective corrective actions and strategies for preventing similar incidents in the future.

RS.AN-06: Actions Performed During an Investigation Are Recorded, and the Records' Integrity and Provenance Are Preserved

Establishing protocols for securing and accurately recording investigative actions forms the bedrock of any robust cybersecurity investigation framework. Such protocols guide investigators, ensuring that every step taken, from initial threat detection to the final analysis, is meticulously documented. This rigorous approach not only facilitates a systematic investigation but also enhances the accountability of the process, ensuring that each action can be audited and scrutinized for its efficacy and compliance with established standards.

The implementation of digital chain-of-custody procedures emerges as a cornerstone in the preservation of evidence integrity. By meticulously tracking the movement and handling of digital evidence from collection to analysis, organizations can safeguard against unauthorized alterations or tampering. These procedures, akin to their physical counterparts, provide a transparent and unbroken lineage of evidence, ensuring that the integrity of the data remains intact throughout the investigative process, thereby bolstering the credibility of the findings.

Utilizing tamper-evident mechanisms plays a pivotal role in ensuring record integrity, serving as a deterrent against unauthorized access or modification of evidence. These mechanisms, from cryptographic hashes to sealable evidence bags for physical components, act as guardians of evidence integrity. By making any attempt at tampering evident, these safeguards reinforce the trustworthiness of the investigative records, thereby enhancing the legal and operational standing of the investigation's outcomes.

The deployment of time-stamping and digital signatures for provenance verification further solidifies the authenticity of investigative records. This technological safeguard ensures that each piece of evidence and corresponding documentation can be traced back to its origin, with the exact time of creation or modification verifiable. Such measures not only fortify the evidence against challenges to its authenticity but also streamline the process of establishing timelines during forensic analysis, an essential aspect of understanding and mitigating cybersecurity incidents.

Training on proper documentation practices and evidence preservation is imperative for maintaining the integrity of the investigative process. Organizations can significantly enhance the reliability and credibility of their investigations by equipping cybersecurity professionals with the knowledge and skills to document their actions and safeguard evidence effectively. This training should cover a broad spectrum, from the basics of secure documentation to advanced techniques in digital evidence management, ensuring that all personnel are proficient in the principles underpinning investigative records' integrity and provenance.

Regular audits of record-keeping systems and compliance processes ensure ongoing adherence to best practices and legal standards. When performed systematically, these audits serve as a health check for the organization's documentation and evidence preservation practices, identifying potential gaps or areas for improvement. By ensuring that record-keeping practices are current and compliant, organizations can mitigate the risk of evidentiary challenges that could undermine the outcomes of cybersecurity investigations.

Selecting secure storage solutions for protecting investigation records is not a decision to be taken lightly. The confidentiality, integrity, and availability of these records hinge upon the storage solutions' robustness. From encrypted digital repositories to physically secure storage facilities, the choice of storage solutions must align with the sensitivity and significance of the information being safeguarded, ensuring that investigation records are protected against unauthorized access, alteration, or destruction.

Integrating documentation practices with legal and regulatory requirements is a nuanced yet nonnegotiable aspect of cybersecurity investigations. This integration ensures that the methods and processes used for documentation and evidence preservation meet and exceed the standards set forth by relevant legal and regulatory bodies. By aligning investigative practices with these requirements, organizations can fortify their legal standing, ensuring that the evidence gathered and the records maintained are admissible and robust in the face of legal scrutiny.

Recommendations

- **Establish a Comprehensive Documentation Policy**: Create and enforce a comprehensive policy that outlines the standards and procedures for documentation and evidence preservation during cybersecurity investigations. This policy should serve as a cornerstone, guiding investigative efforts and ensuring consistency across the organization.
- **Leverage Advanced Digital Forensics Tools**: Invest in and utilize advanced digital forensics tools with built-in chain-of-custody features and tamper-evident capabilities. Such tools can significantly streamline the documentation process, enhancing the integrity and provenance of digital evidence.
- **Conduct Regular Training and Awareness Programs**: Implement ongoing training and awareness programs focused on the latest best practices in documentation and evidence preservation. Ensuring that all personnel involved in cybersecurity investigations are well-informed and up-to-date on these practices is crucial for maintaining the integrity of the investigative process.
- **Perform Periodic Audit and Review of Investigation Records**: Schedule and conduct periodic audits of investigation records and evidence storage facilities to ensure compliance with internal policies and external regulations. These audits can help identify areas for improvement and ensure that documentation and preservation practices remain robust over time.
- **Develop Collaborative Procedures with Legal Teams**: Forge robust collaborative procedures between cybersecurity investigation teams and legal departments to ensure that documentation and evidence-handling practices are aligned with legal and regulatory requirements. This collaboration can provide invaluable guidance on legal considerations and enhance the admissibility of evidence in legal proceedings.

NIST 800-53 Controls

(Refer to our appendix for a comprehensive description of the NIST 800-53 Controls, essential for organizations adhering to strict regulatory standards.)

- AU-07,IR-04,IR-06

Simplified Security Controls (SSC)

This section outlines our SSC, tailored for environments beyond the scope of NIST compliance. These controls offer streamlined, effective security solutions suitable for any organization seeking to enhance its cybersecurity defenses with clear, actionable measures.

Control 1: Secure Incident Documentation Portal

Control Objective: To create a centralized, secure web portal for the recording and management of actions performed during cybersecurity investigations, ensuring the integrity and provenance of investigative records.

Implementation Steps:

1.1 **Portal Development**: Develop a secure web portal that allows for the encrypted entry, storage, and retrieval of incident investigation records. The portal should feature user authentication, role-based access controls, and encryption to protect sensitive information.

1.2 **Implement Documentation Protocols**: Establish and integrate protocols within the portal for the secure and accurate recording of investigative actions, including digital chain-of-custody logs and tamper-evident mechanisms.

1.3 **Train Personnel**: Conduct training sessions for all incident response team members on using the portal effectively, focusing on documentation practices, evidence preservation, and compliance with legal and regulatory requirements.

Expected Outcome: Establishing a secure incident documentation portal centralizes and secures the recording of investigative actions, enhancing the integrity and provenance of records. It facilitates easy access to and management of investigation documentation, streamlines the documentation process, and ensures compliance with legal and regulatory standards.

Control 2: Automated Evidence Preservation System

Control Objective: To automate the process of collecting, storing, and preserving digital evidence, ensuring that the integrity and provenance of the evidence are maintained throughout the investigation process.

Implementation Steps:

2.1 **System Selection and Implementation**: Select and implement an automated evidence preservation system that integrates with existing incident response tools and infrastructure. The system should support digital chain-of-custody procedures and include tamper-evident and time-stamping features.

2.2 **Integration with Forensic Tools**: Integrate the evidence preservation system with forensic analysis tools to automate the collection and preservation of digital evidence, ensuring a seamless flow from evidence collection to analysis.

2.3 **Continuous Training and Policy Update**: Provide ongoing training for the incident response team on using the automated evidence preservation system and regularly update policies to reflect advancements in evidence preservation technologies and practices.

Expected Outcome: Implementing an automated evidence preservation system streamlines the evidence collection and preservation process, reducing the risk of human error and ensuring that the integrity and provenance of digital evidence are maintained. This facilitates more efficient and effective forensic analyses and strengthens the incident investigation process.

Control 3: Comprehensive Audit and Compliance Review Program

Control Objective: To regularly audit record-keeping systems and investigation processes for compliance with established cybersecurity policies, legal standards, and regulatory requirements, thereby enhancing the integrity and accountability of investigative practices.

Implementation Steps:

3.1 **Develop Audit Framework**: Create a comprehensive audit framework encompassing all aspects of incident documentation and evidence handling, including secure recording, chain-of-custody procedures, and secure storage solutions.

3.2 **Schedule Regular Audits**: Establish a schedule for regular audits of the record-keeping systems and investigation processes. An independent internal or external team should conduct these audits to ensure objectivity.

3.3 **Implement Corrective Actions**: Based on audit findings, develop and implement corrective actions to address any identified deficiencies or noncompliance issues. This should also include feedback mechanisms to improve documentation practices and evidence preservation.

Expected Outcome: Establishing a comprehensive audit and compliance review program ensures regular scrutiny of incident documentation and evidence-handling practices, promoting continuous improvement. This program enhances the organization's compliance with legal and regulatory standards, thereby strengthening the integrity and reliability of investigative records and supporting the organization's legal and operational positions in the event of an incident.

RS.AN-07: Incident Data and Metadata Are Collected, and Their Integrity and Provenance Are Preserved

Collecting, preserving, and analyzing incident data and metadata are pivotal in the cybersecurity incident response process. The foundational step in this process involves a comprehensive strategy for extracting data and metadata from systems affected by a cybersecurity incident. This extraction encompasses the retrieval of information necessary for understanding the incident and ensures that the data's integrity is not compromised during the collection phase.

Strategies for comprehensive data and metadata extraction from affected systems are integral to a robust incident response plan. These strategies should be designed to capture a wide array of information, including time stamps, user activities, system changes, and network traffic data, without altering the original data. Implementing such strategies requires a combination of technical tools and procedural protocols, ensuring that every piece of relevant information is collected systematically and securely. This meticulous approach facilitates a deeper incident analysis, enabling responders to reconstruct events accurately and identify the attack vectors used.

Techniques for ensuring data integrity during collection and storage are paramount. As data moves from its original location to storage for analysis, its authenticity must remain unquestionable. Techniques such as write blockers, which prevent any alterations to the storage medium during the data acquisition phase, and immutable storage solutions, where data cannot be altered once written, are crucial. These techniques protect the data against accidental and malicious alterations, thus preserving its integrity for forensic analysis.

The application of secure transfer protocols maintains data provenance by ensuring data is transmitted securely from its source to the analysis environment. Protocols such as Secure Shell (SSH) for file transfers and Transport Layer Security (TLS) for data in transit provide encryption that safeguards the data from interception and tampering. These secure pathways are essential for maintaining the chain of custody, a critical component in the forensic analysis that contributes to the credibility of the investigation findings.

Utilizing cryptographic hashes for data integrity verification is a technical safeguard that provides a fingerprint of the data at the time of collection. This fingerprint, or hash value, can be used to verify that the data has not been altered from the time it was collected, regardless of where it is stored or how it is transferred. Hashing algorithms like SHA-256 are widely recognized for their robustness, ensuring that even the smallest change to the dataset results in a significantly different hash value, thereby signaling potential tampering.

Given the legal and ethical implications, the development of policies for handling sensitive or personal data within the collected incident data is a critical consideration. These policies should outline clear procedures for anonymizing or pseudonymizing personal data to protect privacy, compliance with legal frameworks such as the General Data Protection Regulation (GDPR), and secure handling practices that prevent unauthorized access to sensitive information. Establishing these policies safeguards the organization from legal repercussions and builds trust with stakeholders by demonstrating a commitment to ethical data-handling practices.

Integration of data collection efforts with forensic analysis tools streamlines the investigation process. Forensic tools that can automatically process collected data and metadata to identify patterns, anomalies, and potential indicators of compromise are invaluable. This integration allows for a more efficient analysis, reducing the time to understand the scope and impact of an incident. Moreover, it ensures that the data collected is immediately ready for analysis, minimizing the window of opportunity for attackers to cause further damage.

Establishing access controls for data and metadata repositories is essential to ensure that only authorized individuals can access the information. This measure prevents unauthorized access, alterations, or deletions of critical data, further preserving its integrity and provenance. Access controls and audit logs that record every access or modification attempt create a secure environment for storing sensitive incident-related data.

Recommendations

- **Implement Advanced Data Collection Tools**: Utilize advanced tools and technologies for data and metadata collection that ensure integrity and reduce the risk of alteration during extraction. These tools should support automated, secure, and comprehensive data capture.
- **Adopt Rigorous Data Handling Protocols**: Establish and enforce strict data handling protocols that outline the steps for securing data throughout its lifecycle, from collection to analysis to storage. These protocols should include encryption, secure transfer methods, and access controls.
- **Ensure Compliance with Legal Standards**: Develop and maintain data collection and handling policies that strictly comply with legal and regulatory standards, particularly for sensitive and personal information. Regularly update these policies to reflect changes in laws and regulations.
- **Foster Continuous Training and Awareness**: Conduct regular training sessions for incident response teams and other relevant personnel on the latest data integrity and provenance preservation techniques. Promote an organizational culture that values secure and responsible data handling.
- **Integrate Forensic Analysis Tools with Data Repositories**: Seamlessly integrate forensic analysis tools with data repositories to facilitate immediate and secure analysis of collected data. This integration should support quickly identifying security breaches and enable timely remediation efforts.

NIST 800-53 Controls

(Refer to our appendix for a comprehensive description of the NIST 800-53 Controls, essential for organizations adhering to strict regulatory standards.)

- AU-07,IR-04,IR-06

Simplified Security Controls (SSC)

This section outlines our SSC, tailored for environments beyond the scope of NIST compliance. These controls offer streamlined, effective security solutions suitable for any organization seeking to enhance its cybersecurity defenses with clear, actionable measures.

Control 1: Data Integrity Assurance Protocol

Control Objective: To ensure the integrity and provenance of incident data and metadata from collection to analysis and storage, employing cryptographic hashes and secure transfer protocols.

Implementation Steps:

1.1 **Implementation of Cryptographic Hash Functions**: Immediately upon collection, apply cryptographic hash functions (e.g., SHA-256) to data and metadata to create a unique digital fingerprint, ensuring any unauthorized alterations can be detected.

1.2 **Adoption of Secure Transfer Protocols**: Utilize secure transfer protocols (e.g., SFTP and HTTPS) for transmitting data and metadata between collection points and storage or analysis destinations, preserving data integrity and provenance.

1.3 **Periodic Integrity Checks**: Schedule regular integrity checks on stored data and metadata using the initial cryptographic hashes to verify that data has not been altered post-collection.

Expected Outcome: The Data Integrity Assurance Protocol ensures that all incident-related data and metadata are accurately collected, securely transferred, and remain unaltered during storage, maintaining the integrity and provenance of critical forensic evidence throughout the incident response process.

Control 2: Sensitive Data Handling Framework

Control Objective: To establish comprehensive policies and procedures for the handling, storing, and transferring of sensitive or personal data collected during incident response, ensuring compliance with data protection regulations.

Implementation Steps:

2.1 **Policy Development**: Develop and document policies outlining the procedures for handling sensitive data, including collection, storage, access, transfer, and deletion, ensuring alignment with legal and regulatory requirements.

2.2 **Data Classification and Access Control**: Implement data classification schemes to identify sensitive or personal data and establish strict access controls based on the principle of least privilege, ensuring only authorized personnel can access such data.

2.3 **Training and Awareness**: Conduct regular training sessions for all team members involved in incident response on the importance of sensitive data protection, the details of the handling framework, and the legal consequences of noncompliance.

Expected Outcome: The establishment of a Sensitive Data Handling Framework ensures that all personal and sensitive data collected during incident response are handled in a manner that is secure, compliant with legal standards, and respectful of privacy, thereby reducing the risk of data breaches and enhancing trust in the organization's incident response process.

Control 3: Forensic Readiness and Data Collection Integration

Control Objective: To integrate incident data and metadata collection efforts seamlessly with forensic analysis tools, ensuring efficient and effective analysis while maintaining data integrity and provenance.

Implementation Steps:

3.1 **Integration of Collection and Analysis Tools**: Select and deploy forensic analysis tools compatible with data collection tools, ensuring seamless integration for automatic data ingestion and immediate analysis capability.

3.2 **Establish Standard Operating Procedures (SOPs) for Tool Use**: Develop SOPs for integrated tools, detailing processes for data capture, transfer, analysis, and storage, ensuring consistency and reliability in forensic investigations.

3.3 **Continuous Improvement and Update Process**: Implement a continuous improvement process for the integration framework, incorporating feedback from forensic analysts and incident response teams to optimize data collection and analysis effectiveness and efficiency.

Expected Outcome: The Forensic Readiness and Data Collection Integration control ensures that incident data and metadata are efficiently and effectively analyzed, with a seamless flow from collection to forensic analysis. This integration enhances the speed and accuracy of incident response, enables quicker root cause identification, and strengthens the organization's overall cybersecurity posture.

RS.AN-08: An Incident's Magnitude Is Estimated and Validated

Collecting, preserving, and analyzing incident data and metadata are pivotal in the cybersecurity incident response process. The foundational step in this process involves a comprehensive strategy for extracting data and metadata from systems affected by a cybersecurity incident. This extraction encompasses the retrieval of information necessary for understanding the incident and ensures that the data's integrity is not compromised during the collection phase.

Strategies for comprehensive data and metadata extraction from affected systems are integral to a robust incident response plan. These strategies should be designed to capture a wide array of information, including time stamps, user activities, system changes, and network traffic data, without altering the original data. Implementing such strategies requires a combination of technical tools and procedural protocols, ensuring that every piece of relevant information is collected systematically and securely. This meticulous approach facilitates a deeper incident analysis, enabling responders to reconstruct events accurately and identify the attack vectors used.

Techniques for ensuring data integrity during collection and storage are paramount. As data moves from its original location to storage for analysis, its authenticity must remain unquestionable. Techniques such as write blockers, which prevent any alterations to the storage medium during the data acquisition phase, and immutable storage solutions, where data cannot be altered once written, are crucial. These techniques protect the data against inadvertent and malicious alterations, thus preserving its integrity for forensic analysis.

The application of secure transfer protocols maintains data provenance by ensuring data is transmitted securely from its source to the analysis environment. Protocols such as SSH for file transfers and TLS for data in transit provide encryption that safeguards the data from interception and tampering. These secure pathways are essential for maintaining the chain of custody, a critical component in the forensic analysis that contributes to the credibility of the investigation findings.

Utilizing cryptographic hashes for data integrity verification is a technical safeguard that provides a fingerprint of the data at the time of collection. This fingerprint, or hash value, can be used to verify that the data has not been altered from the time it was collected, regardless of where it is stored or how it is transferred. Hashing algorithms like SHA-256 are widely recognized for their robustness, ensuring that even the smallest change to the dataset results in a significantly different hash value, thereby signaling potential tampering.

Given the legal and ethical implications, the development of policies for handling sensitive or personal data within the collected incident data is a critical consideration. These policies should outline clear procedures for anonymizing or pseudonymizing personal data to protect privacy, compliance with legal frameworks such as the GDPR, and secure handling practices that prevent unauthorized access to sensitive information. Establishing these policies safeguards the organization from legal repercussions and builds trust with stakeholders by demonstrating a commitment to ethical data-handling practices.

Integration of data collection efforts with forensic analysis tools streamlines the investigation process. Forensic tools that can automatically process collected data and metadata to identify patterns, anomalies, and potential indicators of compromise are invaluable. This integration allows for a more efficient analysis, reducing the time to understand the scope and impact of an incident. Moreover, it ensures that the data collected is immediately ready for analysis, minimizing the window of opportunity for attackers to cause further damage.

Establishing access controls for data and metadata repositories is essential to ensure that only authorized individuals can access the information. This measure prevents unauthorized access, alterations, or deletions of critical data, further preserving its integrity and provenance. Access controls and audit logs that record every access or modification attempt create a secure environment for storing sensitive incident-related data.

Recommendations

- **Implement Advanced Data Collection Tools**: Utilize advanced tools and technologies for data and metadata collection that ensure integrity and reduce the risk of alteration during extraction. These tools should support automated, secure, and comprehensive data capture.
- **Adopt Rigorous Data Handling Protocols**: Establish and enforce strict data handling protocols that outline the steps for securing data throughout its lifecycle, from collection to analysis to storage. These protocols should include encryption, secure transfer methods, and access controls.
- **Ensure Compliance with Legal Standards**: Develop and maintain data collection and handling policies that strictly comply with legal and regulatory standards, particularly for sensitive and personal information. Regularly update these policies to reflect changes in laws and regulations.
- **Foster Continuous Training and Awareness**: Conduct regular training sessions for incident response teams and other relevant personnel on the latest data integrity and provenance preservation techniques. Promote an organizational culture that values secure and responsible data handling.
- **Integrate Forensic Analysis Tools with Data Repositories**: Seamlessly integrate forensic analysis tools with data repositories to facilitate immediate and secure analysis of collected data. This integration should support quickly identifying security breaches and enable timely remediation efforts.

NIST 800-53 Controls

(Refer to our appendix for a comprehensive description of the NIST 800-53 Controls, essential for organizations adhering to strict regulatory standards.)

- IR-04,IR-08,RA-03,RA-07

Simplified Security Controls (SSC)

This section outlines our SSC, tailored for environments beyond the scope of NIST compliance. These controls offer streamlined, effective security solutions suitable for any organization seeking to enhance its cybersecurity defenses with clear, actionable measures.

Control 1: Incident Magnitude Assessment Tool Implementation

Control Objective: To accurately determine the magnitude of cybersecurity incidents by implementing tools that estimate the scope and impact on organizational operations and assets.

Implementation Steps:

1.1 **Select and Deploy Assessment Tools**: Choose and implement assessment tools that can analyze incident data to estimate the scope and impact, including factors like affected systems, data breaches, and operational disruptions.

1.2 **Integrate Assessment Tools with Incident Response**: Ensure the chosen assessment tools are fully integrated into the incident response process, allowing real-time magnitude estimation as incidents unfold.

1.3 **Continuous Tool Improvement and Training**: Regularly update the assessment tools to reflect evolving cyber threats and conduct ongoing training for the incident response team on their effective use.

Expected Outcome: The deployment of specialized assessment tools will enable the organization to quickly and accurately determine the magnitude of cybersecurity incidents. This will inform the prioritization of response efforts and resource allocation, leading to more effective incident management and recovery.

Control 2: Cross-Functional Incident Impact Analysis Team

Control Objective: To establish a dedicated team that assesses the immediate and potential long-term impact of cybersecurity incidents across various business functions, aiding in accurately estimating incident magnitude.

Implementation Steps:

2.1 **Team Formation**: Form a cross-functional team with representatives from IT, cybersecurity, business operations, legal, and HR to assess incident impact from multiple perspectives.

2.2 **Develop Impact Assessment Protocols**: Create protocols for the team to follow when analyzing incidents, focusing on qualitative and quantitative measures to estimate impact accurately.

2.3 **Regular Impact Assessment Drills**: Conduct regular drills to ensure the team's readiness and ability to quickly assess the impact of potential incidents, refining protocols as needed.

Expected Outcome: Creating a cross-functional incident impact analysis team will facilitate a comprehensive understanding of an incident's magnitude, considering both technical and business impacts. This approach ensures a well-rounded response strategy that minimizes organizational risk and operational disruptions.

Control 3: Dynamic Incident Response Adjustment Protocol

Control Objective: To dynamically adjust incident response strategies based on real-time assessments of an incident's magnitude, optimizing the response to the specific scope and impact identified.

Implementation Steps:

3.1 **Establish Real-Time Monitoring and Assessment**: Implement real-time monitoring and assessment mechanisms to continuously evaluate the magnitude of an incident as new information becomes available.

3.2 **Develop Flexible Response Plans**: Create flexible incident response plans that can be adapted based on the evolving understanding of an incident's magnitude and impact.

3.3 **Train Response Teams in Dynamic Adjustment**: Train incident response teams in protocols for dynamically adjusting their strategies, emphasizing quick decision-making based on incident magnitude assessments.

Expected Outcome: Implementing a dynamic incident response adjustment protocol allows the organization to respond more effectively to cybersecurity incidents by tailoring its approach to each incident's specific characteristics and impacts. This leads to more efficient use of resources and better mitigation of potential damages.

23

Incident Response Reporting and Communication (RS.CO)

> Seamless communication in the aftermath of a cybersecurity incident isn't just a best practice—it's the linchpin of trust and effectiveness, bridging the gap between swift recovery and sustained resilience, ensuring that every message, every update, and every shared piece of information weaves a stronger fabric of collective defense.

Ensuring timely and efficient communication with internal and external stakeholders following an incident is crucial in a comprehensive cybersecurity strategy. This involves establishing clear notification procedures, identifying key stakeholders, and determining the priority and method of communication. Crafting messages that are both clear and concise minimizes misunderstandings and ensures that all parties are adequately informed. Additionally, aligning with legal and regulatory requirements is essential to maintain compliance and trust. Automated notification systems and thorough staff training support these goals by promoting consistency and understanding.

Similarly, effective information sharing is integral to a robust response strategy. This includes defining what information should be shared and with whom, employing secure mechanisms for information exchange, and balancing the need for transparency with the imperative of confidentiality. Coordination with external partners enhances the breadth and effectiveness of the response, while ongoing training ensures that staff are equipped to handle information responsibly. Ultimately, evaluating the impact of these practices on the response outcomes and continuously seeking improvement based on stakeholder feedback are vital steps for refining the incident response process and strengthening cybersecurity measures.

Author's Note: RS.CO-01 was deprecated from the CSF 2 and is not included in this chapter.

RS.CO-02: Internal and External Stakeholders Are Notified of Incidents

Understanding the importance of notifying stakeholders during a cybersecurity incident is critical for maintaining trust and transparency. When an incident occurs, it is not only the IT department that is affected; the repercussions ripple across the entire organization and beyond. This section delves into the nuanced process of stakeholder notification, which begins with identifying who needs to know what and when. This approach ensures that relevant parties are informed promptly and appropriately, thereby facilitating a coordinated response effort.

A Comprehensive Guide to the NIST Cybersecurity Framework 2.0: Strategies, Implementation, and Best Practice, First Edition. Jason Edwards.
© 2025 John Wiley & Sons Ltd. Published 2025 by John Wiley & Sons Ltd.

Identifying and prioritizing stakeholders for incident communication is a foundational step in the notification process. Different stakeholders may have other information needs and stakes in the incident. For example, internal stakeholders like employees might need immediate safety instructions, while external stakeholders like partners or customers may require assurance and information about how the incident affects them. Prioritizing stakeholders according to their needs, the nature of their relationship with the organization, and the potential impact on them helps streamline the communication process and ensures that critical information reaches the most affected parties first.

Once stakeholders are identified and prioritized, the next step involves the development of notification protocols. This includes establishing clear timelines for when different stakeholders should be notified and determining the most effective channels for communication. For example, immediate internal notifications may occur through direct messages or internal alert systems, while external communications may require more formal channels such as emails or public statements. The protocols should also consider the urgency and sensitivity of the information, ensuring that communications are prompt and secure.

Crafting clear and concise notification messages is vital to avoid misunderstandings and panic. In the heat of a cybersecurity incident, clear communication can reduce confusion and help stakeholders understand their role in the response effort. Messages should be straightforward, avoiding technical jargon that all recipients may not understand. It is also important to provide enough detail to inform stakeholders without overwhelming them or disclosing sensitive information that could exacerbate the situation.

Ensuring compliance with legal and regulatory requirements for notifications is non-negotiable. Many jurisdictions have specific laws governing the disclosure of security incidents, mainly if personal data is involved. Failure to comply with these regulations can result in significant penalties and damage to an organization's reputation. Therefore, understanding and incorporating these requirements into the notification process is crucial.

Automated notification systems can significantly enhance communication efficiency and consistency with stakeholders. Automation ensures that messages are dispatched quickly and uniformly, reducing the chance of human error and ensuring that all affected parties receive the necessary information. These systems can be set up to trigger alerts based on specific criteria, ensuring that the notification process begins as soon as an incident is detected.

Staff training on the notification process and best communication practices is essential for a successful incident response. All employees should understand their roles and responsibilities in the event of an incident, including who to notify and how to communicate effectively. This training should cover the importance of timely reporting, the use of predefined templates or scripts, and the need for sensitivity and discretion in communications.

Regularly updating stakeholders with incident progress and resolutions helps maintain trust and demonstrates transparency. Stakeholders must not be left in the dark after the initial notification. Regular updates can provide reassurance, reduce the spread of rumors, and ensure stakeholders are informed of the steps to resolve the incident and prevent future occurrences.

A post-incident review of the notification process's effectiveness is crucial for continuous improvement. This review should evaluate the timeliness, accuracy, and effectiveness of the communication efforts and identify areas for improvement. Feedback from stakeholders can provide valuable insights into how the process can be refined to meet their needs in future incidents better.

Recommendations

- **Establish a Comprehensive Incident Communication Plan**: Develop a detailed plan that outlines how to notify stakeholders during a cybersecurity incident. This plan should include identifying and prioritizing stakeholders, communication channels, message templates, and compliance requirements.
- **Implement an Automated Notification System**: Utilize technology to streamline the notification process. An automated system can ensure that messages are sent promptly and consistently, reducing the likelihood of human error.
- **Conduct Regular Training and Simulations**: Regular training sessions and simulated incident response exercises can help prepare staff for real-life scenarios. This ensures that everyone understands their roles and responsibilities in the notification process.
- **Maintain Transparency and Regular Updates**: Keep stakeholders informed with regular updates throughout the incident response process. Transparency builds trust and can mitigate the negative impacts of an incident.
- **Review and Refine After Action**: Conduct a thorough review of the notification process following an incident. Gather feedback from stakeholders and use this information to improve future responses. Continuous improvement is key to effective incident communication.

NIST 800-53 Controls

(Refer to our appendix for a comprehensive description of the NIST 800-53 Controls, essential for organizations adhering to strict regulatory standards.)

- IR-04,IR-06,IR-07,SR-03,SR-08

Simplified Security Controls (SSC)

This section outlines our Simplified Security Controls (SSC), tailored for environments beyond the scope of NIST compliance. These controls offer streamlined, effective security solutions suitable for any organization seeking to enhance its cybersecurity defenses with clear, actionable measures.

Control 1: Development of a Comprehensive Stakeholder Notification System

Control Objective: To establish a systematic approach for notifying internal and external stakeholders about cybersecurity incidents, ensuring timely, clear, and compliant communication.

Implementation Steps:

1.1 **Define Stakeholder Groups**: Identify and categorize internal and external stakeholders based on their roles, needs, and levels of impact regarding cybersecurity incidents.

1.2 **Create Notification Protocols**: Develop clear protocols outlining the timelines, channels, and formats for communicating with different stakeholder groups.

1.3 **Implement Legal Compliance Checks**: Incorporate steps to ensure all notifications meet legal and regulatory standards, minimizing the organization's risk exposure.

Expected Outcome: Establishing a comprehensive stakeholder notification system enhances the organization's incident response capabilities. Timely and effective communication mitigates the impact of cybersecurity incidents, maintains trust, and ensures legal compliance.

Control 2: Integration of Automated Notification Mechanisms

Control Objective: To leverage technology to automate the notification process and ensure efficiency, consistency, and scalability in stakeholder communication during cybersecurity incidents.

Implementation Steps:

2.1 **Select and Configure Automation Tools**: Choose suitable automated notification systems and configure them according to the organization's communication protocols and stakeholder groups.

2.2 **Integrate Systems with Incident Response**: Link the notification system with the incident response platform to trigger alerts based on predefined criteria.

2.3 **Conduct Regular Testing and Updates**: Regularly test and update the automated notification system based on evolving communication needs and technological advancements.

Expected Outcome: Integrating automated notification mechanisms streamlines the communication process during incidents, ensuring all stakeholders receive timely and consistent information. This reduces human error and frees up resources for other critical incident response activities.

Control 3: Comprehensive Training Program for Incident Communication

Control Objective: To equip all relevant personnel with the knowledge and skills necessary for effective communication during and after cybersecurity incidents, enhancing the overall response strategy.

Implementation Steps:

3.1 **Develop a Training Curriculum**: Create a comprehensive training program covering incident communication protocols, legal requirements, and best practices.

3.2 **Conduct Regular Training Sessions**: Hold training sessions for employees at all levels, ensuring they understand their roles and responsibilities in the communication process.

3.3 **Evaluate and Refine Training**: Assess the effectiveness of the training program and make necessary adjustments based on feedback and changing requirements.

Expected Outcome: Implementing a comprehensive training program ensures that all personnel are prepared to communicate effectively during cybersecurity incidents. This enhances the organization's incident response efforts and strengthens stakeholder relationships.

RS.CO-03: Information Is Shared with Designated Internal and External Stakeholders

Effective information sharing during and after a cybersecurity incident is critical to the overall response strategy. This section explores the essential practices and considerations for sharing information with internal and external stakeholders. Information sharing is not merely about distributing data; it is about fostering a collaborative environment where stakeholders can contribute to and benefit from shared intelligence. This collaboration can lead to more effective detection, prevention, and response efforts, ultimately reducing the impact of cybersecurity incidents on the organization.

Establishing criteria for what information to share and with whom is the first crucial step in the information-sharing process. Not all information is suitable for all audiences; therefore, it is essential to categorize information based on sensitivity, relevance, and the recipient's role. This categorization helps ensure stakeholders receive pertinent and actionable information without compromising sensitive data or overburdening them with irrelevant details. It also aids in maintaining legal and regulatory compliance, especially when dealing with classified or personal data.

Mechanisms for secure and timely information exchange are fundamental to effective information sharing. Organizations should employ secure communication channels that ensure shared information's confidentiality, integrity, and availability. This could include encrypted emails, secure file transfer protocols, or dedicated information-sharing platforms. Moreover, timeliness is critical; stakeholders should receive information quickly to take necessary actions or make informed decisions.

Balancing transparency with confidentiality in stakeholder communications is a delicate but necessary act. While transparency fosters trust and cooperation, confidentiality protects sensitive information from unauthorized access or disclosure. Organizations must navigate this balance by determining what information can be shared openly and what should be restricted. This balance requires clear policies and guidelines, ensuring that all parties understand their roles and responsibilities in maintaining the confidentiality of sensitive information.

Integration of information sharing with incident management and response processes ensures that information flow is central to the incident response strategy. This integration allows for the immediate and appropriate dissemination of information, enhancing the organization's ability to respond to and recover from incidents. It also facilitates a coordinated response, as stakeholders are informed and engaged in the process, contributing to the overall effectiveness of the response effort.

Coordination with external agencies and industry groups for information sharing can extend the reach and effectiveness of an organization's incident response. Organizations can access a broader range of threat intelligence, best practices, and support resources by collaborating with external entities. This collaboration can be particularly beneficial for addressing complex or large-scale incidents that exceed the capabilities of a single organization.

Documentation of shared information and feedback received is crucial for accountability and continuous improvement. Keeping a record of what information was shared, with whom, and when helps track the flow of information and assess its impact. Additionally, gathering feedback from stakeholders on the usefulness and relevance of the information can provide insights into how to improve future information-sharing practices.

Training on effective and responsible information-sharing techniques is essential for ensuring that all relevant personnel understand how to share information appropriately and securely. This training should cover legal and regulatory requirements, organizational policies, communication protocols, and the ethical considerations of information sharing. Empowering employees with this knowledge ensures that information is shared responsibly and effectively, maximizing its benefits while minimizing risks.

Recommendations

- **Develop and Implement Information Sharing Policies**: Create comprehensive policies that define what information should be shared, with whom, and through which channels. These policies should balance the need for transparency with the need to protect sensitive information and should comply with legal and regulatory requirements.
- **Establish Secure and Efficient Communication Channels**: Implement secure mechanisms for information sharing that ensure the information's confidentiality, integrity, and availability. Regularly review and update these mechanisms to adapt to evolving security threats and technological advancements.
- **Foster Relationships with External Agencies and Industry Groups**: Build and maintain relationships with relevant external entities to enhance collaborative information sharing. Participation in industry forums, working groups, and information-sharing and analysis centers can provide valuable insights and resources.

- **Provide Regular Training and Awareness Programs**: Conduct regular training sessions for all employees on the importance of information sharing and the proper procedures. This training should emphasize the legal, ethical, and operational aspects of sharing information during and after cybersecurity incidents.
- **Evaluate and Refine Information-Sharing Practices**: Regularly assess the effectiveness of information-sharing practices and make improvements based on feedback from stakeholders and changes in the threat landscape. Continuous evaluation and adaptation are key to maintaining an effective information-sharing program.

NIST 800-53 Controls

(Refer to our appendix for a comprehensive description of the NIST 800-53 Controls, essential for organizations adhering to strict regulatory standards.)

- IR-04,IR-06,IR-07,SR-03,SR-08

Simplified Security Controls (SSC)

This section outlines our SSC, tailored for environments beyond the scope of NIST compliance. These controls offer streamlined, effective security solutions suitable for any organization seeking to enhance its cybersecurity defenses with clear, actionable measures.

Control 1: Secure Information-Sharing Platform Implementation

Control Objective: To establish a secure, efficient platform for exchanging information with both internal and external stakeholders, ensuring confidentiality, integrity, and availability of data during and after cybersecurity incidents.

Implementation Steps:

1.1 **Platform Selection**: Evaluate and select an information-sharing platform that meets security requirements, including data encryption and access control features.

1.2 **Integration with Existing Systems**: Ensure the platform integrates seamlessly with current incident management and response systems for streamlined operations.

1.3 **Stakeholder Access Management**: Implement role-based access controls to manage who has access to what information based on the sensitivity of data and the stakeholder's role.

Expected Outcome: Implementing a secure information-sharing platform facilitates a timely and protected exchange of information. This system supports a collaborative response to incidents, enhancing overall security posture and stakeholder trust.

Control 2: Development of Information-Sharing Policies and Procedures

Control Objective: To create comprehensive policies and procedures that guide information sharing, balancing the need for transparency and confidentiality.

Implementation Steps:

2.1 **Policy Formulation**: Draft clear information-sharing policies that outline what information can be shared, with whom, and through which channels, considering legal and regulatory obligations.

2.2 **Procedure Development**: Develop detailed procedures for information sharing, including steps for secure transmission, stakeholder notification, and documentation.

2.3 **Policy and Procedure Review**: Regularly review and update policies and procedures to adapt to new security threats, technological changes, and stakeholder feedback.

Expected Outcome: Developing clear, comprehensive information-sharing policies and procedures ensures that all personnel understand how to share information responsibly. This enhances the organization's ability to respond to incidents effectively while maintaining stakeholder trust and regulatory compliance.

Control 3: Stakeholder Information Sharing Training and Awareness Program

Control Objective: To equip all relevant personnel with the knowledge and skills to share information securely and responsibly, in line with organizational policies and ethical considerations.

Implementation Steps:

3.1 **Training Program Development**: Create a training program that covers legal and regulatory requirements, organizational information-sharing policies, and secure communication protocols.

3.2 **Mandatory Training Sessions**: Conduct mandatory training sessions for all employees involved in incident response and information-sharing processes.

3.3 **Continuous Learning and Feedback Loop**: Establish a feedback mechanism to continuously assess the effectiveness of the training program and make necessary adjustments based on emerging trends and stakeholder feedback.

Expected Outcome: A comprehensive training and awareness program enhances the organization's secure and effective information-sharing capacity. This prepares personnel to handle information responsibly, supporting a coordinated incident response effort and fostering a culture of security awareness.

24

Incident Mitigation (RS.MI)

> In the ballet of cybersecurity incident management, containment and eradication perform as critical movements—where swift, decisive actions isolate danger and purge threats, each step choreographed with precision and insight, ensuring the dance of defense evolves stronger with every performance.

In the critical stages of incident management, containment and eradication are essential. For containment, strategies are designed to quickly limit the spread of an incident within the network, utilizing methods such as network segmentation, the deployment of endpoint detection and response (EDR) tools, and stringent communication controls. These measures, complemented by the coordinated efforts of IT teams for emergency responses and staff training on their specific roles, lay the groundwork for effective incident management. Following containment, the focus shifts to the eradication phase, where the objective is to remove the threat from the system altogether. This involves identifying and eliminating malicious payloads, conducting root cause analyses to prevent future occurrences, and systematically restoring affected systems to secure states. Both phases emphasize the importance of detailed documentation and post-incident analysis to refine protocols and enhance future response capabilities, ensuring a robust defense against cyber threats.

RS.MI-01: Incidents Are Contained

Effective incident containment begins with the development and understanding of comprehensive containment strategies. These are not just reactive measures but proactive plans that are in place before an incident occurs. They outline the steps to be taken immediately after detecting a breach or threat, focusing on limiting its spread and impact. The strategies should be well-communicated to all relevant stakeholders and part of regular training and simulation exercises to ensure readiness.

Immediate actions are crucial in the wake of a security incident. They include cutting off access rights for compromised accounts, isolating affected networks, and securing critical data. These steps must be executed swiftly to prevent the further spread of the incident within the network. Every second counts, and delays can lead to significant data losses or further system compromises. The initial response sets the tone for the subsequent recovery process and can significantly reduce the incident's overall impact.

Deploying network segmentation and isolation techniques is an effective method to contain cybersecurity incidents. Organizations can limit the spread of malicious activities by dividing the network into smaller, manageable segments. If an attacker compromises one segment, the

A Comprehensive Guide to the NIST Cybersecurity Framework 2.0: Strategies, Implementation, and Best Practice, First Edition. Jason Edwards.
© 2025 John Wiley & Sons Ltd. Published 2025 by John Wiley & Sons Ltd.

others remain protected, reducing the overall risk. This approach requires careful planning and implementation, as overly restrictive measures can impede normal business operations. Therefore, balance is key, ensuring security without hindering productivity.

The use of EDR tools is instrumental in the quick containment of incidents. EDR solutions provide real-time monitoring and automatic responses to threats, identifying suspicious activities and isolating affected systems from the network. This immediate response can drastically reduce the time attackers have inside the network, limiting the damage they can inflict. The choice of EDR tools should be aligned with the organization's specific needs and the nature of its IT environment.

Implementation of communication controls is vital to prevent data exfiltration during an incident. These controls include restricting email attachments, disabling file transfers, and monitoring data flows. Ensuring that sensitive information does not leave the organization during a breach is crucial for maintaining trust and compliance. Effective communication controls require a balance between security and operational needs, allowing necessary business processes to continue while protecting critical data.

Coordination with IT teams for emergency patching and configuration changes is crucial in the containment process. Swift action to close security gaps can significantly reduce the impact of the incident. This requires a well-coordinated effort between incident response teams and IT personnel, ensuring that patches and updates are applied promptly and correctly. Such collaboration is essential for maintaining the integrity and availability of IT systems during and after a security incident.

Documentation of containment actions and their effectiveness is an often overlooked but essential component of incident management. Detailed records help in understanding the incident's scope, the effectiveness of the response, and areas for improvement. This documentation is a valuable resource for post-incident analysis and refining future containment strategies. It also aids in regulatory compliance and in demonstrating due diligence to stakeholders.

Training staff on roles and responsibilities during containment efforts is paramount. Well-informed employees can act decisively and appropriately during an incident, reducing the time to containment and recovery. Regular training sessions, with drills and simulations, help ensure everyone knows their part in the containment process. Empowering employees with knowledge and clear guidelines can significantly enhance an organization's cybersecurity posture.

After the containment phase, conducting a post-incident analysis to refine containment protocols is crucial. This analysis should review the effectiveness of the containment strategies, the response times, and the overall handling of the incident. Insights gained from this review can lead to significant improvements in incident response procedures, making them more efficient and effective in the future.

Recommendations

- **Develop a Cross-Departmental Incident Response Team**: Form a dedicated team comprising members from various departments to ensure a coordinated approach to incident containment. This promotes the integration of diverse perspectives and expertise in the development and execution of containment strategies.
- **Establish Clear Communication Channels**: Ensure that there are predefined methods and protocols for internal and external communication during an incident. Effective communication is critical to the successful containment and resolution of cybersecurity events.

- **Conduct Incident Response Drills**: Regularly test and evaluate your incident response plans through drills and simulations. This helps identify weaknesses and areas for improvement, ensuring that your team is well-prepared for actual incidents.
- **Invest in Advanced Detection and Response Tools**: Allocate resources to acquire and integrate advanced EDR tools that align with your organization's needs. These tools are vital for the rapid detection and containment of cybersecurity threats.
- **Create a Continuous Improvement Plan**: Develop a structured approach for updating and refining your incident containment strategies based on lessons learned from past incidents and evolving cybersecurity trends. This should include regular reviews and updates to your incident response plan.

NIST 800-53 Controls

(Refer to our appendix for a comprehensive description of the NIST 800-53 Controls, essential for organizations adhering to strict regulatory standards.)

- IR-04

Simplified Security Controls (SSC)

This section outlines our Simplified Security Controls (SSC), tailored for environments beyond the scope of NIST compliance. These controls offer streamlined, effective security solutions suitable for any organization seeking to enhance its cybersecurity defenses with clear, actionable measures.

Control 1: Development of a Rapid Incident Containment Protocol

Control Objective: To establish a set of rapid response actions to immediately contain cybersecurity incidents, reducing the potential spread within the network and minimizing impact.

Implementation Steps:

1.1 **Protocol Creation**: Develop a comprehensive incident containment protocol that outlines immediate actions, such as cutting off network access for affected systems, deploying emergency patches, and isolating critical network segments.

1.2 **Staff Training**: Conduct regular training sessions for all relevant staff members on their roles and responsibilities during an incident, ensuring they understand and can execute the containment protocol swiftly.

1.3 **Drills and Simulations**: Regularly conduct simulated cyber-attack scenarios to test the effectiveness of the containment protocol and staff readiness, adjusting procedures based on outcomes and feedback.

Expected Outcome: Implementation of the rapid incident containment protocol will enable the organization to limit the spread of incidents within its network quickly. The regular training and drills will ensure staff are prepared and capable of executing the protocol effectively, leading to reduced impact and faster recovery times.

Control 2: Integration of Advanced Endpoint Detection and Response Solutions

Control Objective: To enhance the organization's ability to quickly detect, isolate, and mitigate cyber threats using advanced EDR tools.

Implementation Steps:

2.1 **Solution Selection**: Evaluate and select an advanced EDR solution that fits the organization's specific needs and infrastructure, considering factors such as compatibility, scalability, and feature set.

2.2 **Deployment and Configuration**: Deploy the chosen EDR solution across the organization's endpoints, configuring it according to best practices to ensure comprehensive monitoring and rapid response capabilities.

2.3 **Continuous Monitoring and Improvement**: Establish a continuous monitoring program to oversee the operation of the EDR tools, incorporating regular updates and adjustments based on evolving threats and technological advancements.

Expected Outcome: With advanced EDR tools, the organization will benefit from enhanced detection capabilities and quicker incident response times. This control will lead to a more resilient cybersecurity posture, reducing the likelihood of significant attack damage.

Control 3: Implementation of a Comprehensive Post-Incident Analysis and Documentation Process

Control Objective: To ensure thorough examination and documentation of cybersecurity incidents and containment efforts, fostering continuous improvement in security practices and protocols.

Implementation Steps:

3.1 **Documentation Policy**: Develop and implement a policy requiring detailed documentation of all incident response actions, outcomes, and lessons learned, ensuring consistency and completeness.

3.2 **Analysis Team Formation**: Form a dedicated team responsible for conducting post-incident analyses, identifying causes, evaluating response effectiveness, and recommending improvements.

3.3 **Review and Update Procedures**: Regularly review and update incident containment protocols based on insights gained from post-incident analyses, incorporating new threats and best practices.

Expected Outcome: Establishing a comprehensive post-incident analysis and documentation process will enable the organization to learn from each incident, leading to more robust security measures and effective containment strategies.

RS.MI-02: Incidents Are Eradicated

The process of incident eradication begins with a comprehensive understanding of the incident at hand. This includes identifying the scope, method, and extent of the breach. Eradication processes require a systematic approach to remove all traces of malicious activity from the system. This involves addressing the immediate symptoms and delving deeper to find the root cause of the problem. Understanding the nature of the attack allows for implementing targeted strategies to remove the threat completely and securely.

One of the primary steps in the eradication process is the identification and removal of malicious payloads from infected systems. This step requires detailed analysis and precision to identify and eliminate all malware components. The process involves scanning the systems with updated malware detection tools, removing detected malware, and re-scanning to ensure complete removal. It is crucial to carry out this process systematically to prevent any remnants of the malware from causing further damage.

Techniques for root cause analysis are essential in preventing the recurrence of an incident. This involves thoroughly investigating how the breach occurred, what vulnerabilities were exploited, and why existing defenses failed to prevent it. Understanding the root cause is critical to developing effective measures to prevent similar incidents in the future. This analysis should lead to concrete actions, such as patching vulnerabilities, changing security policies, or improving user education.

Restoring affected systems to a known good state is another vital step in eradicating them. This involves rolling back systems to a state before the compromise occurred, using backups, and reinstalling system components from scratch if necessary. This step ensures that systems are free from any malicious modifications made by the attacker and restores confidence in the integrity of the system and data.

Comprehensive system scans and malware removal tools are integral to the eradication process. These tools should be regularly updated to detect and remove the latest threats. Scans should be thorough and cover all parts of the system, including areas not typically affected by the specific type of malware encountered. This ensures that no part of the threat remains hidden within the system.

Revision and hardening of system configurations are necessary to eliminate vulnerabilities and strengthen the system against future attacks. This includes updating software, changing passwords, disabling unnecessary services, and applying security patches. Each action reduces the attack surface and the likelihood of future breaches. Organizations should adopt a policy of least privilege and ensure that systems are configured according to best security practices.

Verification of eradication is critical and should be performed through follow-up assessments and testing. This step ensures that the threat has been completely removed and that systems are secure. Verification activities can include repeat scans, penetration testing, and monitoring system logs for signs of malicious activity. This verification must be thorough to prevent a recurrence of the incident.

Coordination with security teams to update defensive measures is essential after eradication. Lessons learned from the incident should be used to improve security strategies and responses. This includes updating incident response plans, improving detection capabilities, and refining prevention techniques. Security teams should work together to strengthen all aspects of the organization's security posture in response to the incident.

Documentation of the eradication processes and lessons learned is crucial for improving future response efforts and for compliance purposes. This documentation should include detailed records of the incident, the eradication process, and the steps to prevent future incidents. It serves as a valuable resource for training, helps refine security measures, and provides evidence of compliance with regulations and standards.

Recommendations

- **Implement Regular System Backups**: Establish and maintain a robust backup strategy to ensure that systems can be restored to a known good state after an incident. Regular backups are critical for quickly recovering affected systems without losing critical data.
- **Enhance Malware Detection Capabilities**: Invest in advanced malware detection and removal tools. Regularly update these tools to combat the latest cyber threats and ensure thorough scans are conducted periodically to maintain system integrity.
- **Strengthen Vulnerability Management Programs**: Develop and maintain a comprehensive vulnerability management program. This should include regular vulnerability scanning, timely patching, and hardening system configurations to reduce the risk of future breaches.
- **Conduct Comprehensive Root Cause Analysis**: After an incident, conduct a thorough root cause analysis to understand how the breach occurred and why. Use this information to strengthen security measures and prevent similar incidents.

NIST 800-53 Controls

(Refer to our appendix for a comprehensive description of the NIST 800-53 Controls, essential for organizations adhering to strict regulatory standards.)

- IR-04

Simplified Security Controls (SSC)

This section outlines our SSC, tailored for environments beyond the scope of NIST compliance. These controls offer streamlined, effective security solutions suitable for any organization seeking to enhance its cybersecurity defenses with clear, actionable measures.

Control 1: Enhanced Malware Detection and Removal Protocols

Control Objective: To ensure the thorough and systematic identification and removal of malicious payloads from systems, utilizing advanced detection tools and methodologies.

Implementation Steps:

1.1 **Protocol Development**: Create detailed protocols for using malware detection tools, specifying routine and emergency scanning procedures, malware identification techniques, and removal processes.

1.2 **Tool Selection and Update Procedures**: Select comprehensive malware detection and removal tools capable of identifying the latest threats. Establish a routine for regularly updating these tools to ensure they remain effective against new malware variants.

1.3 **Staff Training and Simulation Drills**: Train IT staff to use these tools and protocols. Conduct regular simulation drills to ensure staff can effectively identify and remove malware under various scenarios.

Expected Outcome: With enhanced malware detection and removal protocols, the organization will significantly reduce the time to identify and eradicate malicious payloads, thus minimizing the impact of incidents. The regular updates and drills will ensure the team remains proficient and prepared.

Control 2: System Configuration and Vulnerability Management Program

Control Objective: To eliminate system vulnerabilities and strengthen security configurations, reducing the likelihood of future breaches and ensuring systems are resilient against attacks.

Implementation Steps:

2.1 **Baseline Configuration Establishment**: Develop and document baseline security configurations for all systems, reflecting the principle of least privilege and best security practices.

2.2 **Regular Vulnerability Scans and Remediation**: Implement a schedule for regular vulnerability scanning and establish procedures for prompt remediation of identified vulnerabilities, including patch management.

2.3 **Configuration Change Management**: Develop a change management process for system configurations that includes review, testing, and approval before implementation to prevent unintended security gaps.

Expected Outcome: Establishing a systematic vulnerability management program and strict configuration management will ensure systems are consistently maintained at their highest security posture, significantly reducing the risk of exploitation.

Control 3: Comprehensive Incident Eradication Tracking and Documentation System

Control Objective: To ensure all steps in the incident eradication process are meticulously documented, enabling effective post-incident analysis and continuous improvement of security measures.

Implementation Steps:

3.1 **Documentation Protocol Establishment**: Develop a standardized process for documenting incident detection, analysis, eradication, and post-incident actions, ensuring all relevant details are recorded.

3.2 **Integration of Documentation into Incident Response Plan**: Ensure that documentation is integral to the incident response plan, with specific roles and responsibilities defined for maintaining records.

3.3 **Review and Learning Process**: Implement a review process for all documented incidents to identify lessons learned and apply these insights to strengthen future incident response and eradication efforts.

Expected Outcome: By establishing a comprehensive documentation system, the organization will have a detailed record of each incident, providing valuable insights for improving future response strategies and compliance with regulatory requirements. This approach will also enhance the organization's ability to conduct thorough post-incident analyses and continuously refine its cybersecurity posture.

25

Incident Recovery Plan Execution (RC.RP)

Transitioning from the throes of incident response to the deliberate steps of recovery encapsulates a pivotal moment where planning, execution, and learning converge, crafting a pathway not just back to operational normalcy but to heightened cybersecurity resilience.

In the complex tapestry of incident response, the transition from immediate reaction to recovery is a critical juncture that requires meticulous planning, execution, and follow-up. The initiation of the recovery portion is predicated on the comprehensive activation of a recovery plan, tailored by lessons gleaned from the incident response phase and anchored by a robust communication strategy. This phase is characterized by seamless coordination between incident response and recovery teams to ensure a smooth transition, underscored by a systematic allocation of resources and personnel. The process mandates continuous monitoring and adjustment to adapt to evolving scenarios and systematic training to enhance team preparedness and awareness. Documentation throughout encapsulates a narrative of actions, decisions, and outcomes, providing a blueprint for compliance, review, and future readiness, laying the groundwork for enhanced resilience and refined operational norms in the aftermath of cybersecurity incidents.

RC.RP-01: The Recovery Portion of the Incident Response Plan Is Executed Once Initiated from the Incident Response Process

The initiation of the recovery process in incident response is a critical step that follows the identification, protection, detection, and response stages within the cybersecurity lifecycle. This section addresses the execution of the recovery portion of the incident response plan. This vital component ensures the organization can return to normal operations while mitigating the impact of the incident. The recovery process involves several key activities, starting with the transition from incident response to recovery, which requires careful planning, communication, and coordination.

Activating the recovery plan is a strategic decision made following the directives of the incident response team. This transition is not merely a procedural step; it embodies the shift from a defensive posture to one focused on restoration and resilience. The recovery plan must be well-integrated with the organization's overall incident response strategy, ensuring a seamless transition. As the recovery phase is initiated, the plan must be activated with clear objectives, predefined roles, and allocated resources to effectively address the aftermath of the security incident.

Coordination between incident response and recovery teams is essential for a seamless transition from dealing with the immediate threats to restoring normal operations. This collaboration ensures

A Comprehensive Guide to the NIST Cybersecurity Framework 2.0: Strategies, Implementation, and Best Practice, First Edition. Jason Edwards.
© 2025 John Wiley & Sons Ltd. Published 2025 by John Wiley & Sons Ltd.

that the recovery actions are informed by the insights gained during the incident response phase, leading to more effective and efficient recovery efforts. The synergy between these teams can significantly reduce downtime and resource allocation mismatches, ensuring that recovery efforts are well-directed and purposeful.

A well-structured communication strategy is crucial for informing all relevant stakeholders about the initiation of the recovery phase. This involves not just internal communication within the organization but also external communication with affected parties, regulators, and possibly the public, depending on the nature of the incident. Clear, concise, and timely information sharing helps manage expectations, reduce uncertainty, and maintain trust during recovery.

Integrating lessons learned during the incident response into recovery actions is critical for enhancing future resilience. This involves analyzing the incident to understand what went wrong, what was done correctly, and how procedures can be improved. Incorporating these lessons into the recovery actions helps address the current incident more effectively and strengthens the organization's preparedness for future incidents.

The allocation of resources and personnel for the recovery execution must be carried out with precision and understanding of the recovery objectives. This includes the deployment of technical resources to restore systems and data, as well as the assignment of personnel to manage recovery tasks. Efficient resource allocation is vital to minimizing downtime and restoring critical business functions as quickly and smoothly as possible.

Documentation of recovery actions is essential for compliance, review, and improvement purposes. This includes recording the decisions, actions, and resources allocated during the recovery phase. Well-documented recovery actions provide a valuable record that can help review the recovery plan's effectiveness, fulfill regulatory requirements, and inform future incident response and recovery efforts.

Training and awareness activities play a significant role in supporting the recovery process. Ensuring that all involved personnel are well-trained and aware of their roles and responsibilities during recovery can significantly enhance the effectiveness and efficiency of the recovery efforts. Additionally, ongoing education and awareness campaigns can help to build a culture of resilience and preparedness within the organization.

Continuous monitoring and adjustment of recovery activities are essential to address evolving threats and unexpected challenges. The recovery phase is not static; it requires ongoing assessment and adaptation to ensure that recovery objectives are being met and that the organization adapts to any new information or circumstances. This dynamic approach helps to ensure that the organization can effectively navigate the complexities of the recovery process and restore normal operations as efficiently as possible.

Recommendations

- **Enhance Coordination Mechanisms**: Strengthen the collaboration between incident response and recovery teams by establishing clear communication channels and protocols. This ensures that valuable insights from the incident response are integrated into the recovery efforts, leading to more effective and cohesive actions.
- **Develop Comprehensive Training Programs**: Create and implement comprehensive training programs focused on recovery processes for all relevant staff. This should include simulations and drills that mimic real-life scenarios, ensuring personnel are well-prepared to execute the recovery plan effectively.

- **Implement Continuous Improvement Practices**: Establish a routine for reviewing and updating the recovery plan based on lessons learned from recent incidents and recovery actions. This continuous improvement approach ensures that the organization remains prepared for evolving cybersecurity threats.
- **Strengthen Resource Management**: Develop a strategy for efficiently allocating resources and personnel during the recovery phase. This should include contingency plans for resource shortages and guidelines for prioritizing recovery efforts based on critical business needs.
- **Enhance Documentation and Compliance Procedures**: Ensure all recovery actions are thoroughly documented and reviewed. This supports compliance with regulatory requirements and provides a valuable resource for post-incident analysis and future planning.

NIST 800-53 Controls

(Refer to our appendix for a comprehensive description of the NIST 800-53 Controls, essential for organizations adhering to strict regulatory standards.)

- CP-10,IR-04,IR-08

Simplified Security Controls (SSC)

This section outlines our Simplified Security Controls (SSC), tailored for environments beyond the scope of NIST compliance. These controls offer streamlined, effective security solutions suitable for any organization seeking to enhance its cybersecurity defenses with clear, actionable measures.

Control 1: Integrated Recovery Planning System

Control Objective: To ensure a cohesive and streamlined transition from incident response to recovery, optimizing resource allocation and minimizing operational downtime.

Implementation Steps:

1.1 **Establish Integrated Teams**: Form integrated teams comprising members from incident response and recovery units to foster collaboration and seamless transition.

1.2 **Develop Transition Protocols**: Create detailed protocols that define the steps for transitioning from incident response to recovery, outlining specific triggers and actions required.

1.3 **Conduct Joint Training Exercises**: Organize regular joint training sessions for incident response and recovery teams to simulate transition scenarios, ensuring familiarity with protocols and enhancing teamwork.

Expected Outcome: Establishing an Integrated Recovery Planning System ensures a seamless transition from incident response to recovery, reducing operational downtime and enhancing the organization's resilience. This team-collaborative approach ensures recovery efforts are well-directed, resource-efficient, and aligned with the organization's security posture.

Control 2: Recovery Communication Plan

Control Objective: To maintain transparency, manage stakeholder expectations, and uphold trust by ensuring clear, consistent, and timely communication throughout the recovery process.

Implementation Steps:

2.1 **Develop a Communication Strategy**: Craft a comprehensive communication plan that outlines the methods, timelines, and responsibilities for communicating with internal and external stakeholders during recovery.

2.2 **Train Communication Teams**: Equip designated communication teams with the necessary skills and information to execute the plan effectively, ensuring they can convey complex information clearly and confidently.

2.3 **Establish Feedback Loops**: Create mechanisms for receiving and integrating stakeholder feedback to adapt real-time recovery efforts and communication strategies.

Expected Outcome: Implementation of a Recovery Communication Plan ensures all stakeholders are informed, expectations are managed, and organizational transparency is maintained during recovery. This fosters an environment of trust and confidence, which is crucial for restoring normal operations and the organization's reputation post-incident.

Control 3: Continuous Recovery Improvement Program

Control Objective: To enhance the effectiveness and efficiency of future recovery efforts by integrating lessons learned and feedback into the recovery process.

Implementation Steps:

3.1 **Capture and Analyze Recovery Data**: Systematically document and review actions, decisions, and outcomes throughout the recovery process for valuable insights.

3.2 **Implement a Feedback Mechanism**: Establish channels for collecting feedback from all involved personnel and stakeholders on the recovery process's effectiveness.

3.3 **Update Recovery Plans Regularly**: Utilize the gathered data and feedback to make informed updates to recovery plans, protocols, and training materials.

Expected Outcome: The Continuous Recovery Improvement Program ensures the recovery process remains dynamic and improves over time, adapting to new threats, technologies, and organizational changes. This leads to more efficient and effective recovery efforts in the face of future incidents, enhancing organizational resilience and preparedness.

RC.RP-02: Recovery Actions Are Selected, Scoped, Prioritized, and Performed

The implementation of recovery actions is an intricate process that requires meticulous planning, selection, scoping, prioritization, and execution. This is essential to mitigate the impact of a cybersecurity incident and restore normal operations promptly and efficiently. The recovery actions form the backbone of an organization's resilience against cyber threats, enabling a structured approach to overcoming the aftermath of an incident. Organizations can significantly reduce the adverse effects of cyber incidents by understanding and applying the right strategies and methodologies.

Selecting appropriate recovery strategies based on the incident's impact is crucial. This decision-making process involves evaluating the nature and extent of the incident to determine the most effective approach for recovery. Factors such as the data type affected, the systems involved, and the overall impact on business operations play a critical role in this context. A well-defined criteria set for selecting recovery strategies ensures that efforts are aligned with the specific needs and priorities of the organization, leading to more effective outcomes.

Scoping recovery efforts is essential to address the affected systems and data precisely. This step involves identifying and delineating the boundaries of the recovery process, focusing on the areas most impacted by the incident. Scoping helps allocate resources efficiently, avoid unnecessary expenditures, and ensure that recovery efforts are concentrated where needed most. By defining the scope clearly, organizations can ensure a targeted and effective approach to recovery, minimizing the spread of damage and accelerating the restoration of services.

Prioritizing recovery tasks is another critical step, ensuring critical functions are restored first. This involves identifying the key services and systems crucial for business operations and prioritizing their recovery above others. Such a prioritized approach helps reduce the overall impact on business continuity and ensures that the most critical components are operational as quickly as possible. Prioritization should be guided by the organization's business impact analysis and continuity plans, ensuring that the most significant areas receive immediate attention.

The execution of recovery actions must focus on efficiency and effectiveness. This involves implementing the planned recovery strategies coordinated and systematically, ensuring that each step is executed correctly and promptly. Utilizing best practices and established procedures can significantly enhance the effectiveness of recovery efforts, minimizing downtime and reducing the likelihood of further issues arising during the recovery process.

The use of project management techniques to track recovery progress is invaluable. These techniques provide a structured framework for organizing, managing, and monitoring the recovery efforts, enabling the timely identification and resolution of any issues. Effective project management ensures recovery tasks are completed within set timelines, resources are utilized efficiently, and stakeholders are informed of the progress.

Adjustments to recovery actions based on real-time feedback and situational changes are essential for a successful recovery process. The dynamic nature of cyber incidents means that situations can evolve rapidly, necessitating adjustments to the recovery plan. Continuous monitoring and feedback mechanisms should be in place to identify such changes and adjust recovery actions accordingly, ensuring the recovery process remains relevant and effective.

Documentation of recovery actions, decision rationale, and outcomes is crucial for accountability, learning, and compliance. This includes detailing the actions taken, the reasons behind each decision, and the results of the recovery efforts. Well-documented recovery actions serve as a valuable resource for post-incident review, regulatory compliance, and continuous improvement processes, enabling organizations to learn from each incident and enhance their resilience against future threats.

Recommendations

- **Develop a Comprehensive Recovery Plan**: Ensure your organization has a comprehensive, well-documented recovery plan. This plan should include clear criteria for selecting recovery strategies, detailed procedures for scoping and prioritizing recovery efforts, and a structured approach for executing recovery actions.
- **Establish Clear Prioritization Guidelines**: Create and implement clear guidelines for prioritizing recovery tasks. These guidelines should be based on your organization's business impact analysis and focus on restoring critical functions and services first to minimize business disruption.
- **Leverage Project Management Practices**: Incorporate project management practices into your recovery process. Use these practices to track progress, manage resources, and ensure that recovery efforts are well-organized and effective.
- **Foster a Culture of Continuous Improvement**: Promote a culture of continuous improvement within your organization. Regularly review and update your recovery plans based on lessons learned from past incidents and changes in your business or technology environment.
- **Invest in Training and Awareness**: Invest in training and awareness programs to ensure that all relevant personnel know the recovery plan and their specific roles. Regular training sessions

and exercises can help prepare your team for efficient and effective action in the event of an incident.

NIST 800-53 Controls

(Refer to our appendix for a comprehensive description of the NIST 800-53 Controls, essential for organizations adhering to strict regulatory standards.)

- CP-10,IR-04,IR-08

Simplified Security Controls (SSC)

This section outlines our SSC, tailored for environments beyond the scope of NIST compliance. These controls offer streamlined, effective security solutions suitable for any organization seeking to enhance its cybersecurity defenses with clear, actionable measures.

Control 1: Recovery Strategy Selection Protocol

Control Objective: To establish a standardized process for selecting the most appropriate recovery strategies based on the specific impact of a cybersecurity incident, ensuring a tailored and effective recovery process.

Implementation Steps:

1.1 **Develop Impact Assessment Criteria**: Create a set of criteria for assessing the impact of cybersecurity incidents on various organizational assets and operations.

1.2 **Strategy Selection Framework**: Develop a framework for selecting recovery strategies based on the assessed impact, considering factors like data sensitivity, affected systems, and business criticality.

1.3 **Regular Review and Update**: Set a schedule for regular reviews and updates to the selection criteria and framework, incorporating feedback from past incidents and changes in business priorities.

Expected Outcome: Implementation of the Recovery Strategy Selection Protocol ensures that recovery efforts are strategically aligned with the specific impacts of incidents, leading to more efficient and effective restoration of operations and minimizing business disruptions.

Control 2: Recovery Project Management System

Control Objective: To implement project management principles and tools to organize, track, and manage recovery tasks effectively, ensuring timely completion and optimal resource allocation.

Implementation Steps:

2.1 **Integration of Project Management Tools**: Adopt and integrate project management software to plan, track, and communicate recovery tasks and progress.

2.2 **Define Recovery Project Parameters**: Set clear timelines, milestones, and responsibilities for each recovery task within the project management tool.

2.3 **Continuous Monitoring and Reporting**: Establish a routine for continuous monitoring of recovery progress and regular reporting to stakeholders to maintain transparency and accountability.

Expected Outcome: Establishing a Recovery Project Management System provides a structured approach to managing recovery efforts, ensuring that tasks are completed within set timelines, resources are utilized efficiently, and stakeholders are kept informed of progress, thereby reducing downtime and enhancing recovery effectiveness.

Control 3: Dynamic Recovery Action Adjustment Mechanism

Control Objective: To ensure the recovery process remains adaptable and responsive to evolving circumstances and feedback, enabling timely adjustments to recovery actions as needed.

Implementation Steps:

3.1 **Establish Feedback Channels**: Create multiple channels for collecting real-time feedback from the recovery team, stakeholders, and automated monitoring systems.

3.2 **Develop an Adjustment Protocol**: Define a clear protocol for assessing feedback and making necessary adjustments to the recovery plan and actions.

3.3 **Regular Training and Simulations**: Conduct regular training sessions and simulation exercises to prepare the recovery team for dynamic adjustments and ensure they understand the adjustment protocol.

Expected Outcome: Implementing a Dynamic Recovery Action Adjustment Mechanism ensures the recovery process can adapt rapidly to changes and feedback, maintaining its relevance and effectiveness in the face of evolving threats and circumstances, thereby improving the organization's resilience and reducing the overall impact of incidents.

RC.RP-03: The Integrity of Backups and Other Restoration Assets Is Verified Before Using Them for Restoration

Verifying the integrity of backups and other restoration assets before their use in restoration processes is critical in ensuring the resilience and security of an organization's data and systems. This chapter delves into the crucial aspects of effectively maintaining, verifying, and utilizing backups and restoration assets. In today's dynamic threat landscape, the ability to recover from incidents quickly and reliably is not just a component of a robust cybersecurity strategy; it is a fundamental business necessity. Understanding and implementing best practices in backup verification can significantly mitigate the risk of data loss and system compromise following a cybersecurity incident.

The first critical step is to introduce robust procedures for regular integrity checks of backups and restoration assets. Regular integrity checks ensure that backup data remains unaltered and free from corruption or tampering. These procedures should be automated, incorporating scheduled checks and alerts to ensure continuous oversight. The frequency of these checks should be determined by the organization's risk assessment, which considers factors such as the value of the data, the likelihood of alteration, and the potential impact of data loss.

Implementation of secure storage and handling practices for backup data is paramount. Backup data should be treated with the same level of security as the original data, involving encryption, controlled access, and secure storage environments. These practices help prevent unauthorized access and ensure backups remain viable and intact when needed. It is essential that organizations not only secure the physical and digital locations of backups but also manage who has access to these assets.

Using cryptographic hashes to verify data integrity, pre-restoration effectively ensures that data has not been altered or corrupted. By comparing the hash of the backup data with the hash of the original data, organizations can confirm the integrity of the backups before they are used. This cryptographic verification safeguards against tampering and corruption, assuring that the restored data replicates the original.

Testing of backup restoration processes to ensure reliability is another crucial step. Regular testing helps identify any issues in the backup and restoration process before an actual need arises. These tests should be comprehensive, covering different scenarios and types of data to ensure that all critical data can be restored accurately and within acceptable recovery time objectives. Testing also helps familiarize the IT team with the restoration process, which can be invaluable during a high-pressure incident response.

Strategies for recovering data without compromising security postures are essential to the recovery process. It involves ensuring that the restoration process does not reintroduce vulnerabilities into the system or overlook breaches that may have caused the initial data loss. Strategies should include checks for malware within backups, restoration to isolated environments for testing before complete restoration, and updating systems to patch vulnerabilities before restoring data.

Documentation of verification processes and results is essential for auditing and improving incident response and recovery procedures. Documentation provides a clear record of what was tested, when, by whom, and the outcomes of those tests. This supports compliance with various regulations and provides invaluable data for post-incident reviews and continuous improvement efforts.

Training on best practices for backup management and data restoration should be a continual process. Regular training sessions help ensure that all staff members understand the importance of backups, know how to handle and store them securely, and are familiar with the restoration process. This training should regularly reflect new threats, technologies, and best practices.

Continuous backup and restoration asset management improvement is necessary to adapt to the evolving cybersecurity landscape. Organizations should regularly review and update their backup and restoration procedures in response to new threats, technological advancements, and lessons learned from past incidents. This ongoing process helps ensure that backup and restoration strategies remain effective and aligned with the organization's cybersecurity posture.

Recommendations

- **Establish Routine Backup Integrity Checks**: Implement automated and regular integrity checks for all backup data. Ensure these checks are part of a routine schedule and are complemented by manual spot-checks to maintain the highest levels of data integrity.
- **Adopt Comprehensive Security Measures for Backups**: Treat backup data with the same level of security as primary data. This includes encrypting backups, ensuring secure storage locations, and managing access control meticulously.
- **Incorporate Regular Testing of Restoration Processes**: Develop a regular schedule for testing backup restoration, simulating various scenarios to ensure all critical data can be effectively restored. This should be part of a broader incident response exercise.
- **Document and Review Backup and Restoration Procedures**: Keep detailed records of backup and restoration processes, including integrity checks and test results. Use this documentation for compliance, auditing, and continuous improvement.
- **Conduct Ongoing Training and Awareness Programs**: Regularly train staff on best practices for backup management and data restoration. This should include updates on new threats and technologies, reinforcing the importance of backups in the organization's cybersecurity strategy.

NIST 800-53 Controls

(Refer to our appendix for a comprehensive description of the NIST 800-53 Controls, essential for organizations adhering to strict regulatory standards.)

- CP-02,CP-04,CP-09

Simplified Security Controls (SSC)

This section outlines our SSC, tailored for environments beyond the scope of NIST compliance. These controls offer streamlined, effective security solutions suitable for any organization seeking to enhance its cybersecurity defenses with clear, actionable measures.

Control 1: Backup Integrity Verification Protocol

Control Objective: To ensure the integrity and authenticity of backups and restoration assets before their use in restoration, minimizing the risk of restoring corrupted or compromised data.

Implementation Steps:

1.1 **Establishment of Integrity Checks**: Develop and implement procedures for regular integrity checks of backups and restoration assets, including using cryptographic hashes to confirm data integrity.

1.2 **Automation of Verification Processes**: Automate the integrity verification process to ensure consistency and frequency. Utilize software tools that can perform scheduled checks and report any discrepancies immediately.

1.3 **Response Plan for Failed Checks**: Create a response plan for scenarios where backups fail integrity checks. This plan should include steps for isolating corrupted backups, notifying relevant personnel, and initiating an investigation into the cause of corruption.

Expected Outcome: The Backup Integrity Verification Protocol ensures that all backups and restoration assets are free from corruption and tampering before use. This increases reliability in the restoration process and enhances the organization's resilience against data loss and cybersecurity incidents.

Control 2: Secure Backup Storage and Handling Framework

Control Objective: To safeguard backup data from unauthorized access and ensure its availability for restoration through secure storage practices and controlled handling procedures.

Implementation Steps:

2.1 **Secure Storage Solutions**: Implement secure storage solutions for backup data, incorporating encryption, access control, and environmental protections to safeguard against unauthorized access and physical damage.

2.2 **Handling and Transportation Protocols**: Establish protocols for safely handling and transporting backup media, ensuring that backups are protected during offsite storage or when being moved between locations.

2.3 **Regular Access Reviews and Audits**: Conduct regular reviews and audits of access logs to backup data, ensuring that only authorized individuals have access and identifying unauthorized access attempts.

Expected Outcome: The Secure Backup Storage and Handling Framework ensures that backup data remains confidential, integral, and available, reducing the risk of data breaches and ensuring backups are ready for use when needed.

Control 3: Backup and Restoration Training Program

Control Objective: To enhance the organization's capability in managing and restoring backups through comprehensive training on best practices and procedures for backup management and data restoration.

Implementation Steps:

3.1 **Development of Training Material**: Develop comprehensive training materials covering best practices, procedures, and tools for effective backup management and data restoration.
3.2 **Mandatory Training Sessions**: Conduct mandatory training sessions for all personnel involved in the backup and restoration process, ensuring they understand their roles and responsibilities.
3.3 **Regular Updates and Refresher Courses**: Regularly update training materials to reflect technological changes, threats, and organizational policies. Conduct refresher courses to ensure that personnel remain knowledgeable and skilled.

Expected Outcome: The Backup and Restoration Training Program ensures that all personnel involved in the backup and restoration processes are competent and knowledgeable. This leads to more effective and secure backup management, reduced risk of errors during the restoration process, and improved overall organizational resilience.

RC.RP-04: Critical Mission Functions and Cybersecurity Risk Management Are Considered to Establish Post-Incident Operational Norms

Reestablishing operations post-incident is a critical phase in the recovery process, which requires careful planning and consideration of the organization's essential functions of mission and cybersecurity risk management strategies. This phase is not simply about returning to business as usual but about seizing the opportunity to strengthen the organization's resilience against future threats. The ability to efficiently and effectively restore operations following an incident indicates a mature cybersecurity posture and is essential for minimizing downtime and financial loss.

Assessing critical mission functions for prioritized restoration is the foundation of effective post-incident recovery. This involves identifying and prioritizing systems and processes essential to the organization's core mission and restoring them first. This prioritization is based on the impact analysis and is crucial for efficient resource allocation. Understanding which functions are critical ensures that the recovery process supports the organization's most important objectives and services.

Integration of cybersecurity risk management in recovery planning is essential to ensure that the lessons learned from the incident are applied to strengthen the organization's security posture. This integration involves analyzing the incident to identify vulnerabilities and failures in the existing security measures and incorporating the necessary changes into the recovery process. By doing so, organizations can restore their operations more securely and enhance their resilience against future incidents.

Developing a strategic plan for operational norms post-incident is necessary to guide the organization through recovery. This plan should outline the steps to restore critical functions, implement improved security measures, and communicate effectively with stakeholders. The plan must also consider the potential for changing threat landscapes and include strategies for adapting to these new challenges.

Consideration of new threat landscapes in post-incident operations is critical. Cybersecurity continuously evolves, and incidents can provide valuable insights into emerging threats and vulnerabilities. Organizations must use these insights to adapt their security strategies and operational practices to protect against future incidents. This might involve adopting new technologies, changing operational procedures, or enhancing employee training programs.

Communication of operational changes to all relevant stakeholders is imperative. This includes employees, customers, partners, and regulators. Effective communication ensures that all parties understand the changes, their reasons, and their roles in the new operational environment. This transparency helps to maintain trust and confidence, which are particularly important following a security incident.

Training and awareness programs on new operational procedures are essential to ensure all employees understand and can effectively implement the changes. These programs should be comprehensive, covering new security policies, procedures, and any new technologies that have been introduced. Training helps to ensure that all staff are aligned with the latest operational norms and are equipped to contribute to the organization's security posture.

Monitoring and evaluation of post-incident operations for effectiveness are crucial for continuous improvement. This involves regularly reviewing the implemented changes to ensure they meet their objectives and adapt as necessary. Monitoring should include measuring the effectiveness of new security measures, the efficiency of restored operations, and the organization's overall resilience to threats.

Recommendations

- **Prioritize Critical Functions in Recovery Planning**: Focus on first identifying and restoring critical mission functions. Ensure these are clearly defined and prioritized within the organization's recovery planning efforts to minimize impact and expedite return to operational norms.
- **Integrate Lessons Learned into Recovery Efforts**: Use the incident as an opportunity to strengthen your cybersecurity framework. Integrate risk management strategies based on lessons learned from the incident into your recovery and future operational plans.
- **Develop Comprehensive Communication Plans**: Establish clear communication strategies to keep all stakeholders informed during the recovery process and about changes to operational norms post-incident. Transparency is key to maintaining trust and ensuring a unified approach to new operational procedures.
- **Implement Ongoing Training and Awareness Initiatives**: Develop comprehensive training programs covering new operational procedures and security measures. Ensure all employees understand their role in safeguarding the organization and are equipped to respond effectively to future threats.
- **Conduct Regular Monitoring and Evaluation**: Establish procedures for ongoing monitoring and evaluation of the effectiveness of post-incident operations. Use these insights to refine and improve operational and security practices continuously.

NIST 800-53 Controls

(Refer to our appendix for a comprehensive description of the NIST 800-53 Controls, essential for organizations adhering to strict regulatory standards.)

- PM-08,PM-09,PM-11,IR-01,IR-08

Simplified Security Controls (SSC)

This section outlines our SSC, tailored for environments beyond the scope of NIST compliance. These controls offer streamlined, effective security solutions suitable for any organization seeking to enhance its cybersecurity defenses with clear, actionable measures.

Control 1: Critical Function Recovery Prioritization Protocol

Control Objective: To ensure systematic and rapid restoration of critical business functions, minimizing operational disruption and aligning recovery efforts with organizational priorities.

Implementation Steps:

1.1 **Critical Function Identification**: Collaborate with department heads to identify and document all critical business functions and the systems they depend on.

1.2 **Impact Analysis and Prioritization**: Conduct a business impact analysis to prioritize critical functions based on their importance to mission fulfillment and operational continuity.

1.3 **Integration into Recovery Plans**: Incorporate the prioritization list into the recovery plans, ensuring that efforts are focused on restoring these functions first.

Expected Outcome: Implementing the Critical Function Recovery Prioritization Protocol ensures that the most crucial business operations are restored first, reducing downtime and financial loss. This structured approach aids in a more strategic and effective recovery, aligning with the organization's key objectives and services.

Control 2: Post-Incident Security Enhancement Program

Control Objective: To strengthen the organization's cybersecurity posture post-incident by integrating lessons learned and addressing identified vulnerabilities.

Implementation Steps:

2.1 **Incident Analysis**: Conduct a thorough analysis of the incident to identify exploited vulnerabilities and security gaps.

2.2 **Security Enhancement Planning**: Develop a plan to address identified issues, including updating security policies, procedures, and systems.

2.3 **Implementation and Review**: Implement the enhancements and periodically review their effectiveness, making adjustments as necessary based on evolving threat landscapes and business needs.

Expected Outcome: The Post-Incident Security Enhancement Program ensures that the organization recovers from the current incident and improves its resilience against future threats. This continual improvement approach leads to a more robust security posture and reduced cyberattack vulnerability.

Control 3: Comprehensive Post-Incident Stakeholder Communication Plan

Control Objective: To maintain transparency and trust by communicating post-incident operational changes and security improvements to all relevant stakeholders.

Implementation Steps:

3.1 **Develop Communication Strategy**: Create a detailed plan outlining what, when, and how information will be communicated to employees, customers, partners, and regulators post-incident.

3.2 **Training for Spokespersons**: Train designated spokespersons in delivering clear, accurate, and consistent messages across different platforms and to various audiences.

3.3 **Continuous Feedback and Adjustment**: Implement mechanisms for gathering stakeholder feedback on the changes and use this to refine ongoing communications and recovery actions.

Expected Outcome: The Comprehensive Post-Incident Stakeholder Communication Plan ensures all parties are well-informed about the incident's impact, recovery efforts, and subsequent operational changes. This fosters an environment of transparency and trust, which is crucial for maintaining stakeholder confidence and support during the recovery phase and beyond.

RC.RP-05: The Integrity of Restored Assets Is Verified, Systems and Services Are Restored, and Normal Operating Status Is Confirmed

Verifying restored assets and confirming operational status is a vital component of the recovery strategy within the NIST Cybersecurity Framework. This phase is crucial in ensuring that the systems and services restored post-incident are secure, intact, and functioning as intended. It is not just about getting systems back online; it is about making sure they are returned to a state free from vulnerabilities and compromises. This step is fundamental in mitigating the risk of repeat incidents and reinforcing the organization's resilience against future cybersecurity challenges.

The processes for comprehensive verification of asset integrity post-restoration involve a series of systematic checks and validations. Each restored asset must undergo rigorous testing to ensure it has not been compromised and functions according to its original specifications and security requirements. This phase should encompass software integrity checks, hardware functionality tests, and security configuration reviews to ensure that all aspects of the asset are operating correctly and securely.

Systematic restoration of systems and services according to priority involves strategically reinstating business operations based on their criticality to the organization's mission. This structured approach ensures that the most crucial services are restored first, minimizing the impact on business continuity and reducing operational risks. It necessitates a well-organized plan aligning with the business impact analysis, ensuring that resources are allocated efficiently and effectively during recovery.

Confirmation of normal operating status through rigorous testing is essential to ascertain that all systems and services are running optimally post-restoration. This involves a series of performance and security tests designed to confirm that the restored systems are functioning correctly and are free from any vulnerabilities that may have been exploited during the incident. This step is crucial in validating that the restoration efforts have been successful and that the business can safely return to standard operational status.

Monitoring tools to ensure the stability of restored operations is critical in maintaining oversight of the newly restored systems. Continuous monitoring allows for the early detection of any irregularities or security breaches, enabling immediate remediation before they can escalate into significant issues. Implementing robust monitoring solutions provides an ongoing assurance of the integrity and security of the operational environment post-incident.

Coordination with business units to validate functional readiness is essential in the restoration process. It involves working closely with different organizational departments to ensure that the restored systems meet their operational needs and security requirements. This collaborative approach helps identify discrepancies or issues from a functional perspective, facilitating a more effective and comprehensive recovery process.

Documentation of restoration and verification activities is vital for future reference and continuous improvement. This involves creating detailed records of the incident response and recovery processes, including actions taken, decisions made, and lessons learned. This documentation is valuable for auditing, compliance, planning, and improving future cybersecurity strategies and responses.

Training on post-restoration assessment and validation techniques is essential for preparing the organization for future incidents. Providing staff with the knowledge and tools they need to assess and validate the integrity of restored systems effectively ensures that the organization is better equipped to handle similar situations in the future. Regular training and drills should keep the response teams well-prepared and up-to-date with the latest restoration and validation methodologies.

Recommendations

- **Establish Robust Verification Processes**: Develop comprehensive processes for verifying the integrity of restored assets, ensuring they cover all critical aspects from security to functionality.
- **Prioritize System Restoration**: Implement a priority-based system restoration strategy that aligns with the organization's business impact analysis to ensure critical services are restored first.
- **Implement Rigorous Testing Regimes**: Conduct thorough testing to confirm the normal operating status of all systems and services post-restoration, ensuring they meet all performance and security standards.
- **Leverage Continuous Monitoring**: Utilize advanced monitoring tools to maintain oversight and ensure the stability of operations post-restoration.
- **Foster Cross-Departmental Collaboration**: Enhance coordination between IT and business units to ensure restored systems meet all functional and security requirements, providing a unified approach to restoration.

NIST 800-53 Controls

(Refer to our appendix for a comprehensive description of the NIST 800-53 Controls, essential for organizations adhering to strict regulatory standards.)

- CP-10

Simplified Security Controls (SSC)

This section outlines our SSC, tailored for environments beyond the scope of NIST compliance. These controls offer streamlined, effective security solutions suitable for any organization seeking to enhance its cybersecurity defenses with clear, actionable measures.

Control 1: Asset Integrity Verification System

Control Objective: To ensure the security and functionality of restored assets by conducting comprehensive integrity checks and validations post-restoration.

Implementation Steps:

1.1 **Develop Verification Procedures**: Establish standardized procedures for conducting software integrity checks, hardware functionality tests, and security configuration reviews on all restored assets.

1.2 **Implement Automated Testing Tools**: Utilize automated testing tools to streamline the verification process, ensuring consistency and thoroughness in evaluating each restored asset.

1.3 **Conduct Regular Training**: Provide regular training sessions for IT and cybersecurity teams on the latest asset verification techniques and tools to maintain high standards of asset integrity verification.

Expected Outcome: The Asset Integrity Verification System ensures that all restored assets are free from vulnerabilities and function according to their original specifications, significantly reducing the risk of repeat incidents and strengthening the organization's cybersecurity defenses.

Control 2: Restoration Priority Management Plan

Control Objective: To minimize operational disruptions and financial loss by restoring critical systems and services in a prioritized and efficient manner.

Implementation Steps:

2.1 **Perform Business Impact Analysis**: Conduct a comprehensive business impact analysis to identify and prioritize critical systems and services based on their significance to the organization's mission.

2.2 **Develop a Restoration Plan**: Create a detailed restoration plan that aligns with the priority levels established in the business impact analysis, outlining specific resources and timelines for each priority level.

2.3 **Monitor and Adjust Priorities**: Continuously monitor the restoration process and adjust priorities as necessary based on evolving business needs and incident dynamics.

Expected Outcome: The Restoration Priority Management Plan ensures a strategic approach to restoring operations, focusing on critical functions first to minimize downtime and ensure a swift return to normal business operations.

Control 3: Continuous Restoration Monitoring Framework

Control Objective: To maintain oversight of restored operations and ensure ongoing stability and security through continuous monitoring and evaluation.

Implementation Steps:

3.1 **Implement Monitoring Tools**: Deploy advanced monitoring tools to continuously track the performance and security of restored systems and services.

3.2 **Establish a Monitoring Team**: Designate a dedicated team responsible for analyzing monitoring data, identifying irregularities, and initiating prompt remediation actions.

3.3 **Develop an Evaluation Protocol**: Create a formal protocol for the regular evaluation of post-incident operations, involving feedback from business units to validate functional readiness and identify areas for improvement.

Expected Outcome: The Continuous Restoration Monitoring Framework provides an ongoing assurance of the stability and security of the operational environment post-incident, enabling early detection of potential issues and facilitating rapid response to maintain operational integrity and security.

RC.RP-06: The End of Incident Recovery Is Declared Based on Criteria, and Incident-Related Documentation Is Completed

Concluding the incident recovery phase is essential to transition from emergency response to normal operations. This phase is critical as it ensures that all aspects of the incident have been addressed and that the organization is prepared to resume standard operational procedures. A structured approach to concluding the recovery phase helps to affirm that the incident has been fully resolved, lessons have been learned, and the organization is better prepared for future challenges.

The development of clear criteria for declaring the end of recovery is crucial. These criteria should be predefined and agreed upon in the incident response plan. Factors might include restoring all services to full functionality, verifying that no threats remain within the system, and confirming that all security measures have been reinstated or enhanced. Establishing these criteria ensures a uniform and objective approach to ending the recovery process.

A comprehensive review to ensure all recovery objectives are met is essential before concluding the incident recovery phase. This review should encompass technical, operational, and business considerations, ensuring that all aspects of the recovery have been completed satisfactorily. It is important that this review is thorough and includes input from all relevant departments to ensure that no elements of the recovery are overlooked.

Finalization and closure of incident-related documentation are crucial steps in the recovery process. This involves completing all reports, logging all actions taken, and ensuring that all documentation is accurate and comprehensive. This documentation is critical for legal, compliance, and auditing purposes and provides a valuable record for future incident response planning and training.

Post-recovery debriefs are essential to capture lessons learned and best practices. These debriefs should involve all key personnel in the incident response and recovery phases. The goal is to identify what was done well, what could be improved, and how the organization can respond to future incidents. This is a vital step in the continuous improvement of cybersecurity practices.

A communication strategy for signaling recovery completion to stakeholders is important. This strategy should outline how and when to communicate at the end of the recovery phase, ensuring that all internal and external stakeholders are appropriately informed. Effective communication helps to restore confidence among customers, employees, and partners and signals the return to normal business operations.

Archiving of incident and recovery documentation for legal and compliance purposes is necessary. This ensures that all information related to the incident is preserved and can be accessed for future reference, legal inquiries, or compliance audits. Proper archiving practices help to protect the organization legally and support continuous improvement efforts.

Evaluation of the entire incident response and recovery process is crucial for enhancing future preparedness. This evaluation should be thorough and cover every aspect of the response and recovery, identifying strengths, weaknesses, and areas for improvement. This reflective process is key to evolving the organization's cybersecurity posture and enhancing its resilience against future threats.

Recommendations

- **Define Clear Recovery Completion Criteria**: Develop and implement objective criteria for concluding the incident recovery phase. Ensure these criteria cover all necessary technical, operational, and business elements.
- **Institute Thorough Review Processes**: Establish comprehensive review processes to ensure all recovery objectives have been met before officially ending the recovery phase. Involve relevant departments to provide a holistic assessment.
- **Enhance Documentation Practices**: Prioritize the completion and accuracy of all incident-related documentation. Ensure all actions, decisions, and outcomes are thoroughly recorded for future reference and compliance.

- **Conduct Effective Post-Recovery Debriefs**: Regularly conduct debrief sessions following recovery to capture and integrate lessons learned into future incident response plans and training programs.
- **Streamline Communication and Archiving**: Develop a robust communication strategy for the recovery phase's conclusion and ensure proper archiving of all incident-related documents for legal, compliance, and educational purposes.

NIST 800-53 Controls

(Refer to our appendix for a comprehensive description of the NIST 800-53 Controls, essential for organizations adhering to strict regulatory standards.)

- IR-04,IR-08

Simplified Security Controls (SSC)

This section outlines our SSC, tailored for environments beyond the scope of NIST compliance. These controls offer streamlined, effective security solutions suitable for any organization seeking to enhance its cybersecurity defenses with clear, actionable measures.

Control 1: Recovery Conclusion Criteria and Approval Process

Control Objective: Establish standardized criteria and an approval process for officially concluding the incident recovery phase, ensuring all recovery objectives have been satisfactorily met and normal operations can safely resume.

Implementation Steps:

1.1 **Develop Criteria**: Define clear, measurable criteria for the conclusion of the recovery phase based on the successful restoration of services, verification of system integrity, and achievement of all recovery objectives.

1.2 **Approval Committee Formation**: Form a recovery conclusion committee comprising members from key departments such as IT, cybersecurity, business operations, and compliance to evaluate whether the recovery criteria have been met.

1.3 **Formal Conclusion Process**: Implement a formal process where the recovery conclusion committee reviews the completed recovery actions against the predefined criteria and provides official approval for concluding the recovery phase.

Expected Outcome: Establishing a Recovery Conclusion Criteria and Approval Process ensures a systematic and validated approach to ending the incident recovery phase. This results in a more secure and controlled transition back to normal operations, minimizing the risk of unresolved issues or premature closure.

Control 2: Post-Incident Review and Improvement Program

Control Objective: To capture lessons learned and best practices from the incident response and recovery process, facilitating continuous improvement and preparation for future incidents.

Implementation Steps:

2.1 **Conduct Debrief Sessions**: Organize post-recovery debrief sessions involving all stakeholders involved in the incident management process to discuss what was successful and what could be improved.

2.2 **Document Findings**: Systematically document the findings and lessons learned from the debrief sessions, highlighting strengths and improvement areas.

2.3 **Update Policies and Procedures**: Review and update incident response and recovery policies, procedures, and training materials based on the documented lessons learned and best practices.

Expected Outcome: Implementing a Post-Incident Review and Improvement Program leads to continuous improvement of the incident response and recovery processes. This fosters a learning culture within the organization, enhancing preparedness and response capabilities for future incidents.

Control 3: Incident Documentation Archiving and Accessibility Protocol

Control Objective: To ensure comprehensive and secure archiving of all incident-related documentation for legal, compliance, and reference purposes while maintaining ease of accessibility for future analysis and audits.

Implementation Steps:

3.1 **Standardize Documentation Practices**: Establish standardized practices for the creation, finalization, and closure of all incident-related documentation, ensuring consistency and completeness.

3.2 **Secure Archiving System**: Implement a secure, centralized archiving system for storing all incident and recovery documentation, ensuring protection against unauthorized access and data loss.

3.3 **Accessibility and Review Mechanism**: Develop a mechanism for controlled access to archived documentation, facilitating ease of retrieval for legal compliance, audits, and future incident response planning.

Expected Outcome: The establishment of an Incident Documentation Archiving and Accessibility Protocol ensures that all documentation related to incident response and recovery is securely stored and readily accessible. This control aids in legal compliance, facilitates future audits and reviews, and enhances the organization's ability to learn from past incidents, improving overall cybersecurity resilience.

26

Incident Recovery Communication (RC.CO)

> Navigating the aftermath of a cybersecurity incident requires more than technical acumen; it demands a mastery of communication, transforming the complex choreography of recovery into a transparent narrative that reassures, informs, and engages stakeholders at every step.

In the intricate journey of incident recovery within cybersecurity, effective communication is a pivotal cornerstone. The process encompasses a broad spectrum of activities, from the initial identification and engagement with key internal and external stakeholders to the meticulous crafting and dissemination of updates regarding recovery progress. At the heart of these endeavors lies developing a comprehensive communication plan tailored to outline messages' specific channels, frequency, and content, ensuring they are both accessible and devoid of technical jargon. This strategy prioritizes the security and reliability of communication channels to safeguard sensitive information and integrates feedback mechanisms, allowing for a dynamic exchange between the organization and its stakeholders. The continuous refinement of these communication practices, underscored by regular training for the recovery team and the iterative review of communication plans, underlines the commitment to transparency and efficacy in restoring operational capabilities while maintaining stakeholder trust and compliance with regulatory standards.

Author's Note: RC.CO-01 and RC.CO-02 were deprecated from the CSF 2 and are not included in this chapter.

RC.CO-03: Recovery Activities and Progress in Restoring Operational Capabilities Are Communicated to Designated Internal and External Stakeholders

Effective communication during recovery activities is pivotal in managing and mitigating the aftermath of a cybersecurity incident. This section explores the strategic approach to engaging with internal and external stakeholders during recovery. A robust recovery plan focuses on restoring operational capabilities and maintaining trust and transparency through effective communication.

The identification of key internal and external stakeholders is the first critical step. Internal stakeholders include employees, management, and the IT department, while external stakeholders may consist of customers, partners, regulators, and possibly the media. Understanding the needs and expectations of each stakeholder group is crucial in tailoring communication strategies that address their concerns and keep them informed about recovery efforts.

A Comprehensive Guide to the NIST Cybersecurity Framework 2.0: Strategies, Implementation, and Best Practice, First Edition. Jason Edwards.
© 2025 John Wiley & Sons Ltd. Published 2025 by John Wiley & Sons Ltd.

A communication plan is essential for outlining how information will be disseminated during recovery. This plan should detail the channels of communication, the frequency of updates, and the type of content to be shared. The objective is to ensure all stakeholders are informed of timely and relevant information regarding the recovery progress.

Crafting clear, concise, and accurate messages is vital when updating stakeholders on recovery progress. The information shared should be straightforward, avoiding technical jargon that may confuse nontechnical stakeholders. Accuracy is paramount to maintain credibility and trust throughout the recovery process.

Ensuring consistency and accuracy of information across all communications helps build and maintain stakeholder trust. Inconsistencies in the information shared can lead to confusion and erode stakeholder confidence in the organization's ability to recover from the incident.

Using secure and reliable communication channels is crucial for protecting sensitive information during recovery. This involves employing encrypted communication tools and secure platforms to safeguard against unauthorized entities' potential interception of communications.

Incorporating feedback mechanisms allows stakeholders to respond or inquire about the recovery progress. This two-way communication fosters a collaborative environment where stakeholders can contribute insights or raise concerns, which may aid recovery.

Training for the recovery team on effective communication strategies equips them with the skills necessary to convey recovery efforts effectively. This training should cover crisis communication, emotional intelligence, and technical communication skills.

Regularly reviewing and adjusting communication plans based on stakeholder needs ensures that the communication strategy remains effective. This iterative process allows the organization to adapt its communication approach to changing circumstances and stakeholder feedback.

Recommendations

- **Develop a Comprehensive Stakeholder Analysis**: Conduct a thorough analysis to identify all potential internal and external stakeholders. This will ensure that no key group is overlooked during the communication process.
- **Customize Communication Plans**: Tailor communication plans to meet different stakeholder groups' specific needs and preferences. This may involve using different channels or varying the level of detail in the information provided.
- **Implement Regular Training Sessions**: Conduct regular training for the recovery team and any other employees involved in communication roles. Focus on improving communication skills, particularly in crises.
- **Establish Secure Communication Protocols**: Prioritize the security of your communication channels to protect sensitive information. Utilize encrypted messaging services and secure email platforms as standard tools for recovery communication.
- **Solicit and Act on Feedback**: Encourage feedback from stakeholders on the effectiveness of your communication efforts. Use this feedback to refine and improve your communication strategy continuously.

NIST 800-53 Controls

(Refer to our appendix for a comprehensive description of the NIST 800-53 Controls, essential for organizations adhering to strict regulatory standards.)

- IR-04,IR-06,SR-08

Simplified Security Controls (SSC)

This section outlines our Simplified Security Controls (SSC), tailored for environments beyond the scope of NIST compliance. These controls offer streamlined, effective security solutions suitable for any organization seeking to enhance its cybersecurity defenses with clear, actionable measures.

Control 1: Development and Implementation of a Comprehensive Stakeholder Communication Plan

Control Objective: To create a structured approach for disseminating information during recovery efforts, ensuring that all relevant internal and external stakeholders are informed with clear, consistent, and accurate updates regarding incident recovery progress.

Implementation Steps:

1.1 **Stakeholder Identification and Analysis**: Conduct a thorough identification and analysis of all potential stakeholders affected by cybersecurity incidents, categorizing them based on their impact and influence over the recovery process.

1.2 **Communication Plan Development**: Develop a comprehensive communication plan that details the communication channels, frequency, content, and protocols for updates. This plan should be tailored to the needs and preferences of the identified stakeholder groups.

1.3 **Plan Execution and Monitoring**: Implement the communication plan, ensuring that updates are delivered as scheduled. Monitor the effectiveness of the communication efforts, making adjustments as necessary to address stakeholder feedback and evolving recovery details.

Expected Outcome: Implementing a comprehensive stakeholder communication plan ensures that all parties affected by a cybersecurity incident are promptly and accurately informed of recovery efforts. This transparency fosters trust, reduces misinformation, and enhances collaborative efforts toward recovery.

Control 2: Establishment of Secure Communication Channels

Control Objective: To safeguard sensitive recovery information from unauthorized access or disclosure during recovery, utilizing secure and encrypted communication channels for all internal and external communications.

Implementation Steps:

2.1 **Assessment of Communication Tools**: Evaluate current communication tools and platforms for security features, including encryption and data protection capabilities, to determine their suitability for sensitive communications.

2.2 **Implementation of Secure Channels**: Select and implement approved communication channels that meet the organization's security requirements. This may include secure messaging apps, encrypted email services, and secure conferencing tools.

2.3 **Training and Awareness**: Conduct training sessions for all employees on the importance of using secure communication channels and provide guidelines on effectively using these tools for internal and external communications.

Expected Outcome: By establishing secure communication channels, the organization protects sensitive recovery information from potential cyber threats, ensuring that communications regarding the recovery process are confidential and secure.

Control 3: Regular Training on Crisis Communication Strategies

Control Objective: To equip the recovery team and relevant stakeholders with the necessary communication skills to effectively manage and convey information during the recovery phase of a cybersecurity incident, emphasizing clarity, empathy, and accuracy.

Implementation Steps:

3.1 **Development of Training Program**: Develop a comprehensive training program on crisis communication strategies tailored to the roles and responsibilities of the recovery team and other key stakeholders involved in communication.

3.2 **Conducting Regular Training Sessions**: Schedule and conduct regular training sessions that cover aspects of crisis communication, including how to craft clear, empathetic, and accurate messages, as well as how to handle inquiries from stakeholders and the media.

3.3 **Continuous Improvement**: Incorporate feedback from training participants and updates in best practices to continuously improve the training content and delivery methods, ensuring the training remains relevant and effective.

Expected Outcome: Regular training on crisis communication strategies enhances the ability of the recovery team and stakeholders to communicate effectively during cybersecurity incidents, reducing misunderstandings and improving the overall management of the incident recovery process.

RC.CO-04: Public Updates on Incident Recovery Are Shared Using Approved Methods and Messaging

Communicating effectively with the public during the incident recovery phase is a delicate balance that organizations must manage. This section outlines the strategies for public communication following a cybersecurity incident, ensuring that information is conveyed accurately, responsibly, and in a manner that maintains public trust.

Establishing robust protocols is the foundation of effective public communication during incident recovery. These protocols define who can speak on behalf of the organization, what can be shared, and how information should be released. They serve as a guideline to prevent misinformation and ensure a unified response.

Coordination with public relations and legal teams is critical to approve messages and methods of communication. This collaboration ensures that all public statements are legally sound, align with the organization's strategic messaging, and address public concerns without compromising the integrity of the recovery efforts.

Selecting the right platforms and channels for communication is essential for broad and effective reach. Organizations must consider a mix of traditional media, social media, and direct communication channels to disseminate recovery updates to different public segments efficiently.

Crafting messages that are informative, reassuring, and free of technical jargon is crucial. The goal is to clarify the incident's impact, what is being done to address it, and how it affects the public without causing unnecessary alarm.

The timing of updates is another important consideration. Organizations should strive to provide timely and relevant information that keeps the public informed without compromising the details of the recovery efforts or enabling further exploitation by malicious actors.

Monitoring public response and media coverage is necessary to adjust messaging if necessary. Public perception can shift rapidly, and staying attuned to these changes allows organizations to respond proactively, correcting misinformation and addressing emerging concerns.

Ensuring consistency in messaging across different platforms and statements is vital for maintaining credibility. Discrepancies in information can lead to confusion and erode public trust. Therefore, all communication should be coherent and aligned, regardless of the medium.

Documentation of public communications is important for compliance, review, and future learning. Keeping records of what was communicated, when, and through which channels can help organizations refine their communication strategies and respond more effectively to similar incidents in the future.

Recommendations

- **Implement a Preapproved Communication Template**: Develop templates for public communication that can be quickly adapted to specific incidents. This ensures prompt and consistent messaging across all communications.
- **Conduct Regular Training with Spokespersons**: Regular training sessions for designated spokespersons, including executives, can prepare them for effective communication during crises, emphasizing clarity, empathy, and transparency.
- **Leverage Social Media Wisely**: Use social media platforms judiciously to communicate updates, recognizing their power to reach broad audiences quickly while managing the spread of misinformation.
- **Engage in Active Listening**: Monitor public feedback actively across all communication channels to gauge the effectiveness of your messaging and adjust strategies as needed to address concerns and misinformation.
- **Review and Learn from Each Incident**: After an incident, thoroughly review the communication efforts to identify strengths and areas for improvement. Use these insights to enhance future communication strategies.

NIST 800-53 Controls

(Refer to our appendix for a comprehensive description of the NIST 800-53 Controls, essential for organizations adhering to strict regulatory standards.)

- CP-02,IR-04

Simplified Security Controls (SSC)

This section outlines our SSC, tailored for environments beyond the scope of NIST compliance. These controls offer streamlined, effective security solutions suitable for any organization seeking to enhance its cybersecurity defenses with clear, actionable measures.

Control 1: Establishment of a Public Communication Protocol for Incident Recovery

Control Objective: To create a standardized approach for communicating with the public during incident recovery phases, ensuring messages are approved, accurate, and consistent across all platforms, thereby maintaining public trust and compliance.

Implementation Steps:

1.1 **Protocol Development**: Draft a comprehensive public communication protocol that outlines who is authorized to speak on behalf of the organization, the approval process for public messages, and guidelines for message content and timing.

1.2 **Coordination with Key Teams**: Ensure the protocol is developed with public relations, legal teams, and cybersecurity response teams to ensure all public communications are legally compliant, strategically sound, and technically accurate.

1.3 **Training and Awareness**: Train all employees, especially those in public-facing roles, on the public communication protocol to ensure consistency and readiness to respond in a unified manner during an incident.

Expected Outcome: Establishing a public communication protocol ensures that all communications during an incident recovery phase are consistent, accurate, and legally compliant, thereby maintaining public trust and organizational credibility.

Control 2: Selection and Integration of Diverse Communication Channels

Control Objective: To ensure broad and effective reach of public updates on incident recovery by integrating a mix of communication channels tailored to diverse audience segments.

Implementation Steps:

2.1 **Channel Identification and Evaluation**: Identify and evaluate a mix of traditional and digital communication channels, including social media, press releases, and direct emails, for their effectiveness in reaching the intended public segments.

2.2 **Integration Plan**: Develop an integration plan for cohesively utilizing these channels, ensuring that messages are adapted to fit the format and audience of each channel without compromising message integrity or consistency.

2.3 **Continuous Monitoring and Adjustment**: Monitor the effectiveness of different channels in real time and adjust the strategy to improve reach and engagement with the public.

Expected Outcome: By selecting and integrating diverse communication channels, the organization can ensure that recovery updates reach a broad audience effectively, facilitating transparency and maintaining public confidence throughout recovery.

Control 3: Real-time Monitoring and Response System for Public Feedback

Control Objective: To establish a system for real-time monitoring of public response and media coverage, enabling quick adjustments to messaging and strategies based on public perception and feedback.

Implementation Steps:

3.1 **Monitoring Tool Implementation**: Implement monitoring tools and services capable of tracking public response and media coverage across various platforms, including social media, news outlets, and forums.

3.2 **Response Team Formation**: Form a dedicated response team trained to analyze feedback, identify potential misinformation, and decide on the necessary adjustments to public communications.

3.3 **Feedback Integration Process**: Develop a process for quickly integrating feedback and insights gained from monitoring into the communication strategy, ensuring that public concerns are addressed proactively.

Expected Outcome: Establishing a real-time monitoring and response system allows the organization to stay ahead of misinformation, adjust messaging as needed, and maintain a positive public image by responding proactively to concerns and feedback during the recovery phase.

A

Appendix A: Glossary of Terms

Access Rights and Credentials Access rights refer to the permissions granted to users or systems to access specific resources, while credentials are the proof of identity used to authenticate a user's access. Managing access rights and credentials is vital in cybersecurity to ensure that only authorized entities can access sensitive information or systems.

Adapting to Legal and Regulatory Requirements Adapting to legal and regulatory requirements involves staying compliant with laws and regulations relevant to cybersecurity, such as general data protection regulation (GDPR) or health insurance portability and accountability act (HIPAA), by adjusting security practices and protocols accordingly to avoid penalties or legal consequences.

Adjusting for Technological Advancements Adjusting for technological advancements involves updating cybersecurity measures to address new threats and vulnerabilities introduced by emerging technologies, ensuring that security protocols protect against evolving cyber risks.

Alignment with Organizational Mission Alignment with the organizational mission entails ensuring that cybersecurity initiatives and practices are congruent with the overarching goals and values of the organization, contributing to its overall objectives and success while safeguarding its assets and reputation.

Basic Cybersecurity Practices Basic cybersecurity practices encompass fundamental security measures and protocols to protect systems, networks, and data from common cyber threats, such as using strong passwords, regularly updating software, and implementing firewalls and antivirus software.

Business Goals Alignment Business goals alignment involves integrating cybersecurity objectives with broader organizational goals to ensure that security measures support and enhance business operations rather than hindering them, fostering a secure and resilient business environment.

Business Strategy Integration Business strategy integration entails incorporating cybersecurity considerations into an organization's overall strategic planning process, ensuring that security initiatives are aligned with business objectives and integrated seamlessly into decision-making processes.

Communication Strategy Communication strategy involves developing a comprehensive plan for effectively disseminating information related to cybersecurity, both internally and externally, to ensure that stakeholders are informed and aware of security risks, policies, and procedures.

Continuity Planning Continuity planning involves developing strategies and procedures to ensure critical business functions can continue uninterrupted during a disruptive incident or disaster, including cyberattacks, enabling swift recovery and minimizing operational downtime.

A Comprehensive Guide to the NIST Cybersecurity Framework 2.0: Strategies, Implementation, and Best Practice,
First Edition. Jason Edwards.
© 2025 John Wiley & Sons Ltd. Published 2025 by John Wiley & Sons Ltd.

Continuous Improvement Models Continuous improvement models involve implementing frameworks such as Plan-Do-Check-Act (PDCA) or Six Sigma to continually assess, refine, and enhance cybersecurity practices, fostering a culture of ongoing improvement and adaptation to evolving threats and challenges.

Continuous Learning and Adaptation Continuous learning and adaptation involve acquiring new knowledge and skills while adjusting cybersecurity strategies and practices to respond to emerging threats and changing circumstances, ensuring that security measures remain effective and current.

Continuous Monitoring Continuous monitoring refers to the real-time or near-real-time surveillance of systems, networks, and activities to detect and respond to cybersecurity threats promptly, helping to minimize the impact of potential security incidents and breaches.

Critical Objectives and Services Critical objectives and services are the key functions, assets, or processes within an organization that are essential for its operations and require heightened protection to ensure their availability, integrity, and confidentiality against cyber threats.

Cross-Functional Teams Cross-functional teams consist of individuals from diverse departments or disciplines working collaboratively on cybersecurity initiatives, leveraging their varied expertise to address complex security challenges effectively and comprehensively.

Cybersecurity Cybersecurity encompasses the technologies, processes, and practices designed to protect systems, networks, and data from unauthorized access, misuse, modification, or destruction, safeguarding against cyber threats and vulnerabilities.

Cybersecurity Competencies in Hiring Cybersecurity competencies in hiring involve identifying and recruiting candidates with the necessary skills, knowledge, and experience to fulfill cybersecurity roles effectively, ensuring that the organization has a capable workforce to address its security needs.

Cybersecurity in HR Practices Cybersecurity in HR practices involves integrating security considerations into human resources processes, such as employee training, background checks, and access control, to mitigate insider threats and enhance overall security posture.

Cybersecurity Policies Cybersecurity policies are formal documents that outline the rules, guidelines, and procedures governing the organization's approach to security, establishing expectations for employee behavior, system usage, and incident response to mitigate risks effectively.

Cybersecurity Policy Development Cybersecurity policy development is creating and implementing comprehensive policies that address various security aspects tailored to the organization's specific needs, risks, and regulatory requirements.

Cybersecurity Posture Cybersecurity posture refers to the overall strength and resilience of an organization's security defenses and practices, reflecting its ability to detect, prevent, and respond to cyber threats effectively.

Cybersecurity Resource Allocation Cybersecurity resource allocation involves distributing resources, such as budget, personnel, and technology, to support cybersecurity initiatives and priorities based on risk assessments and strategic objectives.

Cybersecurity Risk Management Cybersecurity risk management identifies, assesses, prioritizes, and mitigates risks to the organization's systems, networks, and data, aiming to minimize the likelihood and impact of security incidents and breaches.

Cybersecurity Supply Chain Risk Management (C-SCRM) Cybersecurity supply chain risk management (C-SCRM) focuses on identifying and mitigating security risks associated with third-party vendors, suppliers, and partners whose products or services are integral to the organization's operations.

Cybersecurity Training and Awareness Cybersecurity training and awareness programs aim to educate employees and stakeholders about security best practices, policies, and procedures, empowering them to recognize and mitigate cyber threats effectively and reducing the likelihood of security breaches caused by human error or negligence.

Data Analysis Data analysis involves examining and interpreting data to identify trends, patterns, and insights that can inform decision-making and improve cybersecurity strategies, such as detecting anomalies indicative of security breaches or vulnerabilities.

Data Protection Measures Data protection measures encompass a range of safeguards and controls implemented to ensure the confidentiality, integrity, and availability of sensitive data, including encryption, access controls, and data loss prevention technologies.

Data Retention and Destruction Data retention and destruction policies dictate how long data should be stored and when it should be securely disposed of to minimize the risk of unauthorized access or misuse, ensuring compliance with legal and regulatory requirements.

Dependency Mapping and Risk Analysis Dependency mapping and risk analysis involve identifying and assessing the interdependencies between systems, applications, and services to understand potential vulnerabilities and prioritize risk mitigation efforts effectively.

Dynamic Cybersecurity Risk Management Process Dynamic cybersecurity risk management processes adapt and evolve in response to changing threats, vulnerabilities, and organizational requirements, ensuring that risk management strategies remain effective and relevant.

Effective Communication Effective communication involves conveying cybersecurity-related information clearly and accurately to relevant stakeholders, facilitating understanding and collaboration to address security challenges efficiently.

Enforcement Mechanisms Enforcement mechanisms ensure compliance with cybersecurity policies and procedures, often involving disciplinary actions or technological controls to mitigate security risks effectively.

Ethical Cybersecurity Decision-Making Ethical cybersecurity decision-making involves considering moral and professional principles when addressing security issues, ensuring that actions taken are responsible, fair, and aligned with ethical standards.

Gamification in Cybersecurity Training cybersecurity training involves incorporating game elements, such as rewards, challenges, and competition, into educational programs to engage learners and enhance retention of security concepts and practices.

Human Capital Investment in Cybersecurity Human capital investment in cybersecurity entails allocating resources to recruit, train, and retain skilled professionals, recognizing the critical role of personnel in implementing effective security measures and responding to cyber threats.

Incident Response Plans Incident response plans outline the steps to be taken during a cybersecurity incident, detailing roles, responsibilities, and procedures to minimize the impact and facilitate swift recovery.

Incorporating Emerging Threat Intelligence Incorporating emerging threat intelligence involves monitoring and analyzing emerging cyber threats to proactively anticipate and mitigate potential risks, enhancing the organization's resilience against evolving security challenges.

Industry Standards Integration Industry standards integration involves aligning cybersecurity practices with established frameworks and guidelines, such as NIST or ISO standards, to ensure compliance and adopt best practices for security management.

Key Performance Indicators (KPIs) Key performance indicators (KPIs) are metrics used to measure the effectiveness of cybersecurity initiatives and processes, providing insights into performance and enabling informed decision-making.

Key Risk Indicator Key Risk Indicators (KRIs) are metrics used to monitor and assess specific aspects of cybersecurity risk within an organization. These indicators provide early warnings of potential threats or vulnerabilities, enabling proactive risk management and decision-making to mitigate risks effectively. KRIs are essential for identifying emerging risks and maintaining situational awareness to protect critical assets and operations from cyber threats.

Leadership Accountability Leadership accountability refers to the responsibility of organizational leaders for cybersecurity outcomes, fostering a culture of ownership and ensuring that security priorities are prioritized and addressed effectively.

Leadership in Cybersecurity Leadership in cybersecurity involves guiding and directing security efforts at the organizational level, setting strategic objectives, and promoting a security awareness and resilience culture.

Leadership Support for Cybersecurity Leadership support for cybersecurity entails backing security initiatives with resources, advocacy, and visible commitment from top management, reinforcing the importance of cybersecurity across the organization.

Legal and Regulatory Landscapes Legal and regulatory landscapes encompass the laws, regulations, and compliance requirements relevant to cybersecurity, guiding organizations in ensuring legal compliance and managing regulatory risks.

Measurable Objectives Measurable objectives are specific, quantifiable goals to assess progress and success in achieving cybersecurity outcomes, providing performance evaluation and improvement benchmarks.

Mission-Centric Security Training Mission-centric security training provides tailored cybersecurity education that aligns with the organization's mission and operational requirements, enhancing relevance and effectiveness.

Monitoring and Reporting Monitoring and reporting involve continuously observing security-related activities and generating reports to track performance, detect anomalies, and facilitate informed decision-making in cybersecurity management.

Organizational Integrity Organizational integrity refers to the commitment to ethical behavior, transparency, and accountability in cybersecurity practices, fostering trust and credibility internally and externally.

Performance Evaluation Framework A performance evaluation framework provides a structured approach to assessing the effectiveness of cybersecurity initiatives, including criteria, methodologies, and performance indicators.

Performance Management and Cybersecurity Performance management in cybersecurity involves monitoring, evaluating, and optimizing the effectiveness of security measures and personnel to achieve organizational security objectives.

Performance Shortfalls and Successes Performance shortfalls and successes are instances where cybersecurity measures either fail to meet expectations or achieve desired outcomes, providing opportunities for improvement or validation of effectiveness.

Phishing Awareness Phishing awareness programs educate users about the risks of phishing attacks and how to recognize and avoid them, reducing the likelihood of falling victim to phishing scams and compromising security.

Policy Dissemination Policy dissemination involves distributing cybersecurity policies and guidelines to relevant stakeholders, ensuring awareness and compliance with established security protocols.

Positive Risk Management Positive risk management maximizes opportunities while minimizing threats, leveraging risk to drive innovation and growth while maintaining cybersecurity resilience.

Privacy and Civil Liberties Privacy and civil liberties refer to individuals' rights to control their personal information and freedoms, guiding the ethical and legal considerations in cybersecurity practices to respect privacy rights and civil liberties.

Real-World Scenarios in Training Real-world scenarios in training simulate actual cybersecurity incidents or situations, providing learners with practical experience and skills to respond effectively to real-life security challenges.

Regulatory Compliance Regulatory compliance involves adhering to laws, regulations, and industry standards relevant to cybersecurity, ensuring that organizations operate within legal and regulatory frameworks to mitigate risks and avoid penalties.

Risk Appetite and Tolerance Risk appetite and tolerance define the organization's willingness to accept and manage risks, guiding decision-making and risk management strategies to align with business objectives and priorities.

Risk Management Framework A risk management framework provides a structured approach to identifying, assessing, and mitigating cybersecurity risks, guiding organizations in managing risk effectively and consistently.

Risk Management Objectives Risk management objectives articulate the goals and priorities of the organization's risk management efforts, guiding the development and implementation of risk mitigation strategies.

Risk Mitigation Strategies Risk mitigation strategies are measures implemented to reduce the likelihood or impact of cybersecurity risks, such as implementing controls, transferring risk, or accepting risk within defined tolerances.

Risk-Aware Culture A risk-aware culture fosters awareness, accountability, and proactive engagement in managing cybersecurity risks throughout the organization, promoting a collective responsibility for security.

Roles and Responsibilities Roles and responsibilities define the duties and accountabilities of individuals or teams involved in cybersecurity, clarifying expectations and ensuring effective coordination and execution of security tasks.

Scheduled Policy Review Scheduled policy review involves regularly evaluating and updating cybersecurity policies and procedures to ensure relevance, effectiveness, and compliance with evolving threats and requirements.

Secure Offboarding Procedures Secure offboarding procedures ensure that departing employees' access to systems and data is promptly revoked and that sensitive information is protected during the transition, minimizing the risk of insider threats and unauthorized access.

Stakeholder Engagement Stakeholder engagement actively involves relevant parties, such as employees, partners, and customers, in cybersecurity discussions and decision-making processes to foster collaboration and shared responsibility for security.

Stakeholder Feedback Stakeholder feedback solicits input and perspectives from relevant parties on cybersecurity matters, informing decision-making and continuous improvement efforts to address concerns and align with stakeholder needs.

Strategic Asset A strategic asset is a resource or capability contributing significantly to achieving organizational objectives, including cybersecurity initiatives that protect valuable assets and support business goals.

Strategic Planning Strategic planning involves setting long-term goals and objectives for cybersecurity, defining strategies and tactics to achieve them, and allocating resources effectively to support security initiatives.

Strategy Adjustment Process The strategy adjustment process involves periodically reviewing and adapting cybersecurity strategies in response to changes in the threat landscape, technology, or business environment, ensuring ongoing relevance and effectiveness.

Supplier Criticality Supplier criticality assesses the importance of third-party vendors and suppliers to the organization's operations, guiding the prioritization of cybersecurity measures and risk management efforts.

Supply Chain Vulnerabilities Supply chain vulnerabilities are weaknesses or risks within the interconnected network of suppliers, partners, and vendors that can be exploited to compromise the organization's security, requiring proactive mitigation strategies to safeguard against potential threats.

Tailored Training Content Tailored training content is customized to address specific cybersecurity needs, roles, and skill levels within the organization, maximizing relevance and effectiveness in educating personnel on security best practices.

Technological Resources for Cybersecurity Technological resources for cybersecurity encompass tools, software, and hardware solutions designed to detect, prevent, and mitigate cyber threats, including firewalls, intrusion detection systems, encryption software, and security analytics platforms.

Technology for Risk Assessment Technology for risk assessment involves using automated tools and software applications to identify, analyze, and prioritize cybersecurity risks, leveraging data analytics and machine learning algorithms to enhance the accuracy and efficiency of risk management processes.

Technology in Policy Enforcement Technology in policy enforcement refers to using technological controls and mechanisms to ensure compliance with cybersecurity policies and regulations, such as access control systems, encryption, and data loss prevention solutions, to enforce policy guidelines and prevent unauthorized actions.

Training Programs Training programs are structured initiatives to educate employees and stakeholders on cybersecurity best practices, policies, and procedures, enhancing awareness, skills, and capabilities to mitigate security risks effectively and foster a culture of security awareness within the organization.

B

Appendix B: Descriptions of NIST 800-53 Controls

The National Institute of Standards and Technology (NIST) Special Publication 800-53, Revision 5, stands as a cornerstone document for federal agencies and organizations aiming to safeguard their information systems. This publication presents a comprehensive set of security and privacy controls designed to address various threats and technologies, reflecting the latest advancements and challenges in the cybersecurity landscape. Understanding and implementing these controls is crucial for establishing a robust security posture as organizations navigate the complexities of protecting their digital assets. For those seeking to become NIST 800-53 compliant, it is essential to delve into the specifics of each control, assessing how they align with organizational needs and regulatory requirements.

The controls detailed in NIST 800-53 are categorized into families, each addressing a specific security or privacy concern area. These families cover a broad spectrum of topics, from access control (AC) to incident response (IR) and risk assessment (RA) to system and communications protection. By comprehensively applying these controls, companies can create a multilayered defense strategy that mitigates risks, enhances resilience, and ensures their information systems' confidentiality, integrity, and availability. Organizations must thoroughly review each control, tailoring their implementation strategies to the unique aspects of their operational environment and the specific threats they face.

For those looking to achieve compliance with NIST 800-53, it is imperative to consult the entire publication for an in-depth understanding of each control's requirements, objectives, and implementation guidance. The publication provides valuable insights into the rationale behind each control, offering a framework for how they can be effectively integrated into an organization's security protocols. Interested parties can access the complete document and the latest updates on the NIST website at https://csrc.nist.gov/publications/detail/sp/800-53/rev-5/final. Reviewing the full text will ensure that organizations can accurately align their security measures with the standards outlined in the publication.

Moreover, NIST 800-53 controls offer a valuable starting point for organizations not bound by federal compliance requirements but seeking to enhance security measures. The principles and practices outlined in the publication are universally applicable, offering a solid foundation for developing a tailored security strategy. By adopting and adapting these controls, businesses of all sizes and sectors can strengthen their defenses against evolving cybersecurity threats. Whether pursuing compliance or seeking to bolster security, the NIST 800-53 framework provides a comprehensive and adaptable approach to modern cybersecurity challenges.

A Comprehensive Guide to the NIST Cybersecurity Framework 2.0: Strategies, Implementation, and Best Practice, First Edition. Jason Edwards.
© 2025 John Wiley & Sons Ltd. Published 2025 by John Wiley & Sons Ltd.

Access Control Overview

The AC family establishes a comprehensive framework for managing and enforcing access permissions to systems, applications, and data. These controls are designed to protect against unauthorized access, ensuring that only authorized users, processes, or devices have the right to access or manipulate resources according to their roles and responsibilities. By implementing robust AC mechanisms, organizations can safeguard their critical assets, maintain confidentiality, integrity, and availability, and comply with regulatory and security policies. The AC family addresses various aspects of access management, including user authentication, authorization, access enforcement, and monitoring, providing a layered security approach to protect sensitive information from internal and external threats.

AC-1—Policy and Procedures: AC-1 emphasizes developing, documenting, and disseminating AC policies and procedures across an organization. It mandates clear articulation of the policy's objectives, ensuring alignment with legal and regulatory frameworks, and designates officials to oversee this crucial process. Regular reviews and updates are required to keep the policy and procedures relevant and effective. This foundational control is critical to establishing a coherent access management strategy and enhancing security and privacy through structured risk management and collaborative efforts.

AC-2—Account Management: AC-2 outlines the comprehensive framework for managing system accounts, including policy formulation on account types, creation, modification, and termination. It specifies the roles of account managers, criteria for account creation, and processes for monitoring and updating account statuses in alignment with personnel changes. Enhanced features include automated account management, special handling for temporary and emergency accounts, and stringent measures for deactivating inactive or noncompliant accounts. This control is critical for maintaining strict security protocols and ensuring precise control over access privileges.

AC-3—Access Enforcement: AC-3 focuses on enforcing approved authorizations for logical access to information and system resources aligned with AC policies. It differentiates between active entities (users or processes) and passive ones (devices, files, and records) to ensure access is appropriately regulated. This control extends to applications and services, enhancing security and privacy beyond the system level. It supports various enhancements, including dual authorization for heightened security, mandatory and discretionary ACs to restrict or permit information flow based on policy definitions, and role-based AC to assign access based on job functions. Additional provisions ensure secure, controlled access to security-relevant information and enable the dynamic management of access permissions, further solidifying the framework for secure information access and flow within organizations.

AC-4—Information Flow Enforcement: AC-4 mandates the enforcement of approved authorizations controlling the flow of information within and between systems based on defined information flow control policies. It addresses the regulation of information transfer, not just access, focusing on maintaining security and privacy across different domains and adhering to specific data structures and content requirements. Enhancements include using security and privacy attributes for flow control decisions, establishing protected processing domains, and implementing dynamic information flow controls responsive to operational needs. The control also covers encrypted information flow, metadata use, and the employment of hardware mechanisms for one-way flows. By specifying restrictions on embedded data types and enforcing policy filters, AC-4 plays a crucial role in safeguarding against unauthorized or harmful information transfers, supporting the integrity and confidentiality of data as it moves within and beyond organizational boundaries.

AC-5—Separation of Duties: AC-5 focuses on dividing organizational roles and responsibilities to reduce the risk of malicious activity going undetected. This control ensures that no individual controls all aspects of any critical process, requiring collaboration for sensitive tasks and decisions. It helps prevent fraud and errors, as duties and privileges for various operations are distributed among multiple people or systems. This approach is fundamental to internal security policies. It is a critical component of a robust AC strategy, safeguarding against insider threats by limiting the scope of action any person or system can perform without oversight.

AC-6—Least Privilege: AC-6 mandates that users are provided only with the access necessary to perform their duties, minimizing the potential for unauthorized actions or access to sensitive information. This control requires strict enforcement of access limitations for all users, processes, and devices, ensuring that privileges are assigned based on necessity rather than convenience. Regular reviews and adjustments of user privileges are essential to this control, providing access rights stay aligned with job requirements and reducing the risk of security breaches by exploiting excessive privileges.

AC-7—Unsuccessful Logon Attempts: AC-7 aims to protect against unauthorized access by limiting the number of unsuccessful attempts a user can make. After reaching a specified threshold, actions such as account lockout or alert notifications are triggered. This control helps thwart brute-force attacks and promptly identifies and addresses potential intrusions. It is a critical measure in maintaining the integrity and security of user accounts across organizational systems.

AC-8—System Use Notification: AC-8 requires displaying system use notifications on login screens, informing users of the conditions, terms, and policies regarding the use of the system. This notification serves as a legal boundary and awareness tool, informing users of their responsibilities and the consequences of misuse before accessing system resources. It is a preventive measure that helps in legal enforcement and policy compliance, contributing to the organization's security posture.

AC-9—Previous Logon Notification: AC-9 enhances user accountability and awareness by providing users with information about their previous logon sessions upon successful authentication. This control can alert users to the potential unauthorized use of their credentials if the last logon time displayed does not match their expectations. It acts as an early warning system for possible security issues, prompting users to report suspicious activity and thereby aiding in the timely detection of breaches.

AC-10—Concurrent Session Control: AC-10 limits the number of concurrent sessions a user can establish with a system. This control helps mitigate the risk of multiple sessions that could be exploited maliciously. Organizations can reduce the chances of session hijacking and other related security threats by controlling the number of sessions, ensuring a more secure operational environment.

AC-11—Device Lock: AC-11 mandates mechanisms to lock devices after a predefined period of inactivity or under certain conditions, requiring user reauthentication to regain access. This control is essential in preventing unauthorized access to unattended devices and safeguarding sensitive information and system resources. Device lock policies enforce physical security measures for digital assets, playing a crucial role in comprehensive AC strategies.

AC-12—Session Termination: AC-12 ensures user sessions are automatically terminated after predefined conditions of inactivity or elapsed time to mitigate unauthorized access risks. This control is crucial in environments where users may leave their sessions unattended, potentially providing an entry point for unauthorized users. By enforcing session termination policies, organizations can significantly reduce the risk of unauthorized access, ensuring that active sessions are closed promptly when not in use, thereby enhancing the system's security posture.

AC-13—Supervision and Review—Access Control: AC-13 mandates regular supervision, monitoring, and review of ACs and user activities within an organization. This control is designed to ensure that all access attempts, both successful and unsuccessful, are logged, monitored, and reviewed to detect any unauthorized or suspicious activities. By implementing a robust supervision and review process, organizations can quickly identify and mitigate potential security threats, ensuring ACs are effective and comply with organizational policies and standards.

AC-14—Permitted Actions Without Identification or Authentication: AC-14 specifies specific actions that may be performed without requiring user identification or authentication, carefully selected to ensure they do not compromise the system's security. This control acknowledges the necessity for some functionalities to be publicly accessible while ensuring these exceptions do not introduce vulnerabilities. Properly defining and managing these permissions is key to maintaining the balance between usability and security, allowing for necessary operations such as public information access while safeguarding sensitive functions and data.

AC-15—Automated Marking: AC-15 requires the implementation of mechanisms to automatically mark information based on predefined criteria, such as confidentiality levels, facilitating consistent enforcement of ACs, and handling policies across the organization. This automated marking aids in accurately and consistently classifying data, ensuring that information is handled, stored, and transmitted according to its sensitivity level. It streamlines the data classification process and reduces human errors, enhancing the overall information security framework.

AC-16—Security and Privacy Attributes: AC-16 involves assigning, managing, and enforcing security and privacy attributes for information, resources, and users. These attributes are critical in AC decisions, allowing for dynamic and flexible ACs that adapt to changing requirements. By utilizing these attributes, organizations can implement policies such as least privilege and need-to-know more effectively, ensuring that users have access only to the information and resources necessary for their roles.

AC-17—Remote Access: AC-17 sets forth the requirements for secure remote access to organizational systems, emphasizing secure authentication methods, data encryption in transit, and stringent monitoring of remote connections. This control is essential for organizations to ensure that remote access to their networks is controlled, authenticated, and encrypted, reducing the risk of data breaches and unauthorized access. Implementing comprehensive remote access policies and technologies is crucial in today's increasingly distributed work environments.

AC-18—Wireless Access: AC-18 explicitly addresses the security considerations for wireless access to organizational systems, requiring authentication, encryption, and monitoring to protect against the vulnerabilities inherent in wireless communications. This control ensures that wireless networks and connections meet organizational security standards, safeguarding the confidentiality, integrity, and availability of information transmitted over these potentially less secure networks.

AC-19—Access Control for Mobile Devices: AC-19 mandates the establishment of controls to manage the security of mobile devices accessing organizational systems. This control requires implementing policies and technologies to authenticate mobile device users, encrypt sensitive data, and manage their applications. It addresses the unique risks posed by mobile computing, such as loss or theft of devices and the potential for unauthorized access to organizational data. By enforcing strict ACs and security measures for mobile devices, organizations can protect their resources while accommodating the flexibility of mobile computing.

AC-20—Use of External Systems: AC-20 governs the use of external information systems that are not under the organization's control but can access or process its information. This control ensures that organizational data is protected when processed, stored, or transmitted on external systems, requiring that such activities comply with security policies. It emphasizes the need for agreements or contracts that mandate security measures equivalent to the organization's, ensuring data confidentiality, integrity, and availability, regardless of where it resides.

AC-21—Information Sharing: AC-21 involves establishing criteria and procedures for securely sharing information between entities. This control is critical for protecting sensitive information while facilitating information exchange within and between organizations. It requires mechanisms to ensure that information is shared in a manner that maintains its security attributes and is only accessible to authorized parties. Proper implementation of AC-21 supports collaboration and operational objectives while safeguarding against unauthorized disclosure or compromise of information.

AC-22—Publicly Accessible Content: AC-22 mandates reviewing and approving information before making it publicly accessible, ensuring that sensitive or classified information is not inadvertently exposed. This control requires organizations to establish processes for carefully screening content to be published on public platforms, including websites and social media. By controlling what information is made public, organizations can protect sensitive data from unintended disclosure and reduce the risk of security breaches stemming from publicly accessible content.

AC-23—Data Mining Protection: AC-23 requires the implementation of protections against unauthorized data mining activities that could compromise sensitive information. This control is designed to prevent the aggregation, analysis, or combination of data in ways that could reveal sensitive patterns, relationships, or information not apparent from the individual data items. By safeguarding against data mining, organizations can protect the privacy of individuals and the security of organizational data from sophisticated analysis techniques aimed at uncovering confidential information.

AC-24—Access Control Decisions: AC-24 focuses on enforcing access control decisions based on all relevant information, including the context of access requests, the resources accessed, and associated risks. This control ensures that decisions on granting or denying access are made comprehensively, considering a wide range of factors to assess the appropriateness of access accurately. It supports a dynamic and context-aware AC system that can adapt to varying operational needs and security requirements.

AC-25—Reference Monitor: AC-25 emphasizes the need for a reference monitor. This abstract machine mediates all access requests to objects by subjects to ensure that ACs are enforced securely, tamper-resistantly. This control is foundational to safe computing, ensuring AC policies are effectively and consistently applied. The reference monitor enforces the principle of least privilege and ensures that unauthorized access attempts are reliably blocked, providing a critical mechanism for maintaining the security and integrity of the system.

Awareness and Training (AT) Overview

The AT family of controls emphasizes the critical role of human factors in maintaining information security. It focuses on developing, implementing, and maintaining a comprehensive security awareness program that educates all organizational members about their security responsibilities. This program aims to enhance personnel's security knowledge and behavior through targeted training sessions, thereby reducing human error and improving response to security incidents. The AT controls ensure that all users know the potential security risks associated with their activities and the necessary precautions to mitigate them, fostering a culture of security within the organization.

AT-1—Policy and Procedures: AT-1 establishes the foundation for training and awareness activities within an organization by requiring developing, documenting, and disseminating training policies and related procedures. These policies and procedures should clearly define training requirements, ensuring that all personnel know the security risks associated with their activities and understand the relevant policies, standards, and procedures designed to mitigate them. The control emphasizes updating training materials and methods to reflect new threats, technologies, and organizational changes. Regular review and updates to the training policy and procedures help ensure the training program remains effective and relevant, supporting the organization's security posture and compliance efforts.

AT-2—Security Awareness Training: AT-2 mandates that all organizational users, including contractors and third-party users, receive security awareness training that covers the organization's security policies, the acceptable use of information systems, and procedures for responding to security incidents. This control aims to foster a culture of security awareness where users understand their role in protecting the organization's assets and are equipped with the knowledge to act securely in their daily activities. The training is required upon hiring and annually after that, reinforcing security concepts and keeping security at the forefront of users' minds. Effective security awareness training reduces the likelihood of user-induced security breaches and enhances the organization's resilience against cyber threats.

AT-3—Role-Based Training: AT-3 specifies the need to provide targeted, role-based security training to personnel with significant security responsibilities. This control recognizes that different organizational roles have unique security requirements and that tailored training can address these specific needs more effectively. Role-based training ensures that personnel are adequately prepared to securely perform their duties by focusing on each role's responsibilities and system interactions. The objective is to ensure that individuals understand the security implications of their actions and the security procedures specific to their job functions, thereby enhancing the overall security capabilities within the organization.

AT-4—Training Records: AT-4 requires the maintenance (MA) of accurate and comprehensive records of all security training activities conducted by the organization. These records should include details about the training content, the date of training, and the participants. Keeping detailed training records allows organizations to verify that all personnel have received the required training, identify any individuals who may need additional training, and evaluate the effectiveness of the training program over time. Training records are essential for demonstrating compliance with internal policies and external regulatory requirements, and they serve as a valuable resource for planning future training initiatives.

AT-5—Contacts with Security Groups and Associations: AT-5 encourages organizations to actively engage with external security groups and associations to stay informed about the latest security trends, threats, and best practices. This control highlights the value of leveraging external expertise and collaborative efforts to enhance an organization's security posture. By establishing and maintaining contacts with security groups, organizations can gain insights into emerging security issues, access resources for improving security AT, and contribute to the broader security community. Such engagement can also provide opportunities for sharing experiences and strategies, fostering a culture of continuous improvement in security practices.

AT-6—Training Feedback: AT-6 emphasizes collecting participant feedback on the security training and awareness programs. This feedback is instrumental in assessing the effectiveness of the training, identifying areas for improvement, and tailoring future training to meet the needs of the organization and its personnel better. By actively soliciting and analyzing training feedback, organizations can ensure that their training programs remain relevant, engaging, and effective in promoting a secure and aware organizational culture. Implementing changes based on feedback demonstrates the organization's commitment to continuous improvement and can lead to more effective training outcomes.

Audit and Accountability (AU) Overview

The Audit and Accountability (AU) family of controls mandates creating, protecting, and analyzing records to monitor, analyze, investigate, and report on information system activity. These controls are vital for detecting security violations, performance issues, and flaws in system processes. By maintaining comprehensive audit logs and ensuring that activities can be traced to the responsible entity, organizations can enhance their security posture, ensure compliance with legal and regulatory requirements, and support forensic investigations. The AU controls cover aspects such as audit log generation, storage, review, and analysis, providing a framework for accountability and transparency in managing information systems.

AU-1—Policy and Procedures: AU-1 requires organizations to develop, document, and disseminate an AU policy clearly defining audit-related activities and responsibilities. This policy and its accompanying procedures form the backbone of the organization's audit strategy, ensuring that all actions affecting system security are recorded, monitored, and analyzed. By establishing a comprehensive audit policy, organizations can identify unauthorized or malicious activity, support forensic investigations, and ensure compliance with legal and regulatory requirements. Regular review and updates to the audit policy and procedures are necessary to adapt to changes in the operational environment and evolving threats, maintaining the effectiveness of the audit program.

AU-2—Event Logging: AU-2 mandates the logging of specific types of events to create an audit trail that can be used to support security incident detection, analysis, and response efforts. This control specifies the need for detailed logging of user activities, system alerts, and security events to ensure administrators and security personnel have the necessary information to identify and investigate suspicious activities. By defining which events should be logged and ensuring that logs are generated consistently and comprehensively, organizations can enhance their situational awareness and improve their ability to promptly detect and respond to potential security incidents.

AU-3—Content of Audit Records: AU-3 specifies the essential elements that must be included in audit records to provide a comprehensive and actionable account of security-relevant events. This control ensures that audit records contain sufficient detail, including user identification, event type, date and time, and the event's outcome. Such detailed audit records enable organizations to effectively analyze security incidents, track user behavior, and comply with legal and regulatory requirements. By standardizing the content of audit records, organizations can facilitate efficient and thorough analysis, aiding in the detection, understanding, and remediation of potential security issues.

AU-4—Audit Log Storage Capacity: AU-4 emphasizes the importance of allocating adequate storage capacity for audit logs to ensure that logs are retained for a period sufficient to meet operational, legal, and regulatory requirements. This control requires organizations to plan and implement storage solutions that can accommodate the volume of audit data generated, preventing the loss of valuable information due to storage limitations. By proactively managing audit log storage, organizations can ensure the availability of audit records for investigation and analysis purposes, supporting ongoing security monitoring and compliance efforts.

AU-5—Response to Audit Logging Process Failures: AU-5 outlines the procedures to be followed if an audit logging process fails. This control ensures that organizations have mechanisms to alert personnel to such failures and take corrective action, maintaining the continuity and integrity of the audit logging function. Whether it involves notifying system administrators, automatically halting system processes, or implementing alternative logging mechanisms, the goal is to address failures promptly to minimize the risk of lost audit data. Effective management of audit logging process failures is critical for preserving the ability to track and analyze security-related events.

AU-6—Audit Record Review, Analysis, and Reporting: AU-6 mandates the regular review, analysis, and reporting of audit records to identify unauthorized activities, policy violations, and other security-relevant events. This control ensures that organizations actively scrutinize audit logs to detect anomalies, potential security incidents, and compliance issues. By systematically reviewing audit records, organizations can uncover insights into the efficacy of their security controls, identify areas for improvement, and take informed action to enhance their security posture. The analysis and reporting of audit findings are essential components of an effective security program, enabling proactive management of security risks.

AU-7—Audit Record Reduction and Report Generation: AU-7 focuses on processing and reducing audit records to events of interest and generating reports that can assist in responding to audit findings. This control allows organizations to filter and analyze vast amounts of audit data, focusing on the most critical information for security and compliance purposes. By utilizing tools and processes for audit record reduction and report generation, organizations can efficiently identify significant events, trends, and patterns within the audit data, enhancing their ability to manage and respond to security incidents.

AU-8—Time Stamps: AU-8 requires using time stamps in audit records to record the date and time of events accurately. This control ensures that all audit records include synchronized time information, providing a reliable sequence of events crucial for event reconstruction and forensic analysis. Time stamps are critical in investigating security incidents, correlating events across different systems, and establishing the chronology of user actions and system responses. By implementing precise and synchronized time-stamping mechanisms, organizations can enhance the integrity and usefulness of audit records.

AU-9—Protection of Audit Information: AU-9 mandates the protection of audit information to prevent unauthorized access, modification, and deletion. This control is essential for ensuring audit records' confidentiality, integrity, and availability, which contain sensitive information about system operations and security events. Protecting audit information involves implementing ACs, encryption, and integrity-checking mechanisms to safeguard audit logs from tampering or unauthorized disclosure. By securing audit information, organizations can maintain the trustworthiness of their audit logs as a critical resource for security monitoring, IR, and compliance verification.

AU-10—Nonrepudiation: AU-10 ensures that actions taken on information systems cannot be denied by the individuals or entities responsible for those actions. This control involves capturing and retaining evidence that can conclusively prove the occurrence of specific actions and the identities of the individuals or entities involved. Nonrepudiation mechanisms, such as digital signatures and cryptographic techniques, are critical for ensuring accountability, supporting legal and regulatory compliance, and providing irrefutable evidence in the event of disputes or security incidents. Organizations can deter malicious behavior and protect against false denials of involvement in security-relevant activities by implementing nonrepudiation measures.

AU-11—Audit Record Retention: AU-11 requires organizations to retain audit records for a defined period to support accountability and analysis needs. This control ensures that audit records are available sufficiently to permit the reconstruction of events for incident investigation, legal proceedings, and compliance audits. The retention period for audit records should be determined based on legal, regulatory, and operational requirements, ensuring that valuable historical audit data is preserved for as long as necessary. By establishing and enforcing audit record retention policies, organizations can maintain access to critical historical data, supporting ongoing security, compliance, and forensic efforts.

AU-12—Audit Record Generation: AU-12 mandates generating audit records for defined events, providing the information necessary to reconstruct events, detect unauthorized access, and verify compliance. This control requires that systems are configured to automatically generate detailed audit records for specific types of activities, ensuring comprehensive monitoring of security-relevant actions. By systematically generating audit records, organizations can enhance their visibility into system operations, support effective security monitoring, and facilitate the timely detection and response to potential security incidents.

AU-13—Monitoring for Information Disclosure: AU-13 focuses on monitoring and detecting unauthorized information disclosures through audit analysis. This control involves scrutinizing audit records and system activity to identify instances where sensitive information may have been intentionally or unintentionally exposed to unauthorized individuals or entities. By implementing processes for monitoring information disclosure, organizations can quickly identify and respond to potential data leaks or breaches, mitigating the impact of unauthorized information exposure and protecting the confidentiality of sensitive data.

AU-14—Session Audit: AU-14 requires auditing user session activities to provide a comprehensive record of actions performed during a session. This control ensures that detailed audit records are generated for activities within a user session, offering insights into user behavior, system interactions, and potential security violations. Session auditing is vital for identifying unauthorized access, investigating security incidents, and understanding the context of specific actions within a session. By capturing detailed session audit records, organizations can enhance their ability to monitor, analyze, and respond to security events related to user sessions.

AU-15—Alternate Audit Logging Capability: AU-15 mandates the establishment of an alternate audit logging capability to ensure continuity of auditing in the event of primary audit failure. This control provides a backup mechanism for capturing audit records, safeguarding against data loss, and ensuring that critical security events are logged even when the primary auditing system is compromised or unavailable. Organizations can maintain a reliable audit trail by implementing an alternate audit logging capability, supporting ongoing security monitoring and IR efforts, regardless of system failures or disruptions.

AU-16—Cross-organizational Audit Logging: AU-16 addresses the complexities of audit logging in environments involving multiple organizations, requiring the coordination of audit logging practices across organizational boundaries. This control ensures that audit records generated by systems shared among different entities are managed to meet all parties' security and compliance requirements. Cross-organizational audit logging is crucial for maintaining accountability, supporting incident investigation, and ensuring compliance in collaborative environments. Organizations can effectively manage audit information in multiorganizational contexts by establishing agreements and mechanisms for shared audit logging, facilitating cooperation, and mutual security assurances.

Security Assessment and Authorization (CA) Overview

The security assessment and authorization (CA) family of controls establishes a process for assessing security controls in federal information systems and authorizing their operation. This family emphasizes the importance of ongoing assessments to understand security risks and determine if the controls are effectively mitigating those risks. The CA controls provide a structured approach to security certification, system accreditation, and continuous monitoring, ensuring that systems operate at an acceptable level of risk. By implementing these controls, organizations can understand their security posture comprehensively and make informed decisions about system authorization and operation.

CA-1—Policy and Procedures: CA-1 requires organizations to develop, document, and disseminate a security assessment and authorization policy that outlines the process for evaluating the security controls in place and authorizing system operation based on this evaluation. This foundational control ensures that all activities related to assessing security controls and permitting systems are governed by a formal set of guidelines. It mandates regular reviews and updates to the policy and procedures to adapt to technological changes, threats, and organizational objectives. By establishing a structured approach to security assessment and authorization, organizations can ensure consistent application of security practices and make informed decisions about system operations and risks.

CA-2—Security Assessments: CA-2 focuses on conducting assessments of security controls to determine their effectiveness in protecting organizational information systems. This control involves a comprehensive evaluation process, including testing and reviewing security controls to identify weaknesses and noncompliance with organizational security policies. The outcome of these assessments provides essential feedback for improving security posture and forms the basis for making risk-based decisions about system authorization. Regular security assessments are crucial for accurately understanding security risks and ensuring controls function as intended.

CA-3—Information Exchange: CA-3 addresses the security considerations of exchanging information between organizations. This control requires the establishment of agreements that specify the requirements for protecting exchanged information and ensuring that these requirements are consistent with both parties' security policies. It emphasizes the need for secure methods of information exchange, protection mechanisms for the information being exchanged, and regular monitoring to detect and respond to security breaches. Organizations can safeguard sensitive information while facilitating necessary communications and collaboration by managing the risks associated with information exchange.

CA-4—Security Certification: CA-4 mandates a formal evaluation process, known as security certification, to assess an information system's security features and capabilities. This control ensures that systems meet established security requirements before they are authorized for operation. The certification process thoroughly examines the system's security controls, documentation, and risk management practices to verify compliance with organizational security standards. Security certification is a critical step in the authorization process, providing decision-makers with the information needed to assess risks and determine the system's readiness for operation.

CA-5—Plan of Action and Milestones (POA&M): CA-5 requires developing and maintaining a POA&M for addressing weaknesses found during security assessments. This control ensures organizations have a documented strategy for mitigating, accepting, or transferring identified risks. The POA&M is a tool for tracking remediation efforts, assigning responsibilities, and establishing timelines for resolving security vulnerabilities. By systematically managing the resolution of identified weaknesses, organizations can enhance their security posture and demonstrate a proactive approach to managing information system risks.

CA-6—Authorization: CA-6 involves the formal decision by a senior organizational official to authorize the operation of an information system and to explicitly accept the risk to organizational operations, organizational assets, individuals, other organizations, and the nation based on the implementation of an agreed-upon set of security controls. This control ensures that systems are authorized for operation only after a comprehensive assessment of their security posture and that the decision to operate is made with full knowledge of the potential risks. The authorization process is a critical component of risk management, providing a formal mechanism for accepting the security risks associated with system operations.

CA-7—Continuous Monitoring: CA-7 mandates the implementation of a continuous monitoring strategy to maintain situational awareness of the security state of information systems. This control requires organizations to collect security-related information, analyze this information for indications of unauthorized activity or changes in risk, and respond appropriately. Continuous monitoring activities include the regular assessment of security controls, analysis of audit logs, and the detection of security incidents. By actively monitoring their information systems, organizations can detect and respond to security threats more quickly and effectively, maintaining a robust security posture.

CA-8—Penetration Testing: CA-8 requires organizations to conduct periodic penetration testing to identify vulnerabilities and weaknesses in their information systems that adversaries could exploit. This control involves simulating cyberattacks under controlled conditions to evaluate the effectiveness of security controls and identify exploitable vulnerabilities. Penetration testing provides valuable insights into the real-world effectiveness of an organization's security measures and identifies areas where improvements are needed. By regularly testing their defenses, organizations can stay ahead of potential attackers and strengthen their security posture.

CA-9—Internal System Connections: CA-9 addresses the security considerations for connecting multiple information systems within an organization. This control requires the documentation and authorization of internal system connections, ensuring that these connections do not introduce unacceptable risks. It emphasizes the need for securing interfaces between systems, managing ACs for shared resources, and monitoring internal traffic for suspicious activity. By carefully managing internal system connections, organizations can prevent unauthorized access and data leakage between systems, safeguarding sensitive information and critical infrastructure.

Configuration Management (CM) Overview

The Configuration Management (CM) family of controls focuses on establishing and maintaining the integrity of hardware, software, and firmware configurations within an organization's information systems. Organizations can prevent unauthorized changes that might expose systems to vulnerabilities by standardizing and carefully managing system configurations. The CM controls cover areas such as configuration baseline establishment, configuration change control, security impact analysis for changes, and configuration monitoring. Implementing these controls helps ensure that systems are securely configured and remain in a known, trusted state throughout their lifecycle.

CM-1—Policy and Procedures: CM-1 mandates developing, documenting, and disseminating a CM policy and related procedures. This control establishes the foundation for managing changes to system configurations and ensuring that systems are maintained in a secure, authorized state throughout their lifecycle. The policy and procedures define roles and responsibilities for CM, outline the process for making and documenting configuration changes, and set requirements for auditing and reviewing system configurations. By adhering to a formal CM policy, organizations can maintain control over system changes, reduce the risk of unauthorized alterations, and ensure the integrity of their information systems.

CM-2—Baseline Configuration: CM-2 requires organizations to establish and maintain baseline configurations for their information systems. This control involves documenting the authorized configuration of a system at a specific point in time, providing a reference point for future changes and audits. Baseline configurations include details about system components, software versions, network configurations, and security settings. Organizations can quickly detect unauthorized changes by maintaining baseline configurations, restoring systems to a known secure state, and ensuring all systems are configured according to organizational security standards.

CM-3—Configuration Change Control: CM-3 mandates a formal process for managing changes to system configurations. This control ensures that all changes are reviewed, approved, documented, and communicated before implementation. The change control process helps prevent unintended consequences, such as introducing new vulnerabilities or disrupting system operations. By controlling changes through a structured process, organizations can maintain the security and stability of their information systems, ensuring that changes do not adversely affect system security or operational effectiveness.

CM-4—Impact Analyses: CM-4 requires organizations to conduct impact analyses before implementing configuration changes. This control involves assessing the potential effects of proposed changes on system security, functionality, and performance. Impact analyses help identify potential risks associated with changes and inform decision-making about whether, how, and when to proceed. By conducting thorough impact analyses, organizations can avoid the negative consequences of changes, ensuring that modifications to system configurations do not undermine security or operational goals.

CM-5—Access Restrictions for Change: CM-5 focuses on limiting access to change system configurations to authorized personnel only. This control requires organizations to implement mechanisms to prevent unauthorized individuals from changing system configurations. Access restrictions help to protect systems from malicious alterations, reduce the risk of configuration errors, and maintain the integrity and security of information systems. By carefully controlling who can make configuration changes, organizations can prevent unauthorized access and modifications, safeguarding their systems against potential security breaches.

CM-6—Configuration Settings: CM-6 mandates establishing and managing security configuration settings for information technology products employed within information systems. This control involves identifying secure configuration settings based on industry standards, organizational policies, and operational requirements. By applying and enforcing secure configuration settings, organizations can reduce vulnerabilities, prevent unauthorized access, and ensure that systems are configured to operate securely. Regularly reviewing and updating configuration settings in response to new threats and technological changes is essential for maintaining the security posture of information systems.

CM-7—Least Functionality: CM-7 requires that information systems be configured to provide only essential capabilities and functions to fulfill their intended purpose. This control minimizes the system's attack surface by reducing unnecessary services, features, and functionality attackers could exploit. By enforcing the principle of most minor functionality, organizations can mitigate potential vulnerabilities, enhance system security, and limit the potential impact of successful attacks. Implementing this control involves regularly reviewing system configurations and functionalities to ensure that only necessary components are enabled and operational.

CM-8—System Component Inventory: CM-8 mandates the MA of an inventory of information system components that includes up-to-date information on each element. This control is crucial for effective CM, security assessment, and IR activities. An accurate system component inventory allows organizations to track the status of components, identify unauthorized devices, and manage system configurations more effectively. Organizations can ensure that all components are accounted for, configured securely, and monitored for security compliance by keeping a comprehensive inventory.

CM-9—Configuration Management Plan: CM-9 requires organizations to develop, document, maintain, and implement a CM plan. This plan outlines the organization's approach to managing system configurations, including the processes for establishing baselines, making and documenting changes, and monitoring the system for compliance with established configurations. A well-defined CM plan is essential for ensuring system configuration changes are controlled and coordinated, reducing the risk of introducing vulnerabilities or disrupting system operations. The plan should be regularly reviewed and updated to adapt to technological changes, threats, and organizational objectives.

CM-10—Software Usage Restrictions: CM-10 involves implementing restrictions on the use of software within information systems. This control is designed to prevent the installation and execution of unauthorized or potentially harmful software that could compromise system security. By enforcing software usage policies, organizations can control the software landscape on their systems, reducing the risk of malware infection and data breaches. Restrictions may include prohibiting the use of unauthorized software, ensuring the use of licensed and approved software only, and implementing whitelisting or blacklisting approaches to manage software execution.

CM-11—User-installed Software: CM-11 addresses the risks of allowing users to install software on organizational information systems. This control requires organizations to establish policies and procedures for managing user-installed software, ensuring that such software does not introduce vulnerabilities or violate organizational security policies. By controlling software installation by users, organizations can reduce the risk of unauthorized or insecure software affecting system security and integrity. The control may involve prohibiting user-installed software altogether, requiring approval for software installations, or implementing technical controls to enforce software usage policies.

CM-12—Information Location: CM-12 mandates the management of information location within information systems. This control ensures that information is stored, processed, and transmitted only in authorized, secure locations and compliant with organizational policies. By controlling the information area, organizations can protect sensitive data from unauthorized access, prevent data leakage, and ensure compliance with legal and regulatory requirements. Implementing this control requires understanding the flow of information within and across systems and applying appropriate safeguards to control where information resides and moves.

CM-13—Data Action Mapping: CM-13 requires organizations to map the flow of data within their information systems and understand the actions taken on data at various points in the system. This control helps organizations identify potential security risks associated with data processing and transmission, ensuring that data is handled securely throughout its lifecycle. Organizations can implement targeted security controls to protect data from unauthorized access, modification, and disclosure by understanding how data flows and what actions are taken. Data action mapping is essential for developing effective data protection strategies and supporting compliance with privacy and security regulations.

CM-14—Signed Components: CM-14 mandates the use of signed components within information systems to ensure the authenticity and integrity of system components. This control involves verifying digital signatures on software and system components before installation or execution, assuring that the components have not been tampered with and are from a trusted source. By requiring signed components, organizations can prevent the introduction of malicious software, protect against software tampering, and maintain the security and integrity of their information systems.

Implementing this control requires mechanisms for verifying digital signatures and policies for managing signed components throughout their lifecycle.

Contingency Planning (CP) Overview

The Contingency Planning (CP) family of controls provides a framework for preparing for, responding to, and recovering from system disruptions. These controls emphasize the importance of planning for potential security incidents, natural disasters, or system failures that could impact the availability of information systems and data. By developing and implementing comprehensive contingency plans, organizations can ensure the continuity of operations and the protection of critical information assets. The CP controls guide the establishment of emergency response, backup operations, disaster recovery capabilities, and system and data recovery processes, ensuring organizational resilience in the face of disruptions.

CP-1—Policy and Procedures: CP-1 mandates establishing a CP policy and associated procedures to prepare for, respond to, and recover from system disruptions or failures. This control ensures organizations have a formal framework for developing, implementing, and maintaining contingency plans. Such policies and procedures are crucial for minimizing the impact of unexpected events on mission-critical services and data. Regular updates and reviews of the CP policy and procedures help adapt to changes in the operational environment, technology, and threats, ensuring that the organization remains resilient in the face of disruptions.

CP-2—Contingency Plan: CP-2 requires organizations to develop, document, implement, and maintain a contingency plan that outlines procedures for restoring systems and data after a disruption or failure. This control ensures predefined plans for various incidents, enabling a swift and organized response to minimize downtime and data loss. The contingency plan should cover aspects such as backup operations, disaster recovery, and emergency operations and be tailored to the specific needs and risks of the organization. Regular testing and updates to the contingency plan ensure its effectiveness and the organization's preparedness for unforeseen events.

CP-3—Contingency Training: CP-3 emphasizes the need to provide contingency training to personnel with responsibilities outlined in the contingency plan. This control ensures that individuals executing the contingency plan know their roles and are equipped with the knowledge and skills to effectively perform their duties under crisis conditions. Training enhances the overall readiness of the organization to respond to disruptions, supporting a more efficient and effective recovery process. Ongoing training and exercises help maintain a high level of preparedness and ensure that personnel are current on the latest contingency procedures and technologies.

CP-4—Contingency Plan Testing: CP-4 mandates regular contingency plan testing to evaluate its effectiveness and the organization's readiness to execute it under real-world conditions. This control involves conducting exercises that simulate various disruptions, assessing the response mechanisms, and identifying areas for improvement. Testing helps validate the feasibility and effectiveness of recovery strategies, ensuring that the organization can recover critical functions and information systems within acceptable timeframes. Feedback from testing is used to refine the contingency plan, enhancing the organization's resilience to disruptions.

CP-5—Contingency Plan Update: CP-5 requires the periodic review and update of the contingency plan to ensure it remains relevant, effective, and aligned with current organizational objectives, threats, and technologies. This control recognizes that changes in the operational environment, including new business processes, system upgrades, and emerging threats, can impact the adequacy of existing contingency measures. Regular updates to the contingency plan ensure that it evolves to address new challenges and maintains its utility in guiding the organization through recovery efforts after a disruption.

CP-6—Alternate Storage Site: CP-6 focuses on using an alternate storage site to ensure that backups of essential data, software, and system images are securely and geographically separated from the primary site. This control mitigates the risk of simultaneous loss of primary and backup data due to natural disasters, attacks, or other catastrophic events. An alternate storage site enables organizations to restore critical information and systems rapidly, supporting continuity of operations. The choice of an alternate storage site should consider factors such as distance from the primary site, security measures, and accessibility to ensure the protection and availability of backup resources when needed.

CP-7—Alternate Processing Site: CP-7 mandates establishing and using an alternate processing site that enables the organization to continue critical operations when the primary site is unavailable. This control ensures a predefined location equipped with the necessary systems and infrastructure to resume essential functions with minimal downtime. The alternate site should be selected based on physical security, environmental controls, and the ability to support the organization's operational requirements. Regular testing and MA of the alternate processing site are crucial for ensuring its readiness during a primary site disruption.

CP-8—Telecommunications Services: CP-8 addresses the need for redundant telecommunications services to support continuity of operations in the event of a failure or unavailability of primary communications capabilities. This control ensures that organizations have alternative means of communication to maintain connectivity among personnel, customers, and partners during disruptions. Redundant telecommunications services, including backup Internet connections and alternate communication channels, are critical for sustaining organizational activities and coordinating recovery efforts in response to incidents.

CP-9—System Backup: CP-9 requires the regular backup of system data, configurations, and essential information to support system recovery and reconstitution after a disruption. This control involves implementing procedures for creating, testing, and storing backups in a secure and recoverable format. System backups are a key component of an effective contingency plan, enabling organizations to quickly restore lost data and recover from system failures. The backup process should be designed to minimize data loss and ensure the integrity and confidentiality of backup data during storage and restoration.

CP-10—System Recovery and Reconstitution: CP-10 mandates developing and implementing procedures for recovering systems and reconstituting their capabilities to a fully operational state after a disruption. This control ensures that organizations can restore systems and data from backups, repair or replace damaged components, and return to normal operations within acceptable timeframes. System recovery and reconstitution procedures should cover hardware restoration, software reinstallation, data recovery, and security feature configuration. Regular testing of recovery procedures is essential for validating their effectiveness and the organization's readiness for system restoration.

CP-11—Alternate Communications Protocols: CP-11 involves preparing to use alternate communications protocols if primary protocols are compromised or become unavailable. This control ensures that organizations have the flexibility to switch to different methods of communication to maintain connectivity and continue operations during disruptions. Alternate communications protocols may include network protocols, services, or technologies that can be rapidly deployed to restore communications capabilities. Planning for alternate protocols involves assessing compatibility, security, and performance implications to ensure seamless transition and continuity of operations.

CP-12—Safe Mode: CP-12 requires the capability to start systems safely during recovery operations to perform diagnostics, repair functions, or recover from malicious code infections. This control enables organizations to isolate and address issues in a controlled environment, minimizing potential damage and facilitating restoration. Operating in safe mode allows for the execution of essential functions and security measures while limiting the operation of potentially compromised or nonessential components. Implementing procedures for using safe mode and training personnel on its activation and use are critical for practical system recovery efforts.

CP-13—Alternative Security Mechanisms: CP-13 mandates identifying and preparing alternative security mechanisms to ensure the continuous protection of information and systems when primary mechanisms fail or are unavailable. This control recognizes the importance of maintaining a security posture and compliance with organizational policies, even during disruptions. Alternative security mechanisms may include backup devices, software solutions, or procedures that can be quickly implemented to replace or supplement primary security controls. Planning for alternative security mechanisms involves evaluating their effectiveness, compatibility, and integration with existing systems to ensure adequate protection during contingency operations.

Identification and Authentication (IA) Overview

The Identification and Authentication (IA) family of controls centers on establishing processes and technologies to uniquely identify and authenticate users (or processes) accessing organizational information systems. These controls are crucial for ensuring that only authorized individuals and entities can access system resources and that the identities of these users or entities are verified before granting access. The IA controls cover aspects such as user registration, management of identifiers, authenticator management, and authentication mechanisms. Organizations can protect their systems and data from unauthorized access and potential security breaches by implementing strong IA practices.

IA-1—Policy and Procedures: IA-1 requires developing, documenting, and disseminating an IA policy and associated procedures. This foundational control establishes the framework for identifying and authenticating an organization's users, devices, and services. The policy and procedures outline requirements for unique identification, authenticator management, and using cryptographic modules for secure authentication. By maintaining a comprehensive IA policy, organizations can ensure that access to systems and data is appropriately controlled and that users and devices are authenticated securely and reliably.

IA-2—Identification and Authentication (Organizational Users): IA-2 mandates the unique IA of organizational users before allowing access to information systems. This control ensures that each user is accurately identified and authenticated using one or more factors, such as something the user knows (password), something the user has (token), or something the user is (biometric). Implementing strong IA mechanisms is crucial for preventing unauthorized access, tracking user activities, and enforcing ACs based on user identity. The control applies to all users, including employees, contractors, and third-party providers, reinforcing the security of systems and sensitive information.

IA-3—Device Identification and Authentication: IA-3 focuses on the IA of devices before establishing connections to organizational information systems. This control ensures that only authorized devices can access network resources and services, preventing unauthorized devices from potentially introducing security threats. Device IA can be achieved through certificates, preshared keys, or device credentials, providing a basis for implementing device-based ACs. By securing the connections between devices and systems, organizations can enhance their network security posture and protect against unauthorized access and data breaches.

IA-4—Identifier Management: IA-4 mandates the management of user identifiers by ensuring that they are uniquely assigned to individual users. This control involves establishing processes for issuing, maintaining, and revoking identifiers throughout the user's association with the organization. Proper management of identifiers is essential for accurately associating users with their actions, supporting accountability, and facilitating AC decisions. The control ensures that identifiers are protected against misuse and duplication, contributing to the overall security of user IA processes.

IA-5—Authenticator Management: IA-5 requires the effective management of authenticators, such as passwords, tokens, and biometric data, used for user authentication. This control involves establishing practices for creating, distributing, storing, recovering, and revoking authenticators to ensure their security and effectiveness. Managing authenticators properly is crucial for maintaining the integrity of the authentication process, preventing unauthorized access, and protecting user credentials from compromise. The control covers aspects such as authenticator complexity, change policies, and protection mechanisms, enhancing user authentication security within the organization.

IA-6—Authentication Feedback: IA-6 focuses on providing appropriate feedback to users during the authentication process without compromising the security of authentication information. This control ensures that users receive clear, nonrevealing feedback when authentication attempts fail or succeed, helping them correct errors without providing clues that attackers could exploit. Organizations can improve user experience and security awareness by managing the input provided during authentication while protecting sensitive authentication data and mechanisms from potential threats.

IA-7—Cryptographic Module Authentication: IA-7 mandates using cryptographic modules that meet specified standards for authenticating users and devices. This control ensures that cryptographic methods and modules used for authentication are secure, reliable, and compliant with recognized standards, such as FIPS 140. The use of approved cryptographic modules enhances the security of the authentication process, protecting against unauthorized access and ensuring the confidentiality and integrity of authentication data. Implementing cryptographic module authentication involves selecting, configuring, and maintaining cryptographic solutions that support the organization's security requirements and authentication policies.

IA-8—Identification and Authentication (Nonorganizational Users): IA-8 identifies and authenticates users who are not part of the organization but require access to organizational information systems. This control ensures that external users, such as customers or partners, are accurately identified and authenticated before being granted access. By applying robust IA mechanisms to nonorganizational users, organizations can control access to their systems and data, maintaining security and privacy while supporting business operations and collaboration with external entities.

IA-9—Service Identification and Authentication: IA-9 requires identifying and authenticating services that interact with organizational information systems. This control ensures that services like application interfaces and system processes are authenticated before establishing connections, preventing unauthorized services from accessing or exchanging data. Service IA are crucial for securing inter-service communications, protecting against unauthorized data access, and ensuring that only trusted services participate in system operations.

IA-10—Adaptive Authentication: IA-10 involves implementing adaptive authentication mechanisms that adjust authentication requirements based on RAs. This control allows organizations to dynamically vary the level of authentication required for access based on context, such as user location, device security posture, or access patterns. Adaptive authentication enhances security by applying stronger authentication methods in higher-risk scenarios while maintaining usability under normal conditions. Organizations can effectively balance security and convenience by employing adaptive authentication and responding to changing risk levels in real time.

IA-11—Reauthentication: IA-11 mandates the periodic reauthentication of users or when significant events occur, ensuring that access privileges remain valid over time or across changes in user context. This control requires users to reconfirm their identity through the authentication process at defined intervals or following specific events, such as privilege escalation or accessing sensitive resources. Reauthentication helps maintain the security of sessions, preventing unauthorized access resulting from compromised sessions or changes in user roles. Reauthentication mechanisms support continuous verification of user identities, reinforcing AC measures.

IA-12—Identity Proofing: IA-12 focuses on identity proofing to establish the identity of users before issuing credentials and granting access to organizational information systems. This control ensures that the identity of users is verified using reliable, independent sources of information, documents, or biometric data before authentication credentials are issued. Identity proofing is a critical step in the IA process, preventing impersonation and ensuring access is granted only to verified individuals. By conducting thorough identity proofing, organizations can enhance the trustworthiness of user identities and the security of the authentication process.

Incident Response (IR) Overview

The IR family of controls outlines strategies and procedures for addressing and managing the aftermath of security breaches or incidents. The primary goal is to limit damage and reduce recovery time and costs. These controls emphasize the importance of preparation, detection, analysis, containment, eradication, recovery activities, and documenting and reporting incident details. By establishing a robust IR capability, organizations can swiftly respond to incidents, mitigate the impacts on business operations, and prevent future occurrences.

IR-1—Policy and Procedures: IR-1 mandates developing, documenting, and disseminating an IR policy and associated procedures. This foundational control ensures that organizations have a formalized approach to responding to and managing incidents that could compromise information system security. The policy defines the organizational structure for IR, roles and responsibilities, and guidelines for action in response to various types of incidents. Regular updates to the policy and procedures ensure that the IR strategy remains effective and relevant in the face of evolving threats and operational changes, enabling organizations to respond swiftly and effectively to incidents, minimize impact, and support swift recovery.

IR-2—Incident Response Training: IR-2 requires personnel with IR roles and responsibilities to receive appropriate training. This training equips them with the knowledge and skills to effectively detect, respond to, and recover from security incidents. The control emphasizes the importance of regular training and exercises to prepare IR teams for real-world scenarios, ensuring they can apply best practices and organizational procedures under pressure. Continuous education and awareness efforts help maintain a high level of readiness, ensuring that the organization's IR capabilities are robust and effective.

IR-3—Incident Response Testing: IR-3 mandates regular testing of the IR capability to evaluate its effectiveness and the organization's readiness to manage and recover from incidents. This control involves conducting exercises that simulate realistic scenarios, allowing teams to practice their response in a controlled environment. Testing identifies gaps in response plans, training deficiencies, and opportunities for improvement, enhancing the organization's preparedness and resilience. Feedback from testing drives refinements to IR strategies, ensuring that teams are well-equipped to handle incidents efficiently.

IR-4—Incident Handling: IR-4 focuses on establishing an organized approach to handling incidents, including preparation, detection, analysis, containment, eradication, recovery, and post-incident activities. This control ensures that incidents are managed systematically, reducing the impact on the organization and preventing further damage. Effective incident handling procedures facilitate rapid response, minimizing downtime and data loss while preserving evidence for further analysis and legal action. The control underscores the need for clear communication channels, roles, and responsibilities, supporting coordinated and effective incident management.

IR-5—Incident Monitoring: IR-5 requires continuous monitoring of information systems for signs of security incidents. This control ensures that potential security issues are detected promptly, enabling swift action to mitigate threats. Incident monitoring involves collecting and analyzing data from various sources, such as logs, network traffic, and anomaly detection systems, to identify unusual or unauthorized activities indicative of a security incident. By maintaining vigilant monitoring, organizations can identify threats earlier, respond more rapidly, and limit the scope and severity of security incidents.

IR-6—Incident Reporting: IR-6 mandates the timely reporting of security incidents within the organization and to external stakeholders as required. This control ensures that relevant parties are informed of incidents, supporting effective response and recovery efforts. Incident reporting procedures specify what information should be reported, to whom, and within what timeframe, facilitating clear and effective communication during and after an incident. Proper reporting supports organizational learning, compliance with legal and regulatory requirements, and, where appropriate, collaboration with law enforcement and other external entities.

IR-7—Incident Response Assistance: IR-7 provides support and assistance to organizational units in responding to security incidents. This control ensures that specialized knowledge and resources are available for incident detection, analysis, and response efforts. Assistance may include forensic analysis, legal expertise, and communication support, helping to manage incidents effectively and mitigate their impact. Organizations can enhance their overall response capabilities by offering IR assistance, benefiting from specialized skills and expertise.

IR-8—Incident Response Plan: IR-8 requires developing and implementing a comprehensive IR plan that outlines procedures for addressing security incidents. This plan provides a structured approach for the organization during an incident, detailing response phases, roles and responsibilities, communication protocols, and recovery strategies. A well-defined IR plan ensures a coordinated and effective response to incidents, minimizes damage and disruption, and facilitates a swift return to normal operations.

IR-9—Information Spillage Response: IR-9 addresses the procedures for responding to information spillage incidents where sensitive information is inadvertently released or transmitted to an unauthorized environment. This control ensures that spillage incidents are quickly contained, assessed, and remediated to prevent further unauthorized disclosure. Information spillage response includes identifying the extent of the spillage, notifying affected parties, removing the spilled information from unauthorized locations, and implementing measures to prevent recurrence. Effective management of information spillage incidents protects sensitive information and maintains the organization's integrity and compliance with security policies and regulations.

IR-10—Integrated Information Security Analysis Team: IR-10 mandates the formation of an integrated team responsible for conducting security analysis across the organization. This team combines expertise from various domains, such as IR, threat intelligence, and security operations, to comprehensively analyze security incidents and trends. The integrated approach enhances the organization's ability to detect, understand, and respond to complex security threats, leveraging diverse skills and perspectives. Collaboration and information sharing within the team support a more effective and holistic security analysis, contributing to improved security posture and IR capabilities.

Maintenance (MA) Overview

The MA family of controls performs periodic and timely information system MA. This includes ensuring that systems are serviced and repaired in a manner that protects the integrity and security of organizational data. The controls cover MA tools, activities, and personnel, emphasizing the importance of authorized MA activities and protecting MA-related information. Implementing these controls helps organizations ensure that their information systems operate effectively and securely over time.

MA-1—Policy and Procedures: MA-1 requires developing, documenting, and disseminating a system MA policy and associated procedures. This control establishes a framework for performing MA on information systems, ensuring that MA activities are carried out in a controlled, secure, and efficient manner. The policy and procedures cover scheduling MA, authorizing MA personnel, protecting sensitive data during MA, and documenting MA activities. By adhering to a formal system MA policy, organizations can minimize the risks associated with MA activities, including potential disruptions to operations and unauthorized access to sensitive information.

MA-2—Controlled Maintenance: MA-2 mandates that MA activities on information systems be monitored and controlled to protect the integrity and security of the systems. This control requires organizations to authorize, schedule, and oversee MA activities, ensuring that they are performed by qualified personnel and per organizational policies. Controlled MA procedures help prevent unauthorized modifications to systems, detect potential security vulnerabilities introduced during MA, and maintain a record of MA activities for accountability and review. Implementing controlled MA practices supports information systems' ongoing security and operational reliability.

MA-3—Maintenance Tools: MA-3 focuses on managing and controlling tools used for system MA to prevent unauthorized access and ensure the integrity of the systems. This control requires organizations to approve, control, and monitor MA tools, especially those that could be exploited to compromise system security. MA tools should be securely stored when not in use, checked for malicious code, and utilized in a manner that minimizes potential security risks. By managing MA tools effectively, organizations can reduce the likelihood of introducing vulnerabilities during MA activities and protect their systems against possible threats.

MA-4—Nonlocal Maintenance: MA-4 addresses the security risks associated with performing MA activities from external locations. This control requires organizations to establish safeguards for nonlocal MA, such as using encrypted channels for remote access, documenting nonlocal MA activities, and ensuring that MA sessions are monitored. By implementing strict controls for nonlocal MA, organizations can mitigate the risks of unauthorized access, data leakage, and potential disruptions, ensuring that remote MA activities are conducted securely and with minimal impact on system security.

MA-5—Maintenance Personnel: MA-5 mandates that MA personnel be appropriately authorized, trained, and supervised to ensure the security and integrity of information systems during MA activities. This control involves verifying the trustworthiness of MA personnel, ensuring they have the necessary clearances and are aware of their security responsibilities. Organizations should also ensure that MA personnel follow established procedures and monitor their activities to prevent unauthorized actions. Organizations can control MA activities by managing MA personnel effectively, reducing the risk of inadvertent or malicious damage to systems.

MA-6—Timely Maintenance: MA-6 emphasizes the importance of performing MA on information systems in a timely manner to address vulnerabilities and ensure the continued security and functionality of the systems. This control requires organizations to establish routine MA schedules and prioritize corrective MA actions based on RAs. Timely MA activities, including applying patches and updates, help protect systems against known vulnerabilities and ensure that security controls remain effective. By prioritizing and scheduling MA activities, organizations can minimize potential security risks and maintain the operational readiness of their information systems.

MA-7—Field Maintenance: MA-7 addresses the challenges and security risks associated with performing MA activities in field or nonorganizational locations. This control requires organizations to implement additional safeguards when conducting MA outside of controlled environments, such as verifying the identity of personnel, ensuring the physical security of MA activities, and protecting sensitive information during MA. Field MA procedures should be designed to maintain the security and integrity of information systems, even in less secure or uncontrolled environments. Organizations can mitigate the risks associated with remote or onsite MA activities by establishing specific controls for field MA.

Media Protection (MP) Overview

The Media Protection (MP) family of controls addresses digital and nondigital MP throughout its lifecycle, from creation to disposal. This includes measures to safeguard information system media, both paper and electronic, against unauthorized access, disclosure, alteration, and destruction. Controls in this family encompass media access, marking, storage, transport, and sanitization practices, ensuring that sensitive data is adequately protected on all media types where it is stored, processed, or transmitted.

MP-1—Policy and Procedures: MP-1 mandates developing, documenting, and disseminating a MP policy and associated procedures. This control establishes the framework for protecting digital and physical media that contain sensitive information, ensuring that media are handled, stored, transported, and disposed of securely. The MP policy defines roles and responsibilities, authorized media use, labeling requirements, and procedures for sanitizing and disposing of media. By adhering to a formal MP policy, organizations can prevent unauthorized access to sensitive information, mitigate the risk of data breaches, and ensure compliance with legal and regulatory requirements for information protection.

MP-2—Media Access: MP-2 requires that access to digital and physical media containing sensitive information be restricted to authorized individuals. This control ensures that media are protected against unauthorized access, theft, and misuse, safeguarding the confidentiality, integrity, and availability of the information they contain. Organizations should implement physical and logical ACs, maintain an inventory of media, and monitor access to media to detect and respond to unauthorized access attempts. By controlling access to media, organizations can reduce the risk of unauthorized disclosure of sensitive information and maintain the trustworthiness of their information systems.

MP-3—Media Marking: MP-3 mandates marking digital and physical media with appropriate security labels to indicate the sensitivity of the information they contain. This control facilitates the correct handling of media based on classification, supporting compliance with organizational policies and legal requirements for information protection. Media marking helps personnel identify sensitive information, apply appropriate handling procedures, and prevent unauthorized access or disclosure. By implementing media marking practices, organizations can enhance information security, support accountability, and ensure that sensitive information is managed per its classification level.

MP-4—Media Storage: MP-4 focuses on securely storing digital and physical media to protect them from unauthorized access, damage, and environmental hazards. This control requires organizations to implement physical and ecological protections for media storage areas, such as ACs, fire suppression systems, and environmental controls. Secure storage practices help ensure the preservation and integrity of media and the information they contain, supporting the availability of information and the continuity of operations. By managing media storage securely, organizations can protect sensitive information from physical and environmental risks, maintaining the confidentiality and integrity of their data.

MP-5—Media Transport: MP-5 mandates the secure transport of digital and physical media to prevent loss, theft, and unauthorized access during transportation. This control involves implementing encryption, secure containers, and chain of custody documentation to safeguard media during transit. Secure transport practices are essential for maintaining the confidentiality and integrity of sensitive information when it is moved between locations, whether within the organization or to external partners. By ensuring the secure transport of media, organizations can mitigate the risks associated with transporting sensitive information and maintain control over their data.

MP-6—Media Sanitization: MP-6 requires the sanitization of digital and physical media before disposal or reuse to ensure that sensitive information is irrecoverably removed. This control involves selecting and applying appropriate sanitization techniques, such as degaussing, shredding, or overwriting, based on the classification of the information and the media type. Media sanitization prevents unauthorized access to sensitive information after media are no longer needed, supporting data privacy and security. By implementing effective media sanitization practices, organizations can protect sensitive information throughout the media lifecycle, reducing the risk of data breaches and ensuring compliance with data protection regulations.

MP-7—Media Use: MP-7 restricts digital and physical media use to authorized individuals and processes to prevent unauthorized access and data breaches. This control requires organizations to implement policies and procedures for authorizing, monitoring, and controlling media use within information systems. By managing media use effectively, organizations can ensure that sensitive information is accessed and processed securely, protecting confidentiality, integrity, and availability. Restrictions on media use help mitigate the risks associated with removable media, such as USB drives and DVDs, which can be easily lost or used to introduce malware into systems.

MP-8—Media Downgrading: MP-8 addresses the procedures for downgrading the classification level of digital and physical media, ensuring that information is protected appropriately throughout its lifecycle. This control involves assessing the sensitivity of the information, applying necessary sanitization processes, and remarking media to reflect the new classification level. Media downgrading is performed when information no longer requires a higher level of protection, allowing for broader distribution or disposal of media. By establishing controls for media downgrading, organizations can maintain the security of information as its classification level changes, ensuring that data is handled and protected in accordance with its current sensitivity.

Physical and Environmental Protection (PE) Overview

The Physical and Environmental Protection (PE) family of controls is designed to prevent unauthorized physical access, damage, and interference to information systems and facilities. These controls address physical security measures such as AC mechanisms, surveillance, and environmental protection against natural and human-made disasters. Implementing PE controls ensures the safeguarding of physical resources and information systems, contributing to the organization's overall security posture.

PE-1—Policy and Procedures: PE-1 requires organizations to develop, document, and disseminate policies and procedures for physical and environmental protection. This foundational control establishes the framework for protecting the organization's facilities and information systems from physical and environmental threats. The policy and procedures cover physical ACs, monitoring and surveillance, environmental controls, and emergency procedures. Regular review and updates ensure that the physical security measures remain effective in light of changing threats and operational requirements. By maintaining a comprehensive physical and environmental protection policy, organizations can safeguard their assets, personnel, and data from physical threats and ensure the resilience of their operations.

PE-2—Physical Access Authorizations: PE-2 mandates the management of physical access authorizations to facilities and areas containing information systems. This control ensures access is granted based on a valid need to access the physical premises, aligning with the individual's role and responsibilities. Authorization processes include issuing access badges, maintaining access lists, and regularly reviewing and revoking access as necessary. Organizations can prevent unauthorized access to critical information systems and protect against potential physical threats by controlling who is authorized to enter sensitive areas.

PE-3—Physical Access Control: PE-3 focuses on implementing appropriate controls to manage and monitor physical access to facilities and information system resources. This control involves deploying barriers, locks, guards, and surveillance systems to deter, detect, and respond to unauthorized physical access. Physical ACs are critical for ensuring the security of information systems, protecting assets from theft, vandalism, and environmental hazards. Effective implementation of physical AC measures supports the organization's overall security posture and operational integrity.

PE-4—Access Control for Transmission: PE-4 addresses the protection of information transmission within physical boundaries to prevent unauthorized access, interception, or tampering. This control ensures that data transmissions, especially sensitive information, are protected through encryption and secure cabling solutions. By safeguarding information during transmission, organizations can maintain data confidentiality, integrity, and availability, mitigating the risk of data breaches and ensuring compliance with security policies and standards.

PE-5—Access Control for Output Devices: PE-5 requires the implementation of ACs for output devices to prevent unauthorized access to information. This control includes securing printers, displays, and other devices that output sensitive data, ensuring only authorized individuals can retrieve or view the information. Managing access to output devices helps prevent data leakage and protects sensitive information from being accessed or disclosed without authorization, supporting the confidentiality and integrity of organizational data.

PE-6—Monitoring Physical Access: PE-6 mandates monitoring physical access to facilities and information system resources to detect and respond to unauthorized access attempts. This control involves using surveillance cameras, entry and exit logs, and alarm systems to monitor and record access activities. Monitoring physical access provides valuable information for security investigations, deters unauthorized access, and helps ensure the safety and security of the physical environment. By actively monitoring access, organizations can quickly identify and respond to security incidents, enhancing their physical security measures.

PE-7—Visitor Control: PE-7 focuses on managing visitors' access to facilities and information system resources. This control requires organizations to implement procedures for authorizing, monitoring, and controlling visitor access, including escorting visitors and maintaining visitor logs. Visitor control measures help ensure that visitors do not compromise the security of information systems or access sensitive areas without proper authorization. Effective visitor management supports the organization's security policies and protects against unauthorized physical access.

PE-8—Visitor Access Records: PE-8 mandates the MA of access records for visitors to facilities containing information systems. This control ensures that detailed information about visitors, including their identity, the purpose of their visit, and the areas accessed, is documented and retained. Visitor access records support accountability, facilitate security investigations, and help ensure access to sensitive areas is appropriately managed. Organizations can audit and review visitor access by keeping accurate and comprehensive visitor access records and enhancing physical security controls.

PE-9—Power Equipment and Cabling: PE-9 requires protecting power equipment and cabling from damage and unauthorized access. This control ensures that power supplies and cabling, critical for the operation of information systems, are securely housed and protected against physical threats. Safeguarding power equipment and cabling helps prevent disruptions to information systems, supports the continuity of operations, and reduces the risk of damage or interference that could impact system availability and integrity.

PE-10—Emergency Shutoff: PE-10 mandates emergency shutoff procedures and mechanisms for power supplies and information systems to protect against fires, explosions, and other emergencies. This control includes marked and accessible emergency shutoff switches and procedures for safely shutting down systems in an emergency. Emergency shutoff capabilities are essential for minimizing damage to systems and facilities, protecting personnel, and ensuring rapid emergency response, enhancing the organization's resilience and safety measures.

PE-11—Emergency Power: PE-11 requires the availability of emergency power systems to ensure the continued operation of critical information systems during power outages. This control involves the implementation of uninterruptible power supplies, generators, and other emergency power solutions to provide backup power to essential systems. Emergency power capabilities are crucial for maintaining system availability, supporting mission-critical operations, and ensuring the safety of personnel during power disruptions. By planning for and providing emergency power, organizations can enhance their preparedness for power outages and ensure the continuity of key functions.

PE-12—Emergency Lighting: PE-12 mandates the provision of emergency lighting in facilities containing information systems to ensure personnel safety and systems' security during power outages or emergencies. This control requires the installation of reliable emergency lighting systems that activate automatically during power failures, illuminating exit routes and critical areas. Emergency lighting supports evacuation efforts, facilitates access to emergency equipment, and helps prevent accidents during power outages, contributing to the overall safety and security of the facility.

PE-13—Fire Protection: PE-13 focuses on implementing fire protection measures to safeguard information systems and facilities from fire damage. This control includes installing fire detection systems, fire suppression equipment, and fire barriers and developing fire response procedures. Fire protection measures are essential for preventing the loss of life, minimizing damage to information systems and facilities, and ensuring the continuity of operations in the event of a fire. By adopting comprehensive fire protection strategies, organizations can mitigate the risk of fire-related disruptions and enhance their resilience to fire emergencies.

PE-14—Environmental Controls: PE-14 requires the implementation of environmental controls to protect information systems from damage due to environmental hazards. This control involves using temperature and humidity controls, water detection systems, and air filtration mechanisms to maintain optimal environmental conditions for information systems. Environmental controls help prevent equipment failures, data loss, and system disruptions caused by adverse environmental conditions, supporting the availability and reliability of information systems. By managing the physical environment, organizations can ensure the longevity and performance of their information systems, reducing the risk of environmentally induced damage.

PE-15—Water Damage Protection: PE-15 mandates measures to protect information systems and facilities from water damage. This control includes implementing waterproofing solutions, drainage systems, and leak detection equipment to prevent water ingress and mitigate the impact of water-related incidents. Water damage protection is crucial for preventing the loss of equipment, data, and operational capability due to flooding, leaks, and other water-related hazards. By adopting effective water damage protection strategies, organizations can safeguard their assets and maintain continuity of operations in the face of water-related risks.

PE-16—Delivery and Removal: PE-16 focuses on securing information system components' delivery and removal process to prevent unauthorized access, tampering, or theft. This control requires organizations to establish procedures for verifying the legitimacy of deliveries and removals, documenting the movement of components, and ensuring the security of components during transit. Secure delivery and removal practices are essential for maintaining the integrity of information systems, controlling access to sensitive components, and protecting against the introduction of unauthorized or compromised equipment. By managing the delivery and removal process, organizations can mitigate the risks associated with the movement of system components and ensure the security of their information systems.

PE-17—Alternate Work Site: PE-17 mandates the provision of alternate work sites for personnel to ensure continuity of operations if primary work locations are unavailable. This control involves identifying suitable alternate locations, equipping them with necessary information systems and security measures, and developing procedures for relocating personnel and operations. Alternate work sites are critical for maintaining organizational functions during disruptions, disasters, or emergencies that render primary sites inoperable. By planning for alternate work sites, organizations can enhance their resilience, ensuring that critical operations can continue despite adverse conditions.

PE-18—Location of System Components: PE-18 requires organizations to consider the physical location of information system components when making decisions about system housing, architecture, and operational processes. This control involves evaluating risks associated with the physical environment, proximity to hazards, and accessibility to unauthorized individuals. Strategic placement of system components can mitigate physical security risks, enhance system protection, and support the organization's overall security posture. By carefully selecting the location of system components, organizations can reduce the likelihood of physical damage, unauthorized access, and environmental threats.

PE-19—Information Leakage: PE-19 addresses the need to protect against unintentional information leakage through electromagnetic, acoustic, or other emissions from information systems. This control involves implementing measures to detect, measure, and mitigate emanations that could be intercepted and used to gain unauthorized access to sensitive information.

Protecting against information leakage is crucial for maintaining the confidentiality of sensitive data, especially in environments with a high risk of eavesdropping or interception. By adopting countermeasures against information leakage, organizations can safeguard their information against unintended disclosures and enhance the security of their communications and data processing activities.

PE-20—Asset Monitoring and Tracking: PE-20 mandates the monitoring and tracking of physical assets to prevent loss, theft, and unauthorized access. This control requires organizations to maintain an inventory of physical assets, implement tracking mechanisms, and regularly audit asset location and status. Asset monitoring and tracking support accountability, enable rapid response to incidents involving physical assets, and enhance the overall security of organizational resources. By keeping close tabs on the location and condition of physical assets, organizations can ensure that critical resources are protected and available for operational needs.

PE-21—Electromagnetic Pulse Protection: PE-21 focuses on implementing protections against the effects of electromagnetic pulse (EMP) and other electromagnetic interference (EMI) that could potentially disrupt or damage information systems. This control involves assessing the risk of EMP/EMI events, implementing shielding and surge protection measures, and developing response strategies to mitigate the impact on system operations. Protecting against EMP and EMI is essential for maintaining the availability and integrity of information systems, especially in environments susceptible to electromagnetic disturbances. By adopting measures to counteract EMP and EMI, organizations can safeguard their systems against electromagnetic threats and ensure the continuity of critical functions.

PE-22—Component Marking: PE-22 requires marking system components to identify sensitive or critical assets and support asset management and security processes. This control involves labeling components with identifiers, classification levels, and handling instructions to facilitate the proper management, protection, and disposal of assets. Component marking enhances the traceability of assets, supports compliance with security policies, and helps prevent unauthorized access or misuse of sensitive components. By implementing systematic component marking practices, organizations can improve the security and accountability of their information system assets, ensuring they are managed appropriately throughout their lifecycle.

PE-23—Facility Location: PE-23 mandates the consideration of security implications when selecting the location of facilities housing information systems. This control involves evaluating potential threats and vulnerabilities associated with the physical environment, including natural disasters, proximity to high-risk areas, and accessibility to unauthorized individuals. Strategic facility location decisions can enhance the physical security of information systems, reduce exposure to environmental and man-made risks, and support the overall resilience of the organization. By carefully selecting facility locations, organizations can mitigate physical security threats and ensure the protection and availability of critical information systems.

Planning (PL) Overview

The Planning (PL) family of controls focuses on developing, documenting, and maintaining security and privacy plans. These plans describe the security controls in place or planned for information systems and articulate policies and strategies for protecting data confidentiality, integrity, and availability. The PL controls ensure that security and privacy considerations are integrated into the technology and decision-making processes at all organizational levels, facilitating a structured approach to risk management.

PL-1—Policy and Procedures: PL-1 requires developing, documenting, and disseminating a security and privacy planning policy and associated procedures. This foundational control establishes the framework for developing, implementing, and maintaining security and privacy plans that protect information systems and sensitive data. The policy and procedures outline requirements for planning activities, roles, and responsibilities, as well as integrating security and privacy considerations into organizational processes. Regular review and updates ensure that planning practices remain effective and aligned with evolving operational needs, regulatory requirements, and threat landscapes. Organizations can ensure a systematic and coordinated approach to managing information system security and privacy risks by adhering to a formal security and privacy planning policy.

PL-2—System Security and Privacy Plans: PL-2 mandates developing and maintaining system security and privacy plans for information systems. These plans document the security and privacy requirements, controls, and risk management strategies tailored to the system's operational context and the data it processes. The system security and privacy plans serve as comprehensive guides for implementing and managing security and privacy controls, ensuring that protection measures are integrated throughout the system lifecycle. Regular updates to the plans ensure they reflect changes in the system environment, technology, or operational practices. Organizations can effectively manage risks by maintaining up-to-date system security and privacy plans and ensuring compliance with organizational policies and regulatory requirements.

PL-3—System Security Plan Update: PL-3 focuses on the periodic review and update of the system security plan to ensure it remains relevant and effective in addressing the security needs of the information system. This control recognizes that changes in technology, threats, operational environments, and organizational priorities can impact the security posture of information systems. Regular updates to the system security plan ensure that it accurately reflects the current operational context, incorporates lessons learned, and addresses new or evolving threats. By keeping the system security plan up-to-date, organizations can ensure that security measures are aligned with current requirements and effectively protect the system and its data.

PL-4—Rules of Behavior: PL-4 requires establishing rules of behavior for individuals accessing organizational information systems. These rules outline users' responsibilities and expected behavior, emphasizing the importance of complying with security policies, protecting sensitive information, and reporting security incidents. The rules of behavior help foster a culture of security awareness and responsibility, ensuring that users understand their role in safeguarding organizational assets. By clearly communicating rules of behavior and obtaining acknowledgment from users, organizations can enhance compliance with security policies and reduce the risk of security violations.

PL-5—Privacy Impact Assessment: PL-5 mandates the conduct of privacy impact assessments (PIAs) for information systems to evaluate the privacy implications of system operations on individuals' privacy rights and data. PIAs help identify and mitigate privacy risks associated with collecting, storing, processing, and disseminating personal information. By systematically assessing privacy impacts, organizations can implement measures to protect individuals' privacy, ensure compliance with privacy laws and regulations, and build trust with users and stakeholders. Regular updates to PIAs in response to system changes or new privacy requirements ensure ongoing protection of privacy and the responsible management of personal information.

PL-6—Security-related Activity Planning: PL-6 focuses on integrating security-related activities into organizational planning processes. This control ensures that security considerations are incorporated into strategic and operational planning, project management, and decision-making activities. By planning for security-related activities, organizations can allocate resources effectively, address security requirements proactively, and ensure security measures align with organizational objectives. Integrating security into planning processes supports a holistic approach to risk management, enhancing the organization's resilience and ability to respond to evolving security challenges.

PL-7—Concept of Operations: PL-7 requires the development of a concept of operations (CONOPS) document that describes how information systems are intended to operate within the organizational context. The CONOPS includes information on system functions, data flows, user interactions, and security considerations, providing a comprehensive overview of system operations. By documenting the concept of operations, organizations can ensure a shared understanding among stakeholders, support system design and development decisions, and identify security requirements and risks associated with operational processes. The CONOPS serves as a valuable tool for planning, implementation, and communication, enhancing the effectiveness of system operations and security measures.

PL-8—Security and Privacy Architectures: PL-8 mandates developing and maintaining security and privacy architectures for the organization and its information systems. These architectures provide a structured framework for integrating security and privacy controls into the system and organizational environments. By defining security and privacy architectures, organizations can ensure that protection measures are systematically designed, implemented, and aligned with business objectives and regulatory requirements. The architectures support a coherent and consistent approach to managing security and privacy risks, facilitating effective mitigation and compliance efforts.

PL-9—Central Management: PL-9 focuses on the central management of security and privacy functions to ensure consistent and effective implementation of policies, procedures, and controls across the organization. Central management enables coordinated oversight of security and privacy activities, standardized practices, and the efficient allocation of resources to address security and privacy risks. By adopting a centralized approach to managing security and privacy, organizations can enhance their ability to respond to threats, ensure compliance with policies and regulations, and foster a unified security and privacy culture.

PL-10—Baseline Selection: PL-10 requires the selection of security and privacy control baselines as the foundation for protecting information systems and managing risks. These baselines consist of minimum controls tailored to the system's risk environment, supporting compliance with organizational policies and regulatory requirements. The selection of appropriate baselines is crucial for establishing a solid security and privacy posture, guiding the implementation of additional controls as needed, and ensuring that systems are protected against known threats. Organizations can effectively address their specific security and privacy requirements and cost-effectively manage risks by selecting and customizing control baselines.

PL-11—Baseline Tailoring: PL-11 involves tailoring security and privacy control baselines to fit the specific needs, conditions, and risk environment of the organization and its information systems. This control recognizes that the one-size-fits-all approach to baselines may not adequately address the unique requirements of each system. Tailoring involves adding, removing, or modifying controls to align with the system's functionality, technology, and risk factors. By customizing control baselines, organizations can ensure that security and privacy measures are practical and efficient, providing adequate protection without imposing unnecessary burdens or constraints on system operations.

Program Management (PM) Overview

The Program Management (PM) family of controls addresses establishing and managing a comprehensive Information Security Program. This includes the development of policies, the assignment of roles and responsibilities, and the allocation of resources to manage and protect information assets. PM controls guide the security program's strategic planning, implementation, assessment, and continuous improvement, ensuring alignment with organizational goals and compliance with regulatory requirements.

PM-1—Information Security Program Plan: PM-1 mandates the development, documentation, implementation, and MA of an organization-wide information security program plan. This plan outlines the organization's strategy for protecting its information systems and managing cybersecurity risks. It includes goals, objectives, roles and responsibilities, resources, and measures for assessing the program's effectiveness. The information security program plan ensures a coordinated, comprehensive approach to securing the organization's digital assets. Regular reviews and updates to the plan ensure it evolves to address new threats, technologies, and business objectives, aligning with the organization's risk management strategy.

PM-2—Information Security Program Leadership Role: PM-2 requires the designation of a senior official responsible for the organization's information security program. This role involves overseeing the security program's development, implementation, and continuous improvement, ensuring that it effectively addresses the organization's cybersecurity risks. The senior official coordinates security efforts across the organization, allocates resources, and ensures compliance with legal and regulatory requirements. The leadership role is critical for establishing clear cybersecurity accountability, fostering a security culture, and ensuring that security considerations are integrated into organizational decision-making processes.

PM-3—Information Security and Privacy Resources: PM-3 allocates appropriate resources to support the organization's information security and privacy programs. This control ensures that sufficient personnel, technology, and financial resources are dedicated to implementing and maintaining effective security and privacy measures. Resource allocation is based on RAs to protect critical assets and information while supporting the organization's mission and business objectives. Organizations can enhance their resilience to cyber threats and protect sensitive information by prioritizing security and privacy resource needs.

PM-4—Plan of Action and Milestones Process: PM-4 mandates establishing a process for creating, maintaining, and tracking a POA&M for addressing weaknesses in information systems. The POA&M is a key tool for documenting security deficiencies, proposed remediation actions, resources required, and timelines for completion. This process facilitates systematic risk management, enabling organizations to prioritize and address security vulnerabilities in a timely manner. Regular reviews and updates to the POA&M ensure that remediation efforts are on track and that emerging vulnerabilities are promptly addressed, supporting continuous security posture improvement.

PM-5—System Inventory: PM-5 requires organizations to maintain an inventory of information systems and their interconnections. This inventory supports effective risk management by providing a comprehensive view of the organization's IT assets, including hardware, software, and network components. The system inventory aids in identifying critical systems, assessing vulnerabilities, and prioritizing security controls. Regular updates to the inventory ensure that it reflects changes in the IT environment, supporting ongoing security and privacy risk management efforts. An accurate and up-to-date inventory is essential for IR, disaster recovery planning, and compliance with security policies and standards.

PM-6—Measures of Performance: PM-6 focuses on developing and implementing measures to evaluate the effectiveness of the information security program. These performance measures provide insight into how well security policies, procedures, and controls are functioning to protect the organization's information systems and data. By establishing clear, measurable criteria for security performance, organizations can identify areas of strength and weakness, inform decision-making, and allocate resources effectively. Regular monitoring and reporting of performance measures support continuous improvement of the security program, enhancing the organization's resilience to cyber threats.

PM-7—Enterprise Architecture: PM-7 mandates the integration of security and privacy considerations into the organization's enterprise architecture. This control ensures that security and privacy are foundational elements of the organization's IT infrastructure, systems, and processes. By aligning security and privacy with the enterprise architecture, organizations can ensure that protective measures are consistently applied across all IT assets, supporting a unified defense posture. The enterprise architecture is a strategic framework for managing complexity, facilitating innovation, and ensuring security and privacy are embedded in the technology landscape.

PM-8—Critical Infrastructure Plan: PM-8 requires organizations to develop and maintain a plan for protecting critical infrastructure components essential to their operations. This plan outlines strategies for ensuring the resilience and security of systems and assets deemed vital for national security, economic stability, or public health and safety. By identifying critical infrastructure and implementing targeted protections, organizations can mitigate risks to these essential services and functions, ensuring their availability and integrity even in the face of cyber threats or physical disasters. The critical infrastructure plan is a crucial component of national and organizational efforts to safeguard vital assets and maintain continuity of operations.

PM-9—Risk Management Strategy: PM-9 mandates the development, implementation, and continuous improvement of a risk management strategy for the organization. This strategy outlines the organization's approach to identifying, assessing, and responding to cybersecurity risks, balancing the need to protect information assets with the organization's mission and business objectives. The risk management strategy provides a framework for making informed decisions about security investments, controls, and risk tolerance levels. Organizations can effectively manage cybersecurity risks, protect sensitive information, and support achieving strategic goals by adopting a comprehensive risk management strategy.

PM-10—Authorization Process: PM-10 focuses on establishing a formal process for authorizing the operation of information systems and accepting the associated security risks. This authorization process involves assessing the security controls in place, determining the residual risk to organizational operations and assets, and making an informed decision to authorize system operation based on this assessment. The process ensures that senior officials are accountable for the security risks associated with their systems and that decisions to operate systems are made with a clear understanding of those risks. By implementing a rigorous authorization process, organizations can ensure that security considerations are integral to system operations and that risks are managed appropriately.

PM-11—Mission and Business Process Definition: PM-11 requires organizations to define their mission and business processes, including identifying information flows and the information systems supporting them. This control supports effective risk management by ensuring that security and privacy considerations are integrated into the organization's operational processes. By understanding the mission, business processes, and associated information flows, organizations can identify critical assets, assess risks, and implement appropriate security controls to protect against threats. This alignment of security with business processes helps ensure that protective measures support the organization's objectives and operational needs.

PM-12—Insider Threat Program: PM-12 mandates establishing an insider threat program to detect, prevent, and respond to threats posed by individuals with authorized access to the organization's information systems and data. This program involves implementing measures to identify potential insider threats, monitor for suspicious activities, and mitigate risks associated with insider actions. An effective insider threat program is critical for protecting sensitive information from unauthorized disclosure, modification, or destruction by trusted insiders. By addressing insider threats proactively, organizations can enhance their security posture and protect against potentially damaging actions by employees, contractors, or other insiders.

PM-13—Security and Privacy Workforce: PM-13 focuses on developing and maintaining a skilled workforce capable of implementing the organization's security and privacy policies and procedures. This control requires organizations to identify workforce needs, provide appropriate training and education, and ensure that personnel can effectively perform their security and privacy responsibilities. Investing in the security and privacy workforce enhances the organization's ability to protect its information assets, comply with regulatory requirements, and adapt to evolving threats. By prioritizing workforce development, organizations can build a culture of security and privacy awareness, supporting the effective management of cybersecurity risks.

PM-14—Testing, Training, and Monitoring: PM-14 mandates regular testing, training, and monitoring activities to evaluate the effectiveness of security and privacy controls and ensure compliance with organizational policies. This control involves assessing vulnerabilities, training personnel on security and privacy practices, and continuously monitoring systems and networks for signs of security incidents. By implementing a comprehensive program of testing, training, and monitoring, organizations can detect and address security weaknesses, enhance the skills and awareness of their workforce, and maintain a vigilant posture against cyber threats.

PM-15—Security and Privacy Groups and Associations: PM-15 encourages organizations to participate in security and privacy groups and associations to share information, collaborate on best practices, and stay informed about emerging threats and technologies. This control recognizes the value of engaging with external communities to enhance the organization's security and privacy knowledge base. Participation in groups and associations facilitates the exchange of threat intelligence, supports cooperative efforts to address common challenges, and promotes a collaborative approach to cybersecurity and privacy. By leveraging external expertise and resources, organizations can improve their ability to protect against and respond to cyber threats.

PM-16—Threat Awareness Program: PM-16 requires organizations to establish a threat awareness program to educate personnel about current and potential cybersecurity threats. This program involves disseminating threat information, conducting awareness activities, and promoting employee vigilance. A well-implemented threat awareness program enhances the organization's defense against cyberattacks by ensuring that personnel know the tactics, techniques, and procedures used by adversaries. By fostering a culture of threat awareness, organizations can improve their detection and response capabilities, reducing the likelihood and impact of security incidents.

PM-17—Protecting Controlled Unclassified Information on External Systems: PM-17 mandates measures to protect CUI when it is processed, stored, or transmitted on external systems not operated by the organization. This control requires organizations to ensure that external partners and service providers implement appropriate security controls to safeguard

CUI in accordance with federal standards and regulations. Protecting CUI on external systems is essential for maintaining the confidentiality and integrity of sensitive information and ensuring compliance with legal and regulatory requirements. By establishing agreements and monitoring compliance, organizations can extend their security posture to external environments, mitigating risks associated with handling CUI.

PM-18—Privacy Program Plan: PM-18 mandates the development, documentation, implementation, and MA of an organization-wide privacy program plan. This plan outlines the organization's strategy for protecting personal information and managing privacy risks per legal, regulatory, and policy requirements. The privacy program plan includes goals, objectives, roles and responsibilities, resources, and measures for assessing the program's effectiveness. By maintaining a comprehensive privacy program plan, organizations can ensure a coordinated, thorough approach to privacy, enhancing the protection of individuals' privacy rights and supporting compliance with privacy obligations.

PM-19—Privacy Program Leadership Role: PM-19 requires the designation of a senior official to be responsible for the organization's privacy program. This role involves overseeing the privacy program's development, implementation, and continuous improvement, ensuring it effectively addresses the organization's privacy risks and compliance requirements. The senior official is tasked with coordinating privacy efforts across the organization, allocating resources, and ensuring adherence to privacy laws and policies. The leadership role is critical for establishing clear accountability for privacy, fostering a culture of privacy awareness, and ensuring that privacy considerations are integrated into organizational processes.

PM-20—Dissemination of Privacy Program Information: PM-20 focuses on disseminating privacy program information to personnel and other relevant stakeholders. This control ensures that individuals are informed about the organization's privacy policies, procedures, and practices and their roles and responsibilities in protecting personal information. Effective communication of privacy program information supports awareness, compliance, and accountability, enhancing the organization's ability to protect individuals' privacy. By making privacy program information widely available and understandable, organizations can foster a culture of privacy awareness and ensure that personnel are equipped to contribute to protecting personal information.

PM-21—Accounting of Disclosures: PM-21 mandates the MA of an accounting of disclosures of personally identifiable information (PII) to external entities. This control requires organizations to document instances where PII is disclosed outside the organization, including the purpose and recipient of the information. Accounting of disclosures supports transparency and accountability in handling personal information, enabling individuals to understand how their information is shared and used. By maintaining accurate records of disclosures, organizations can respond to individuals' requests for information about sharing their PII, supporting privacy rights, and compliance with legal and regulatory requirements.

PM-22—Personally Identifiable Information Quality Management: PM-22 focuses on managing the quality of PII to ensure its accuracy, relevance, timeliness, and completeness. This control involves implementing processes for collecting, maintaining, and using PII to ensure the information is suitable for its intended purposes. PII quality management is essential for making informed decisions, providing accurate services, and protecting individuals' rights. By prioritizing the quality of PII, organizations can enhance trust, reduce errors, and support effective privacy practices.

PM-23—Data Governance Body: PM-23 mandates the establishment of a data governance body to oversee the organization's data management practices, including handling personal information. This control involves creating a cross-functional team responsible for setting data governance policies, establishing data standards, and ensuring compliance with legal and policy requirements. The data governance body plays a crucial role in ensuring that data is managed responsibly, securely, and in a manner that respects privacy. The data governance body supports the organization's objectives and compliance efforts by providing oversight and guidance on data management practices.

PM-24—Data Integrity Board: PM-24 requires the formation of a data integrity board to oversee the use of PII in a manner that ensures its accuracy and integrity. This board is responsible for reviewing and approving proposals for using PII, ensuring that data practices adhere to privacy principles and organizational policies. The data integrity board plays a key role in protecting individuals' privacy by ensuring that PII is used responsibly and that data quality is maintained. By establishing a data integrity board, organizations can enhance their privacy governance structure, supporting the ethical and lawful use of personal information.

PM-25—Minimization of Personally Identifiable Information Used in Testing, Training, and Research: PM-25 focuses on minimizing the use of PII in testing, training, and research activities. This control requires organizations to implement measures to limit the use of PII to the minimum necessary to achieve the intended purpose. Anonymized or pseudonymized data should be used to reduce privacy risks. Minimizing the use of PII in these activities helps protect individuals' privacy and reduce the risk of data breaches. Organizations can demonstrate their commitment to privacy protection by adopting data minimization principles while still supporting testing, training, and research needs.

PM-26—Complaint Management: PM-26 mandates the establishment of mechanisms for managing privacy complaints from individuals regarding handling their PII. This control involves creating procedures for receiving, investigating, and resolving complaints promptly and fairly. Effective complaint management supports individuals' privacy rights, enhances accountability, and provides valuable feedback for improving privacy practices. By proactively addressing privacy complaints, organizations can build trust with individuals, demonstrate their commitment to privacy, and identify opportunities to enhance their privacy program.

PM-27—Privacy Reporting: PM-27 requires organizations to report privacy-related activities, incidents, and compliance status to senior management and other relevant stakeholders. This control involves regularly preparing and disseminating reports detailing the organization's privacy practices, any incidents involving PII, and efforts to address privacy risks. Privacy reporting supports transparency, accountability, and informed decision-making, enabling organizations to demonstrate their commitment to protecting personal information. Organizations can enhance their privacy governance and respond effectively to privacy challenges by maintaining a robust privacy reporting process.

PM-28—Risk Framing: PM-28 focuses on establishing a risk-framing approach that defines the context, assumptions, constraints, risk tolerance, and priorities for managing cybersecurity risks. This control provides a foundation for the organization's risk management program, ensuring that risk decisions are made consistently and aligned with organizational objectives. Risk framing supports a structured approach to identifying, assessing, and responding to risks, facilitating effective risk management practices. By clearly defining how risks are framed, organizations can ensure that risk management efforts are aligned with their strategic goals and risk appetite.

PM-29—Risk Management Program Leadership Roles: PM-29 mandates the designation of senior officials responsible for leading the organization's risk management program. These roles involve overseeing the risk management strategy's development, implementation, and continuous improvement, ensuring that it effectively addresses cybersecurity and privacy risks. The leadership roles are critical for establishing clear accountability for risk management, coordinating efforts across the organization, and ensuring that risk management practices are integrated into organizational processes. By assigning senior officials to lead the risk management program, organizations can enhance their ability to manage risks in a coordinated, strategic manner.

PM-30—Supply Chain Risk Management Strategy: PM-30 requires developing and implementing a strategy for managing risks associated with the supply chain. This control involves identifying and assessing risks posed by suppliers, contractors, and service providers and implementing measures to mitigate those risks. Supply chain risk management is essential for ensuring the security and resilience of the organization's operations, protecting against threats that may arise from third-party relationships. By adopting a comprehensive supply chain risk management strategy, organizations can protect against disruptions, ensure product and service integrity, and maintain stakeholder trust.

PM-31—Continuous Monitoring Strategy: PM-31 mandates the establishment of a continuous monitoring strategy to assess the security and privacy posture of the organization's information systems on an ongoing basis. This control involves implementing tools, techniques, and processes to continuously assess the effectiveness of security and privacy controls, identify vulnerabilities, and detect anomalies and threats. A continuous monitoring strategy supports proactive risk management, enabling organizations to respond rapidly to changes in the threat landscape and operational environment. Organizations can ensure the ongoing protection of their information systems and data by maintaining continuous oversight of their security and privacy posture.

PM-32—Purposing: PM-32 focuses on ensuring that the collection, use, and dissemination of PII is limited to specific, authorized purposes. This control requires organizations to clearly define the purposes for which PII is collected and used and to restrict the use of PII to those purposes. Purposing supports privacy principles of data minimization and purpose limitation, protecting individuals' privacy by preventing the misuse of their information. By adhering to purposing requirements, organizations can enhance individuals' trust, ensure compliance with privacy regulations, and support the responsible management of personal information.

Personnel Security (PS) Overview

The Personnel Security (PS) family of controls establishes security measures that organizations must implement to ensure that individuals who access their information systems and environments are trustworthy and meet established security criteria. This includes controls for personnel screening, employment termination processes, and ensuring that personnel with access to sensitive information are adequately trained and monitored. By implementing PS controls, organizations can reduce the risk of insider threats and ensure that their workforce contributes to the security of their information systems.

PS-1—Policy and Procedures: PS-1 mandates developing, documenting, and disseminating PS policies and procedures. This foundational control establishes the framework for screening, hiring, and managing personnel in a manner that minimizes security risks to the organization. The policies and procedures cover background checks, access agreements, and actions responding to security violations. By maintaining comprehensive PS policies, organizations can ensure that individuals with access to sensitive information and systems are trustworthy and aware of their security responsibilities, enhancing the protection of organizational assets.

PS-2—Position Risk Designation: PS-2 requires organizations to designate the level of risk associated with positions based on their access to sensitive information and systems. This control involves assessing the potential impact of unauthorized access, disclosure, or misuse of information by individuals in specific positions and applying appropriate screening and management practices based on the risk level. Position risk designation supports targeted PS measures, ensuring that individuals in high-risk positions are subject to more rigorous scrutiny. By categorizing positions according to risk, organizations can allocate security resources effectively and reduce the risk of insider threats.

PS-3—Personnel Screening: PS-3 focuses on screening individuals before granting them access to sensitive information and systems. This control requires organizations to conduct background checks and other screening activities to assess candidates' trustworthiness and suitability for positions involving access to critical assets. Personnel screening is a key component of PS, helping prevent unauthorized access, data breaches, and other security incidents. By implementing thorough screening processes, organizations can ensure that individuals accessing sensitive information are reliable and pose a minimal security risk.

PS-4—Personnel Termination: PS-4 mandates procedures for managing the termination of personnel, ensuring that access to information systems and facilities is revoked and organizational assets are secured. This control involves conducting exit interviews, retrieving organizational property, and disabling access credentials promptly upon an individual's departure. Effective management of personnel termination reduces the risk of unauthorized access and protects against potential security threats from former employees. By implementing systematic termination procedures, organizations can safeguard their information assets and maintain the integrity of their security posture.

PS-5—Personnel Transfer: PS-5 requires organizations to manage personnel transfer between positions or departments to address associated security risks. This control involves reassessing the suitability and trustworthiness of individuals for new roles, updating access authorizations, and providing appropriate security awareness training. Managing personnel transfers effectively is essential for ensuring that individuals have access only to information and systems relevant to their new roles, minimizing the risk of unauthorized access or data leakage. By addressing the security implications of personnel transfers, organizations can maintain control over access to sensitive assets and ensure ongoing compliance with security policies.

PS-6—Access Agreements: PS-6 mandates that individuals with access to sensitive information and systems sign access agreements acknowledging their security responsibilities. This control ensures that personnel know the rules of behavior for system usage, the protection of sensitive information, and the consequences of security violations. Access agreements reinforce the importance of security awareness and accountability, supporting compliance with organizational policies and enhancing the protection of information assets. By requiring access agreements, organizations can clarify expectations for secure behavior and strengthen their security culture.

PS-7—External Personnel Security: PS-7 focuses on the security requirements for external personnel, such as contractors and service providers, who have access to the organization's information systems and facilities. This control involves establishing contract security clauses, conducting background checks, and providing security awareness training to ensure that external personnel adhere to the organization's security policies. Managing the security of external personnel is crucial for preventing data breaches, unauthorized access, and other security risks associated with third-party access. Organizations can extend their security perimeter and safeguard their information assets against external threats by implementing stringent security measures for external personnel.

PS-8—Personnel Sanctions: PS-8 mandates sanctions for personnel who violate security policies or procedures. This control involves defining disciplinary actions for security violations, ensuring that individuals are held accountable for their actions, and deterring future violations. Applying sanctions for security breaches reinforces the seriousness of security policies, supports a culture of accountability, and helps maintain the integrity of the organization's security posture. Organizations can address security incidents effectively by enforcing personnel sanctions and promoting compliance with security practices.

PS-9—Position Descriptions: PS-9 requires organizations to develop and maintain position descriptions that include security roles and responsibilities. This control ensures that individuals know their specific security obligations, supporting accountability and effective performance of security-related tasks. Position descriptions are a foundation for security awareness, training, and performance management, aligning individual roles with the organization's security objectives. By clearly defining security roles and responsibilities in position descriptions, organizations can ensure that personnel are equipped to protect information assets and the overall security posture.

Protective Technology (PT) Overview

The Protective Technology (PT) family of controls aims to ensure that the security and resilience of systems are maintained by protecting against malicious code, providing the means for ongoing monitoring, and implementing protective measures at the information system boundary. PT controls focus on technologies that help enforce security policies through mechanisms such as firewalls, intrusion detection systems, and data loss prevention solutions, enhancing the overall defense-in-depth strategy of an organization.

PT-1—Policy and Procedures: PT-1 mandates developing, documenting, and disseminating privacy policies and procedures. This foundational control establishes the framework for managing PII in accordance with legal, regulatory, and policy requirements. The privacy policies and procedures cover aspects such as the collection, use, dissemination, and protection of PII, as well as supporting compliance and protecting individuals' privacy rights. By maintaining comprehensive privacy policies, organizations can ensure a coordinated, comprehensive approach to privacy, enhancing the protection of personal information and supporting compliance with privacy obligations.

PT-2—Authority to Process Personally Identifiable Information: PT-2 requires organizations to establish and document the authority for processing PII. This control involves identifying the legal, regulatory, or policy basis for collecting, using, and sharing PII, ensuring that privacy practices are grounded in legitimate authority. Documenting the authority to process PII supports transparency, accountability, and compliance with privacy laws and regulations. By clearly establishing the basis for processing PII, organizations can ensure that personal information is handled responsibly and per applicable requirements.

PT-3—Personally Identifiable Information Processing Purposes: PT-3 mandates that organizations define and document the specific purposes for processing PII. This control requires that the collection and use of PII be limited to clearly defined purposes, supporting the principles of purpose limitation and data minimization. Documenting processing purposes helps ensure that PII is used in a manner consistent with individuals' expectations and privacy rights. By specifying the purposes for processing PII, organizations can enhance privacy protections, reduce the risk of unauthorized use, and support compliance with privacy regulations.

PT-4—Consent: PT-4 focuses on obtaining consent from individuals to process their PII. This control involves establishing mechanisms for acquiring and documenting individuals' consent, ensuring that consent is informed, voluntary, and specific to the processing activities. Obtaining consent supports individuals' autonomy over their personal information, enhances transparency, and ensures compliance with privacy laws that require consent for certain types of data processing.

By managing consent effectively, organizations can build trust with individuals, protect privacy rights, and mitigate legal and reputational risks associated with processing PII.

PT-5—Privacy Notice: PT-5 mandates the provision of privacy notices to individuals, informing them about the organization's privacy practices related to the processing of PII. This control requires that privacy notices be clear, concise, and accessible, covering aspects such as the purposes of processing, the types of PII collected, and individuals' rights regarding their information. Privacy notices enhance transparency, support informed decision-making by individuals, and ensure compliance with legal and regulatory requirements for privacy disclosures. By providing comprehensive privacy notices, organizations can demonstrate their commitment to privacy protection and foster trust with individuals whose information they process.

PT-6—System of Records Notice: PT-6 focuses on publishing a System of Records Notice (SORN) for systems that contain PII and are subject to privacy regulations. This control involves documenting and publicly disclosing the existence of systems of records, the types of PII collected, the purposes of collection, and the conditions under which PII may be disclosed. The SORN supports transparency and compliance with privacy laws, providing individuals with information about the organization's handling of their personal information. By maintaining and publishing SORNs, organizations can ensure their practices are open to scrutiny and accountable for protecting personal information.

PT-7—Specific Categories of Personally Identifiable Information: PT-7 requires organizations to identify and document specific categories of PII they collect, process, store, or transmit. This control supports the implementation of tailored privacy protections for different types of PII, recognizing that some categories of information may require enhanced safeguards due to their sensitivity. By categorizing PII, organizations can apply appropriate privacy controls, minimize data collection to what is necessary, and ensure that sensitive information is protected per its risk profile. Identifying specific categories of PII helps organizations manage privacy risks effectively and comply with privacy regulations that distinguish between types of personal information.

PT-8—Computer Matching Requirements: PT-8 mandates compliance with legal and regulatory requirements for computer matching activities involving PII. This control consists of ensuring that computer matching programs, which compare PII across different data sources to identify matches, are conducted in a manner that protects individuals' privacy rights and data accuracy. Compliance with computer matching requirements helps prevent unauthorized disclosures, supports data accuracy, and ensures that matching activities are conducted for legitimate purposes. By adhering to computer-matching regulations, organizations can safeguard individuals' privacy while utilizing data-matching techniques for operational or compliance purposes.

Risk Assessment (RA) Overview

The RA family of controls mandates the assessment of security risks associated with the operation and use of information systems. This includes identifying, evaluating, and documenting risks, threats, and vulnerabilities to organizational operations, assets, individuals, and other entities. RA controls guide organizations in understanding their risk posture and making informed decisions regarding implementing appropriate security measures to mitigate identified risks.

RA-1—Policy and Procedures: RA-1 mandates establishing, documenting, and regularly reviewing policies and procedures for conducting RAs within an organization. This foundational control ensures that RAs are carried out in a structured, consistent, and effective manner, enabling organizations to identify, evaluate, and address risks to their operations and assets. The policies and procedures cover the scope, frequency, and methodologies for RAs, ensuring that all relevant threats, vulnerabilities, and impacts are considered. By adhering to well-defined RA policies, organizations can better understand their risk environment, make informed security decisions, and allocate resources effectively to manage risk.

RA-2—Security Categorization: RA-2 requires organizations to categorize their information systems and the information processed, stored, and transmitted by those systems according to the potential impact on the organization should certain events occur. This categorization is based on the level of confidentiality, integrity, and availability that each system requires. By categorizing systems and data, organizations can prioritize their security efforts and apply appropriate controls to protect against threats. This process helps ensure high-value assets receive the most robust protection measures, aligning security practices with the organization's risk management strategy.

RA-3—Risk Assessment: RA-3 focuses on conducting comprehensive RAs to identify vulnerabilities, threat sources, and potential impacts on the organization's operations, assets, individuals, other organizations, and the Nation. This control involves analyzing the likelihood of threat exploitation and the resulting impact to determine the risk level. RAs are crucial for understanding the organization's threat landscape, informing risk response decisions, and implementing effective security controls. Regularly updated RAs ensure that emerging threats and changing organizational contexts are accounted for, supporting dynamic and responsive risk management practices.

RA-4—Risk Assessment Update: RA-4 mandates the periodic update of RAs to reflect changes in the threat environment, vulnerabilities, impact assessments, or other factors affecting organizational risk. This control ensures that the organization understands its current and accurate risk posture, enabling timely adjustments to risk management strategies and security controls. Updates to RAs are critical for maintaining the relevance of the organization's risk management processes and facilitating proactive adjustments to security practices in response to new information or changes in the operational environment.

RA-5—Vulnerability Monitoring and Scanning: RA-5 requires organizations to implement ongoing monitoring and scanning processes to identify weaknesses in information systems that adversaries could exploit. This control involves using automated tools and techniques to regularly scan systems and networks for vulnerabilities, assess the severity of findings, and prioritize remediation efforts. Effective vulnerability management is essential for reducing the attack surface, enhancing the organization's defense against cyber threats, and maintaining the security and resilience of information systems. Organizations can prevent exploitation and protect their critical assets by actively identifying and addressing vulnerabilities.

RA-6—Technical Surveillance Countermeasures Survey: RA-6 focuses on conducting surveys to detect unauthorized surveillance devices and hazards and identify vulnerabilities that could be exploited through technical surveillance. This control aims to protect sensitive information from eavesdropping or interception by adversaries, ensuring the confidentiality and integrity of communications and data. Technical surveillance countermeasures (TSCM) surveys are part of a comprehensive security program, helping organizations to safeguard against espionage, information theft, and unauthorized monitoring. By regularly conducting TSCM surveys, organizations can detect and mitigate technical surveillance threats, enhancing privacy and security for their operations and sensitive activities.

RA-7—Risk Response: RA-7 mandates organizations to develop and implement strategies for responding to identified risks based on the results of RAs. This control involves selecting appropriate risk response measures such as risk avoidance, mitigation, transfer, or acceptance and implementing actions to address the risks in alignment with the organization's risk tolerance. Effective risk response ensures that risks are managed proactively, reducing their potential impact on the organization's objectives and operations. By carefully planning and executing risk response actions, organizations can enhance their resilience to adverse events and ensure their information systems' continued security and availability.

RA-8—Privacy Impact Assessments: RA-8 requires organizations to conduct PIAs for systems, projects, or technologies that process PII. PIAs help identify and mitigate privacy risks associated with the collection, use, and dissemination of PII, ensuring that privacy considerations are integrated into the design and operation of information systems. By conducting PIAs, organizations can enhance transparency, build trust with individuals, and ensure compliance with privacy laws and regulations. PIAs are essential for protecting individuals' privacy rights and managing privacy risks proactively and responsibly.

RA-9—Criticality Analysis: RA-9 focuses on conducting criticality analyses to determine the importance of specific information systems, components, or services to the organization's mission, business functions, or continuity of operations. This control helps organizations identify and prioritize critical assets that require enhanced protection due to their significance. Criticality analysis supports effective resource allocation, ensuring the most vital assets are secured against threats and vulnerabilities. By understanding the criticality of their assets, organizations can tailor their security strategies to ensure the resilience and availability of essential functions, even under adverse conditions.

RA-10—Threat Hunting: RA-10 mandates the active and systematic search for compromise or malicious activity indicators within the organization's information systems and networks. This proactive approach goes beyond automated monitoring and detection to identify threats that may have evaded existing security measures. Threat hunting involves analyzing data from various sources, using advanced analytical techniques, and applying expert knowledge to uncover subtle or sophisticated threats. By hunting threats, organizations can detect and respond to advanced persistent threats (APTs) and other sophisticated attacks, enhancing their overall security posture and reducing the risk of significant breaches.

System and Services Acquisition (SA) Overview

The system and services acquisition (SA) family of controls addresses the risks of acquiring information systems and services. This includes ensuring that security requirements are integrated into the procurement process and that suppliers and products meet these requirements. SA controls cover aspects such as supplier RA, contract management, and oversight of suppliers, ensuring that the security implications of system and service acquisitions are properly managed.

SA-1—Policy and Procedures: SA-1 mandates developing, documenting, disseminating, and reviewing security and privacy policies and procedures for system and service acquisition. This foundational control ensures that acquisitions are performed in a manner that maintains security and privacy, integrating these considerations into the procurement process. The policies and procedures cover all stages of the acquisition process, from planning and solicitation to contract award and management, ensuring that suppliers specify and meet security and privacy requirements. By adhering to formal acquisition policies, organizations can manage the risks associated with external system services and products, ensuring they meet established security and privacy standards before integration into the operational environment.

SA-2—Allocation of Resources: SA-2 requires organizations to allocate sufficient resources to acquire information systems and services, ensuring that security and privacy considerations are adequately funded. This control involves planning for the financial, human, and technical resources needed to implement security and privacy requirements throughout the system development lifecycle (SDLC). By prioritizing resource allocation for security and privacy, organizations can ensure that systems and services acquired can protect sensitive information and meet regulatory compliance requirements, thus supporting the organization's overall risk management strategy.

SA-3—System Development Lifecycle: SA-3 focuses on integrating security and privacy considerations throughout the system development lifecycle. This control ensures that security and privacy are addressed during all phases of the SDLC, from initial requirements analysis through design, development, testing, deployment, and MA. Incorporating security and privacy from the outset and throughout system development helps prevent costly rework, reduces vulnerabilities, and ensures that systems are designed to operate securely and respect privacy by default. This holistic approach to security and privacy enhances the resilience and trustworthiness of information systems and services.

SA-4—Acquisition Process: SA-4 mandates establishing a process to ensure that security and privacy requirements are included in contracts for acquiring information systems and services. This control involves specifying security and privacy criteria, evaluating potential suppliers' ability to meet these criteria, and integrating security and privacy provisions into acquisition contracts. By formalizing security and privacy requirements in the acquisition process, organizations can ensure that external system services and products conform to organizational standards, reducing the risk of introducing vulnerabilities through third-party components or services.

SA-5—System Documentation: SA-5 requires the provision and MA of system documentation that accurately reflects information systems' current security and privacy posture. This control ensures that documentation includes detailed descriptions of the system architecture, components, security controls, and privacy measures. Well-maintained system documentation supports effective system management, audit, and compliance efforts, enabling personnel to understand systems' security and privacy features and facilitating the identification and resolution of issues. By keeping system documentation current, organizations can enhance their ability to manage security and privacy risks effectively.

SA-6—Software Usage Restrictions: SA-6 mandates the establishment of policies and procedures governing the use of software within the organization. This control ensures that only authorized software is used on organizational systems, reducing the risk of malware, unauthorized data access, and other security threats associated with unauthorized software. Policies and procedures typically include software whitelisting, licensing compliance, and prohibiting unauthorized installation. By enforcing software usage restrictions, organizations can maintain control over the software environment, enhancing the security and integrity of their information systems.

SA-7—User-installed Software: SA-7 focuses on preventing users from installing unauthorized software. This control involves implementing technical measures and policies to restrict users' ability to install software without proper authorization, reducing the risk of security vulnerabilities introduced by unauthorized or malicious software. By controlling user-installed software, organizations can protect against malware infections, data breaches, and other security incidents resulting from using unapproved software, maintaining the security posture of their information systems.

SA-8—Security and Privacy Engineering Principles: SA-8 requires the application of security and privacy engineering principles in developing information systems. This control ensures that systems are designed and built with security and privacy as integral components, following best practices and standards for secure and privacy-enhancing engineering. By embedding security and privacy into the system design and architecture, organizations can build systems that are inherently more resistant to threats, support compliance with legal and regulatory requirements, and effectively protect individuals' privacy rights.

SA-9—External System Services: SA-9 mandates establishing requirements and monitoring mechanisms for external system services. This control ensures that third-party providers adhere to the organization's security and privacy requirements, protecting the confidentiality, integrity, and availability of information processed by external services. Through contracts, service level agreements (SLAs), and ongoing monitoring, organizations can manage the risks associated with outsourcing, ensuring that external services do not compromise the organization's security posture or privacy obligations.

SA-10—Developer Configuration Management: SA-10 focuses on implementing configuration management practices by developers throughout the system development life cycle. This control ensures that changes to system components are tracked, reviewed, and approved, maintaining the integrity of the system and its security and privacy posture. Effective developer configuration management supports the reliable operation of systems, facilitates identifying and correcting security vulnerabilities, and helps ensure that privacy protections are consistently applied. By managing changes systematically, organizations can reduce the risk of unauthorized modifications and ensure that systems remain in a known, secure state.

SA-11—Developer Testing and Evaluation: SA-11 mandates comprehensive testing and evaluation of security and privacy features by developers. This control systematically assesses security and privacy mechanisms implemented in information systems, ensuring they function as intended and meet organizational requirements. Developer testing and evaluation are critical for identifying and addressing vulnerabilities and deficiencies early in the system development process, reducing the risk of security incidents and privacy breaches. By conducting thorough testing and evaluation, organizations can enhance their information systems' reliability, security, and privacy.

SA-12—Supply Chain Protection: SA-12 focuses on protecting the organization's supply chain from security threats and vulnerabilities. This control involves assessing the security practices of suppliers and contractors, implementing measures to mitigate supply chain risks, and ensuring the integrity of products and services acquired. Supply chain protection is essential for preventing the introduction of compromised components, counterfeit materials, and other security threats that can undermine the organization's information security posture. By managing supply chain risks proactively, organizations can enhance the security and resilience of their information systems and services.

SA-13—Trustworthiness: SA-13 mandates assessing and enhancing the trustworthiness of information systems and components. This control involves evaluating the security and privacy properties of systems and components to ensure they can protect sensitive information and support critical functions reliably. Trustworthiness encompasses security, privacy, reliability, and resilience and is critical for systems that process sensitive data or support essential services. By focusing on trustworthiness, organizations can ensure that their information systems are robust against threats and capable of maintaining their security and privacy posture under adverse conditions.

SA-14—Criticality Analysis: SA-14 requires organizations to conduct a criticality analysis for information systems and components to determine their importance to the organization's mission, business functions, or continuity of operations. This analysis helps identify systems and components critical to the organization's operations and requires enhanced protection measures. Criticality analysis supports effective risk management by prioritizing resources and security efforts toward protecting the most vital assets. By understanding the criticality of their systems and components, organizations can tailor their security strategies to ensure the resilience and availability of essential functions, even under adverse conditions.

SA-15—Development Process, Standards, and Tools: SA-15 focuses on adopting secure development processes, standards, and tools. This control requires organizations to utilize best practices, methodologies, and tools that support the safe development of information systems. Implementing secure development processes helps ensure systems are designed and built with security considerations from the outset, reducing vulnerabilities and enhancing the overall security posture. Organizations can produce more secure and reliable systems by adhering to recognized development standards and using tools that facilitate secure coding practices, protecting against potential security breaches and attacks.

SA-16—Developer-provided Training: SA-16 mandates that developers receive training on secure coding practices, security testing techniques, and privacy principles. This control ensures that developers have the knowledge and skills to build secure and privacy-respecting information systems. Developer training is critical for reducing coding errors that can lead to vulnerabilities and ensuring that security and privacy are integrated into the development process. By investing in developer training, organizations can enhance their information systems security and privacy capabilities, supporting protecting of sensitive information and compliance with regulatory requirements.

SA-17—Developer Security and Privacy Architecture and Design: SA-17 requires incorporating security and privacy considerations into the architecture and design of information systems. This control involves planning and documenting how security and privacy controls are integrated into the system architecture, ensuring that protective measures are built into the system from the ground up. A well-defined security and privacy architecture supports the effective implementation of controls, enhances system resilience, and facilitates compliance with security and privacy requirements. By focusing on security and privacy architecture and design, organizations can build information systems that are inherently more secure and privacy-enhancing, reducing the risk of vulnerabilities and breaches.

SA-18—Tamper Resistance and Detection: SA-18 focuses on implementing measures to resist and detect tampering with information systems and components. This control involves designing systems and components to prevent unauthorized access, modification, or destruction and alert the organization to tampering. Tamper resistance and detection are critical for protecting sensitive information and ensuring the integrity and availability of systems. By adopting tamper-resistant designs and detection mechanisms, organizations can enhance the security of their information systems, deter attackers, and quickly respond to security incidents.

SA-19—Component Authenticity: SA-19 mandates measures to ensure the authenticity of system components. This control verifies that hardware, software, and firmware components are genuine and have not been tampered with or replaced by counterfeit versions. Ensuring component authenticity is essential for maintaining the integrity of information systems, preventing security vulnerabilities, and protecting against supply chain attacks. By verifying the authenticity of components, organizations can reduce the risk of introducing compromised elements into their systems, supporting the overall security and reliability of their information infrastructure.

SA-20—Customized Development of Critical Components: SA-20 focuses on the customized development of critical system components when commercially available products do not meet security requirements. This control requires organizations to develop or commission the development of custom components that adhere to specific security and privacy standards, ensuring that critical systems are adequately protected. Custom development allows for greater control over the security features of components, addressing unique organizational needs and enhancing the protection of sensitive information. By investing in the customized development of critical components, organizations can ensure that their information systems meet stringent security and privacy requirements, supporting their risk management strategy.

SA-21—Developer Screening: SA-21 mandates screening developers involved in developing information systems and components. This control consists of conducting background checks and assessing the trustworthiness of individuals with access to sensitive information during development. Screening developers is critical for reducing insider threats, preventing unauthorized access or disclosure of sensitive information, and ensuring that systems are developed by personnel who adhere to security and privacy standards. By implementing developer screening practices, organizations can enhance the protection of the development process and the trustworthiness of their information systems.

SA-22—Unsupported System Components: SA-22 focuses on managing and mitigating risks associated with using unsupported system components. This control involves identifying components no longer supported by vendors, assessing the risks posed by their continued use, and implementing measures to mitigate them. Unsupported components may lack security updates and patches, exposing organizations to vulnerabilities and attacks. By addressing the risks associated with unsupported components, organizations can maintain the security and integrity of their information systems, ensuring that they are protected against known threats and vulnerabilities.

SA-23—Specialization: SA-23 requires organizations to consider specialization in acquiring, developing, and implementing information systems and services. This control involves leveraging specialized knowledge, skills, and capabilities to address specific security and privacy requirements. Specialization can enhance the effectiveness of security and privacy measures, enabling organizations to address complex challenges and meet specialized needs. By adopting specialized solutions and expertise, organizations can enhance their security posture and privacy protections, tailoring their approach to the unique requirements of their operational environment and risk landscape.

System and Communications Protection (SC) Overview

The system and communications protection (SC) family of controls provides measures to safeguard information in transit and at rest. This includes protecting the confidentiality and integrity of data through encryption, implementing network security controls to guard against unauthorized access, and ensuring secure communication protocols. SC controls are essential for maintaining a secure data flow within and across organizational boundaries, protecting against eavesdropping, data interception, and data manipulation.

SC-1 Policy and Procedures: SC-1 addresses establishing, documenting, and disseminating security policies and procedures for managing and protecting system communications. This control requires organizations to develop, publish, and maintain policies and procedures consistent with laws, regulations, and strategies for system communication protection. By defining and implementing these policies and procedures, organizations ensure that security considerations are integrated into the communication processes, promoting a secure information technology environment.

SC-2 Separation of System and User Functionality: SC-2 ensures that systems separate user functionality from system functionality to minimize the potential for unauthorized access to sensitive system operations. This control mandates mechanisms that distinguish between user and system activities, preventing users from executing actions that could compromise system security. By implementing this separation, organizations protect critical system functions from unauthorized use and potential exploitation.

SC-3 Security Function Isolation: SC-3 focuses on isolating security from nonsecurity functions to reduce the risk of unauthorized access and modification. This control involves implementing mechanisms that segregate security-related activities within the system, ensuring that these functions are separate from general system operations. Isolation helps maintain the integrity and reliability of security functions, safeguarding them from unauthorized interference.

SC-4 Information in Shared System Resources: SC-4 addresses protecting information stored in shared system resources against unauthorized access and disclosure. This control requires mechanisms to prevent users from accessing data without authorization, especially when resources are shared among multiple users or processes. By securing shared information, organizations can prevent data breaches and ensure that sensitive information remains confidential.

SC-5 Denial-of-service Protection: SC-5 aims to protect information systems from denial-of-service (DoS) attacks that could impair system availability. This control involves implementing measures to detect, respond to, and recover from DoS attacks. By mitigating the effects of these attacks, organizations can maintain system availability and ensure that critical services remain accessible to authorized users.

SC-6 Resource Availability: SC-6 ensures that system resources are available to meet operational and mission requirements. This control focuses on implementing measures to manage and protect system resources against overutilization, degradation, or failure. By maintaining resource availability, organizations can support continuous operation and delivery of services, even in the face of challenges to system performance.

SC-7 Boundary Protection: SC-7 emphasizes establishing boundaries to protect system components within those boundaries from unauthorized access and exfiltration of data. This control involves deploying firewalls, gateways, and encrypted tunnels to monitor and control communications at external and internal boundaries. Boundary protection helps to safeguard sensitive information and system resources from external threats and internal leaks.

SC-8 Transmission Confidentiality and Integrity: SC-8 focuses on protecting the confidentiality and integrity of information during transmission over networks. This control requires using cryptographic measures to encrypt data in transit, ensuring that unauthorized parties cannot intercept or tamper with information. By enforcing transmission confidentiality and integrity, organizations can prevent eavesdropping and data manipulation, securing communications across networks.

SC-9 Transmission Confidentiality: SC-9 is dedicated to ensuring the confidentiality of information as it is transmitted across various networks. This control mandates using robust encryption standards or similar protective measures to shield data during transit, thus preventing unauthorized disclosure. By focusing on the confidentiality of transmissions, organizations can safeguard sensitive information from being intercepted or accessed by unauthorized entities, ensuring that private communications remain secure. This protection is crucial in maintaining the integrity and privacy of data as it moves through potentially insecure or public networks.

SC-10 Network Disconnect: SC-10 emphasizes the need for systems to have the capability to quickly disconnect or isolate from network connectivity in the face of security threats or breaches. This control involves mechanisms to immediately sever network connections to prevent the spread of malicious activity or compromise further system resources. Organizations can effectively contain security incidents by enabling a rapid response through network disconnection, minimizing damage, and protecting sensitive information. Implementing this control helps ensure operational continuity by allowing for controlled isolation and recovery of affected systems.

SC-11 Trusted Path: SC-11 ensures establishing a secure, trusted path for communication between the user and the system, particularly for transmitting sensitive information like authentication credentials. This control requires mechanisms to protect the integrity and confidentiality of data as it travels between endpoints, preventing interception or manipulation by unauthorized parties. By maintaining a trusted path, organizations can safeguard user interactions with systems, ensuring that sensitive operations, such as logging in or transmitting personal information, are securely managed and protected against eavesdropping or tampering.

SC-12 Cryptographic Key Establishment and Management: SC-12 focuses on the secure establishment and management of cryptographic keys used to protect information. This control requires implementing policies and mechanisms for generating, distributing, storing, rotating, and revoking cryptographic keys. Effective key management ensures that keys are protected against unauthorized disclosure and use, maintaining the confidentiality and integrity of encrypted data. By securing the lifecycle of cryptographic keys, organizations can ensure the robustness of encryption practices, supporting secure communications and data protection efforts.

SC-13 Cryptographic Protection: SC-13 mandates using cryptographic measures to protect information confidentiality, integrity, and authenticity. This control involves applying cryptographic algorithms to data at rest and in transit, ensuring that information is accessible only to authorized parties. Organizations can prevent unauthorized access to sensitive information by employing cryptographic protection, ensuring that data remains secure against interception, alteration, or impersonation attacks. Cryptographic protection is a cornerstone of modern security strategies, providing a foundational layer of defense for digital assets.

SC-14 Public Access Protections: SC-14 addresses the unique security challenges of public access systems. This control involves implementing measures to protect the system against unauthorized access, misuse, and abuse while ensuring that public users can access the information or services intended for them. By carefully managing public access, organizations can maintain the security of their systems without compromising the availability and integrity of publicly accessible resources. Public access protections are essential for services like public websites, kiosks, and information portals, where balancing accessibility with security is critical.

SC-15 Collaborative Computing Devices and Applications: SC-15 focuses on securing devices and applications that enable collaborative computing, such as video conferencing tools and shared workspaces. This control requires implementing security measures to protect against unauthorized access and data leakage within collaborative environments. Organizations can facilitate teamwork and information sharing by securing collaborative computing resources while safeguarding sensitive data and communications. Effective security practices for collaborative tools are crucial in today's interconnected work environments, where remote and distributed teams rely heavily on digital collaboration platforms.

SC-16 Transmission of Security and Privacy Attributes: SC-16 mandates the secure transmission of security and privacy attributes associated with information to ensure that policies governing access and handling are maintained across different systems. This control requires mechanisms to embed or associate these attributes with data as it moves within and between systems, ensuring that information is processed and stored according to its designated security requirements. By effectively transmitting security and privacy attributes, organizations can maintain the integrity and confidentiality of information across diverse environments, supporting consistent data protection practices.

SC-17 Public Key Infrastructure Certificates: SC-17 focuses on using Public Key Infrastructure (PKI) certificates to support strong authentication, confidentiality, and data integrity. This control involves the management of digital certificates, which are used to verify the identity of entities and encrypt data. Organizations can establish a trust framework by implementing PKI certificates, ensuring that communications and transactions are securely authenticated and protected. PKI certificates are fundamental in securing web services, email, and other applications that require reliable identity verification and encrypted communications.

SC-18 Mobile Code: SC-18 addresses the security challenges mobile code poses, such as Java applets or JavaScript, which can be executed in web browsers or other environments. This control mandates the establishment of policies and procedures to manage the use of mobile code and to protect against unauthorized code execution that could compromise system integrity or confidentiality. By controlling and securing mobile code, organizations can prevent the execution of malicious or untrusted scripts, safeguarding their systems from potential vulnerabilities introduced by such code. Effective mobile code management is essential in reducing the attack surface of applications and ensuring the security of web-based services.

SC-19 Voice Over Internet Protocol (VoIP): SC-19 is dedicated to securing Voice Over Internet Protocol (VoIP) technologies, which allow for the transmission of voice communications over IP networks. This control requires implementing security measures to protect VoIP communications from interception, eavesdropping, and tampering. By securing VoIP systems, organizations can ensure the confidentiality and integrity of voice communications, protect sensitive information, and maintain the privacy of conversations. VoIP security is critical in preventing unauthorized access and ensuring that voice data is transmitted securely over potentially unsecured networks.

SC-20 Secure Name/Address Resolution Service (Authoritative Source): SC-20 focuses on ensuring the security and integrity of name and address resolution services provided by authoritative sources. This control involves protecting the mechanisms that translate domain names to IP addresses (such as DNS) to prevent attacks like DNS poisoning, where attackers redirect traffic to malicious sites. By securing authoritative name/address resolution services, organizations can prevent attackers from compromising the integrity of these critical network services, ensuring that users are directed to the correct and intended IP addresses.

SC-21 Secure Name/Address Resolution Service (Recursive or Caching Resolver): SC-21 emphasizes the security of recursive or caching services involved in name/address resolution processes. This control requires protecting these services from poisoning and spoofing attacks, ensuring that users are directed to the correct IP addresses. By securing the resolution process, organizations can prevent attackers from redirecting traffic to malicious sites, safeguarding the integrity of web communications, and protecting users from phishing and malware distribution networks.

SC-22 Architecture and Provisioning for Name/Address Resolution Service: SC-22 focuses on the secure architecture and provisioning of name/address resolution services. This control mandates the design and deployment of these services to enhance security and resilience against attacks. It ensures redundancy, secure configuration, and regular updates to protect against vulnerabilities. By adopting a robust architectural approach, organizations can ensure the reliability and security of domain name resolution services, which is crucial for the seamless operation of internet-dependent applications and services.

SC-23 Session Authenticity: SC-23 is dedicated to maintaining the authenticity of user sessions. This control involves implementing mechanisms to verify the identity of users throughout their sessions, protecting against session hijacking and impersonation attacks. By ensuring session authenticity, organizations can prevent unauthorized access to user accounts and sensitive information, preserving the security and integrity of user interactions with systems.

SC-24 Fail in Known State: SC-24 requires systems to be designed to fail in a secure, known state that minimizes risk and potential damage. This control involves implementing fail-safe measures that activate in the event of system failures, ensuring that systems default to a state that preserves security. By preparing for failures in this manner, organizations can protect against unexpected security vulnerabilities that may arise during system outages or malfunctions.

SC-25 Thin Nodes: SC-25 focuses on deploying thin clients or nodes that rely on centralized data processing and storage services. This control involves minimizing the functionality and data stored on end-user devices to reduce the risk of data theft and malware infections. By using thin nodes, organizations can centralize security controls and data protection measures, simplifying management and enhancing overall system security.

SC-26 Decoys: SC-26 involves using decoy systems, files, or resources to detect and analyze unauthorized access attempts. This control includes deploying honeypots or similar deceptive technologies to lure attackers, allowing organizations to monitor intruder activities, identify attack vectors, and improve defensive strategies. Organizations can enhance their threat detection capabilities by employing decoys and gaining valuable insights into potential security breaches.

SC-27 Platform-Independent Applications: SC-27 encourages developing and using applications independent of specific platforms. This control supports software deployment across various operating systems and devices without significant modifications. By fostering platform independence, organizations can ensure software portability and reduce vulnerabilities associated with platform-specific dependencies, improving the resilience and flexibility of application deployment.

SC-28 Protection of Information at Rest: SC-28 mandates the protection of information stored on systems to prevent unauthorized access and disclosure. This control requires the encryption of sensitive data at rest, along with the implementation of ACs and physical security measures. By securing data at rest, organizations can safeguard sensitive information from being compromised, ensuring confidentiality and compliance with data protection regulations.

SC-29 Heterogeneity: SC-29 emphasizes using diverse technologies and platforms within an organization's IT environment. This control involves deliberately implementing varied systems and software to reduce the risk of widespread vulnerabilities. By fostering a heterogeneous environment, organizations can prevent attackers from exploiting a single vulnerability across the entire network, enhancing overall system resilience.

SC-30 Concealment and Misdirection: SC-30 uses concealment and misdirection techniques to protect sensitive information and systems. This control creates ambiguity and deception to confuse and deter attackers, such as obfuscating system information or deploying fake network services. By employing concealment and misdirection, organizations can complicate attackers' efforts, making it more challenging for them to identify valuable targets or navigate the system effectively.

SC-31 Covert Channel Analysis: SC-31 requires the analysis and mitigation of covert channels that could be used to bypass security controls. This control involves identifying and addressing unintended communication paths within systems that may allow data leakage or unauthorized communication. By conducting covert channel analysis, organizations can uncover and secure potential avenues of data exfiltration, enhancing the confidentiality and integrity of their systems.

SC-32 System Partitioning: SC-32 mandates partitioning systems into separate components or domains to limit access and reduce the risk of unauthorized interaction. This control involves creating logical or physical separations to isolate sensitive functions and data, thereby minimizing the potential impact of a breach. By implementing system partitioning, organizations can enhance security through compartmentalization, making it more difficult for attackers to access or compromise the entire system.

SC-33 Transmission Preparation Integrity: SC-33 focuses on ensuring the integrity of information as it is prepared for transmission. This control requires the verification of data integrity before and during its transmission using mechanisms such as checksums or digital signatures. By ensuring transmission preparation integrity, organizations can protect against the tampering or corruption of data in transit, maintaining the accuracy and reliability of information exchanged between systems.

SC-34 Nonmodifiable Executable Programs: SC-34 emphasizes protecting executable programs from unauthorized modification. This control involves securing executable code against tampering, using measures such as code signing and integrity verification. Organizations can prevent malicious alterations that compromise system functionality or security by ensuring that executable programs are nonmodifiable.

SC-35 External Malicious Code Identification: SC-35 mandates identifying and preventing external malicious code from affecting the system. This control requires the implementation of mechanisms such as antivirus software and intrusion detection systems to detect and block malware. By focusing on external malicious code identification, organizations can protect their systems from a wide range of threats, including viruses, worms, and trojan horses, thereby maintaining the integrity and availability of their IT resources.

SC-36 Distributed Processing and Storage: SC-36 strategically uses distributed processing and storage solutions to enhance system resilience and performance. This control encourages the dispersion of computational and storage capabilities across multiple locations or systems, reducing the risk of a single point of failure. Organizations can improve system availability, facilitate scalability, and enhance disaster recovery capabilities by adopting distributed processing and storage.

SC-37 Out-of-band Channels: SC-37 requires the use of out-of-band channels for the secure transmission of sensitive information or control signals. This control involves establishing alternative communication paths separate from the primary network, providing a safe means of communication for critical operations or emergency responses. By utilizing out-of-band channels, organizations can ensure the integrity and confidentiality of crucial communications, especially in network compromise or disruption.

SC-38 Operations Security: SC-38 focuses on protecting information involved in an organization's operational processes and procedures. This control requires measures to safeguard sensitive operational details from exposure or unauthorized access, thereby preventing potential exploitation by adversaries. By emphasizing operations security, organizations can protect the confidentiality of their operational practices and maintain the security of their processes.

SC-39 Process Isolation: SC-39 mandates isolating processes within a system to prevent unauthorized interaction and access. This control involves implementing mechanisms to ensure that individual processes operate independently and without undue influence on each other, protecting against process-based attacks. Organizations can minimize the risk of malicious processes affecting system integrity or security by enforcing process isolation.

SC-40 Wireless Link Protection: SC-40 emphasizes the security of wireless communications to protect against interception and unauthorized access. This control requires the implementation of encryption and secure authentication methods for wireless connections, safeguarding data transmitted over airwaves. By focusing on wireless link protection, organizations can prevent eavesdropping and ensure the confidentiality and integrity of wireless communications.

SC-41 Port and I/O Device Access: SC-41 addresses the security of ports and input/output devices to prevent unauthorized access and data leakage. This control involves restricting access to physical and logical ports and monitoring the use of removable devices. By securing port and I/O device access, organizations can protect against data exfiltration, unauthorized device connections, and other security risks associated with peripheral interfaces.

SC-42 Sensor Capability and Data: SC-42 focuses on the security and integrity of sensor-derived data and the capabilities of sensor systems. This control involves ensuring that sensors operate as intended, are protected from tampering, and that their data is accurate and reliable. By securing sensor capabilities and data, organizations can trust the inputs used for decision-making processes, ensuring that sensor information remains dependable for operational and security purposes.

SC-43 Usage Restrictions: SC-43 mandates the implementation of usage restrictions for systems and applications to prevent unauthorized or harmful activities. This control involves defining and enforcing policies that limit how and by whom systems can be used, including restrictions on software installation and execution. By imposing usage restrictions, organizations can minimize the risk of security breaches and ensure that systems are used per organizational policies and security requirements.

SC-44 Detonation Chambers: SC-44 involves using controlled environments, known as detonation chambers, to execute and analyze suspicious code safely. This control enables organizations to inspect potentially malicious software and email attachments in an isolated environment, preventing harm to the main system. By employing detonation chambers, organizations can better understand suspicious code behavior, aiding in developing defenses and mitigations against emerging threats.

SC-45 System Time Synchronization: SC-45 emphasizes the importance of synchronizing system time across all organizational systems and devices. This control requires the implementation of protocols to ensure accurate and consistent timekeeping, supporting time-dependent security mechanisms like logging, authentication, and digital signatures. By maintaining system time synchronization, organizations can ensure the reliability of time stamps for security and operational purposes, facilitating IR and system management.

SC-46 Cross Domain Policy Enforcement: SC-46 focuses on enforcing security policies across different organizational security domains. This control involves implementing measures to control the flow of information between domains with varying security levels, preventing unauthorized access or data leakage. By enforcing cross-domain policies, organizations can manage and secure the exchange of information, ensuring that data is shared in a controlled and secure manner per organizational security policies.

SC-47 Alternate Communications Paths: SC-47 mandates the availability of alternate communications paths to ensure continued operations during primary communication failures. This control involves planning and implementing backup communication methods that can be activated during outages or attacks on primary channels. Organizations can maintain operational continuity and resilience by establishing alternate communications paths, ensuring that critical communications can proceed even under adverse conditions.

SC-48 Sensor Relocation: SC-48 addresses the ability to relocate sensors as operational needs or threat landscapes change. This control involves flexibility in the deployment of sensing devices, allowing for the adaptation of security monitoring to shifting priorities or emerging threats. By enabling sensor relocation, organizations can optimize their situational awareness and defensive posture, ensuring monitoring efforts align with current security requirements.

SC-49 Hardware-Enforced Separation and Policy Enforcement: SC-49 emphasizes using hardware mechanisms to enforce separation and policy enforcement within systems. This control involves leveraging hardware-based security features to create isolated environments and enforce security policies at the hardware level, providing robust protection against software attacks. By utilizing hardware-enforced separation and policy enforcement, organizations can achieve a higher level of security, protecting critical system components even in the face of sophisticated software threats.

SC-50 Software-Enforced Separation and Policy Enforcement: SC-50 focuses on using software mechanisms to enforce separation and policy compliance within systems. This control requires software-based solutions to isolate applications and data and enforce organizational security policies. Organizations can create controlled environments for sensitive operations by adopting software-enforced separation and policy enforcement, mitigating the risk of unauthorized access and ensuring compliance with security policies.

SC-51 Hardware-Based Protection: SC-51 mandates using hardware-based security protections to defend against threats and vulnerabilities. This control involves leveraging hardware features such as Trusted Platform Modules (TPMs) and hardware security modules (HSMs) to enhance the security of systems. By employing hardware-based protections, organizations can secure cryptographic keys, authenticate devices, and ensure the integrity of their systems, providing a solid foundation for overall security posture.

System and Information Integrity (SI) Overview

The system and information integrity (SI) family of controls ensures that systems and data are accurate, trustworthy, and free from unauthorized modification. This includes implementing measures to detect and protect against malware, managing system flaws, and ensuring the integrity of information processing. SI controls are critical for preventing corruption of data and system functions, ensuring that information remains reliable and systems operate as intended.

SI-1 Policy and Procedures: SI-1 emphasizes establishing, documenting, and maintaining robust security policies and procedures. These guide the organization's approach to managing and protecting its information systems and data. By clearly defining roles, responsibilities, and processes for security, organizations can ensure a consistent and effective approach to maintaining system integrity, protecting against unauthorized access, and responding to security incidents. Regularly reviewing and updating these policies and procedures ensures they remain relevant in the face of evolving threats and changing organizational structures.

SI-2 Flaw Remediation: SI-2 focuses on identifying, evaluating, and correcting software and system flaws. Organizations must implement processes to regularly scan for vulnerabilities, assess the risk they pose, and prioritize their remediation based on this RA. Timely patching or mitigating identified vulnerabilities is critical to protecting the organization from potential exploits that could lead to unauthorized access, data loss, or service disruptions. This control underscores the ongoing nature of security management and the necessity for a proactive stance on system vulnerabilities.

SI-3 Malicious Code Protection: SI-3 mandates the deployment of measures to protect information systems against malware. This includes antivirus software, intrusion detection and prevention systems, and other malware defenses. Organizations must consistently update these measures to combat new and evolving threats. Malicious code protection is a foundational aspect of system security, safeguarding critical assets from unauthorized access, disruption, or destruction caused by viruses, worms, trojans, and other forms of malware.

SI-4 System Monitoring: SI-4 requires continuous monitoring of information systems to detect unauthorized access, modifications, and other security-relevant events. This involves collecting, analyzing, and acting upon security logs and alerts to identify and respond to potential security incidents swiftly. Effective system monitoring is vital for the early detection of threats, enabling organizations to mitigate risks before they can cause significant damage. This control is essential for maintaining situational awareness and ensuring the ongoing security of organizational assets.

SI-5 Security Alerts, Advisories, and Directives: SI-5 involves establishing processes to receive, assess, and respond to security alerts, advisories, and directives. This control ensures that organizations remain informed about emerging threats, vulnerabilities, and recommended protective measures. By actively engaging with external security communities and vendors, organizations can update their defenses, patch vulnerable systems, and adjust security postures in response to new information, maintaining resilience against evolving threats.

SI-6 Security and Privacy Function Verification: SI-6 emphasizes verifying security and privacy controls to ensure they function correctly and effectively. This includes regular testing and evaluations to validate the operational effectiveness of security measures and privacy safeguards. Verification activities help identify potential weaknesses or failures in security and privacy functions, enabling timely remediation and ensuring that protective measures are correctly implemented and maintained across the organization's information systems.

SI-7 Software, Firmware, and Information Integrity: SI-7 is centered on ensuring the integrity of software, firmware, and information. Organizations must implement controls to detect and prevent unauthorized modifications, which could compromise system functionality or security. This includes using cryptographic checksums, digital signatures, and integrity monitoring tools. Maintaining integrity is crucial for preventing malicious tampering and ensuring that systems and data remain accurate and trustworthy.

SI-8 Spam Protection: SI-8 mandates measures to protect against spam, which can be a vector for malware and phishing attacks. Organizations should implement email filtering technologies and user training to detect and manage spam effectively. By reducing the volume of spam, organizations can decrease the risk of malware infection and improve productivity by limiting the amount of irrelevant or malicious content users must contend with.

SI-9 Information Input Restrictions: SI-9 restricts information input to systems, ensuring that only authorized and validated data is processed. This control helps prevent malicious data from entering the system, which could lead to security vulnerabilities or system compromises. Implementing strict input controls, including authentication and authorization mechanisms, helps safeguard system integrity by controlling the sources and types of data allowed.

SI-10 Information Input Validation: SI-10 requires validating information input into systems to prevent attacks such as SQL injection, cross-site scripting, and buffer overflows. Organizations must ensure all input data is rigorously checked for correctness, legitimacy, and security before processing. By validating input, organizations can protect their systems from attacks that exploit input vulnerabilities, maintaining the security and stability of their information processing environments.

SI-11 Error Handling: SI-11 mandates that systems implement proper error handling mechanisms to prevent leakage of sensitive information through error messages and to ensure system stability. Effective error handling includes suppressing detailed system information in user-facing messages, logging errors for system administrator review, and maintaining system operation even when mistakes occur. This control is crucial for avoiding inadvertent disclosures of sensitive information and ensuring that unauthorized individuals cannot exploit error information to gain insights into system architectures or vulnerabilities.

SI-12 Information Management and Retention: SI-12 focuses on adequately managing and retaining information to ensure that data is kept secure throughout its lifecycle and disposed of securely when no longer needed. This involves implementing policies and procedures for data classification, storage, access, and disposal. Effective information management and retention practices help organizations comply with legal and regulatory requirements, minimize risks associated with data breaches, and ensure that valuable information is protected and available when needed.

SI-13 Predictable Failure Prevention: SI-13 involves implementing measures to prevent system failures that could be predicted based on known vulnerabilities, configuration errors, or system loads. This control requires regular system assessments, stress testing, and capacity planning to identify and mitigate potential failure points before they result in system outages or compromises.

Organizations can proactively address predictable failures and enhance system reliability, availability, and security.

SI-14 Nonpersistence: SI-14 mandates using nonpersistent computing environments to minimize the risk of data leakage and malware persistence. This control involves deploying technologies such as virtualization and containerization that can be reset or reimaged easily to a known secure state. Nonpersistent environments help ensure that any malicious changes or data remnants are eliminated after a session ends, thereby enhancing the security of the computing environment.

SI-15 Information Output Filtering: SI-15 requires filtering information output from systems to prevent unauthorized disclosure of sensitive data. This control involves implementing mechanisms to inspect and sanitize data being exported from the system to ensure it does not contain sensitive information that should not be disclosed. By filtering output information, organizations can protect sensitive data from accidental or intentional leakage, maintaining confidentiality and compliance with privacy regulations.

SI-16 Memory Protection: SI-16 focuses on protecting system memory from unauthorized access and manipulation. This control involves implementing address space layout randomization (ASLR) measures and nonexecutable memory protections to prevent exploits such as buffer overflows and code injection attacks. Memory protection mechanisms are vital for securing operating systems and applications against attempts to execute malicious code or access sensitive information stored in memory.

SI-17 Fail-safe Procedures: SI-17 mandates establishing fail-safe procedures to ensure that systems default to a secure state in case of a failure or compromise. This control requires the design of systems and processes that, in the face of unexpected conditions or breaches, will revert to a state that minimizes risk and protects sensitive data. Implementing fail-safe procedures prevents data loss or exposure and maintains system integrity under adverse conditions.

SI-18 Personally Identifiable Information Quality Operations: SI-18 emphasizes the importance of maintaining the quality, accuracy, and relevance of PII throughout its lifecycle. This control involves implementing processes to ensure that PII is regularly updated, corrected, and validated to support privacy and security objectives. High-quality PII operations help organizations meet legal and regulatory requirements, enhance customer trust, and prevent errors or fraud associated with incorrect or outdated information.

SI-19 De-identification: SI-19 focuses on de-identifying sensitive information to reduce privacy risks. This control involves removing, encrypting, or anonymizing PII to prevent data association with specific individuals. By de-identifying data, organizations can use or share information for analysis, research, or other purposes while minimizing privacy concerns and complying with data protection regulations.

SI-20 Tainting: SI-20 involves tracking the flow of sensitive information within systems to prevent unauthorized or unintended use. This control requires the implementation of mechanisms to "taint" sensitive data, marking it for extraordinary handling and preventing its exposure through outputs or in insecure contexts. Tainting helps ensure that data privacy and security policies are enforced throughout data processing and storage, protecting sensitive information from leakage or misuse.

SI-21 Information Refresh: SI-21 mandates regular updates or information refreshes to ensure accuracy, relevance, and security. This control involves replacing obsolete information with current data and revalidating or re-encrypting stored data to maintain its confidentiality and integrity. Regular information refreshes help organizations keep pace with changes, support decision-making with accurate data, and enhance security by renewing protective measures.

SI-22 Information Diversity: SI-22 uses diverse information sources and formats to enhance decision-making and resilience. This control encourages the integration of varied data types and sources to prevent over-reliance on single points of failure or biased information. By promoting information diversity, organizations can improve analytical robustness, foster innovation, and enhance the reliability of insights derived from data analysis.

SI-23 Information Fragmentation: SI-23 involves dividing sensitive information into smaller, nonsensitive fragments that do not reveal the original data when isolated. This control requires mechanisms for securely splitting, storing, and reassembling information to protect against unauthorized access or disclosure. Information fragmentation enhances data security by ensuring that compromised fragments do not expose the complete dataset or sensitive details, thereby protecting information integrity and confidentiality.

Supply Chain Risk Management (SR) Overview

The supply chain risk management (SR) family of controls addresses the security risks associated with the supply chain and acquiring products and services. This includes assessing and managing risks to the organization's supply chain, ensuring the security and resilience of suppliers, and protecting against tampering and counterfeit products. SR controls are essential for ensuring that threats within the supply chain do not compromise the security of the organization's information systems and the integrity of its products and services.

SR-1 Policy and Procedures: SR-1 emphasizes developing, documenting, and maintaining policies and procedures specific to supply chain risk management. This control ensures organizations establish a coherent approach to identifying, assessing, and mitigating risks throughout the supply chain. By formalizing supply chain security practices, organizations can enhance the resilience of their supply chains, protect against disruptions, and ensure the integrity and security of goods and services acquired.

SR-2 Supply Chain Risk Management Plan: SR-2 mandates the creation of a comprehensive supply chain risk management plan that outlines strategies and practices for managing risks associated with suppliers and third-party vendors. This plan should include RA methodologies, mitigation strategies, and processes for continuously monitoring supply chain threats. An effective supply chain risk management plan enables organizations to proactively address vulnerabilities and threats, ensuring the security and reliability of their supply chain operations.

SR-3 Supply Chain Controls and Processes: SR-3 focuses on implementing controls and processes to manage and mitigate supply chain risks. This involves establishing security requirements for suppliers, conducting RAs, and monitoring supplier performance. By implementing robust supply chain controls and processes, organizations can protect against potential vulnerabilities introduced through third-party products or services, ensuring the security and integrity of their supply chain.

SR-4 Provenance: SR-4 mandates verification of the provenance of products and components to ensure they are sourced from reputable and secure sources. This control involves tracking the origin and history of items within the supply chain to prevent the introduction of counterfeit or compromised components. By verifying provenance, organizations can ensure the authenticity and security of their supply chain, protecting against the risks associated with counterfeit, tampered, or otherwise malicious products.

SR-5 Acquisition Strategies, Tools, and Methods: SR-5 emphasizes the need for secure acquisition strategies, tools, and methods to safeguard the supply chain. This control requires adopting practices that ensure the security and integrity of products and services at the point of purchase or acquisition. By employing secure acquisition strategies, organizations can reduce the risk of introducing vulnerabilities through purchased software, hardware, or services, enhancing the overall security of their supply chain.

SR-6 Supplier Assessments and Reviews: SR-6 focuses on the regular assessment and review of suppliers to evaluate their compliance with security requirements and ability to manage risks. This control thoroughly evaluates suppliers' security practices, policies, and performance. Regular supplier assessments and reviews help organizations identify and mitigate potential supply chain risks, ensuring that suppliers meet the necessary security standards and contribute to the organization's overall security posture.

SR-7 Supply Chain Operations Security: SR-7 mandates implementing security measures throughout supply chain operations to protect against tampering, theft, and other security threats. This control involves securing transportation, storage, and handling processes to ensure the integrity and security of products as they move through the supply chain. By enhancing supply chain operations security, organizations can safeguard critical assets and sensitive information, reducing the risk of security breaches and operational disruptions.

SR-8 Notification Agreements: SR-8 requires the establishment of notification agreements with suppliers and partners to ensure timely communication of security issues and incidents that could impact the supply chain. This control involves formalizing mechanisms for sharing information about vulnerabilities, breaches, or other security concerns. Notification agreements enable organizations to respond swiftly to supply chain threats, mitigating potential impacts on their operations and security posture.

SR-9 Tamper Resistance and Detection: SR-9 focuses on using tamper-resistant and tamper-detection mechanisms to protect products and components within the supply chain. This control involves implementing physical and logical measures to prevent unauthorized access, modification, or substitution of items. Tamper resistance and detection mechanisms help ensure the integrity and authenticity of products, reducing the risk of security breaches resulting from tampered or counterfeit components.

SR-10 Inspection of Systems or Components: SR-10 mandates the thorough inspection of systems or components to identify tampering, defects, or noncompliance with security requirements. This control requires establishing procedures for inspecting received products for signs of tampering or compromise. By conducting detailed inspections, organizations can detect and address security issues before systems or components are integrated into their environment, enhancing the security of their supply chain.

SR-11 Component Authenticity: SR-11 emphasizes verifying the authenticity of components to prevent the introduction of counterfeit or compromised items into the organization's systems. This control involves implementing measures to authenticate the source and integrity of components used in products or systems. Verifying component authenticity helps protect against the risks associated with counterfeit parts, which may introduce vulnerabilities or reduce the reliability of systems.

SR-12 Component Disposal: SR-12 focuses on the secure disposal of components to prevent unauthorized access to sensitive information or technology. This control requires implementing processes for the safe and environmentally responsible disposal or recycling of hardware and electronic components. Secure component disposal ensures that sensitive data is properly destroyed or sanitized, reducing the risk of data breaches and protecting intellectual property even at the end of the component's lifecycle.

Index of 800-53 Controls used in the CSF

This index is a comprehensive reference for the NIST SP 800-53 controls discussed in this book, each linked to sections of the NIST Cybersecurity Framework (CSF). The index is organized numerically and includes page numbers for quick navigation. The selection of controls is deliberate, focusing on those relevant to a wide range of organizations while recognizing that some are specific to federal systems. This curated approach ensures that the book provides actionable insights without overwhelming detail.

A Comprehensive Guide to the NIST Cybersecurity Framework 2.0: Strategies, Implementation, and Best Practice, First Edition. Jason Edwards.
© 2025 John Wiley & Sons Ltd. Published 2025 by John Wiley & Sons Ltd.

i

m

p